The Blackwell Guide to the

Philosophy of Science

—— Blackwell Philosophy Guides ——

Series Editor: Steven M. Cahn, City University of New York Graduate School

Written by an international assembly of distinguished philosophers, the *Blackwell Philosophy Guides* create a groundbreaking student resource – a complete critical survey of the central themes and issues of philosophy today. Focusing and advancing key arguments throughout, each essay incorporates essential background material serving to clarify the history and logic of the relevant topic. Accordingly, these volumes will be a valuable resource for a broad range of students and readers, including professional philosophers.

The Blackwell Guide to the
Philosophy of Science

Edited by

Peter Machamer and Michael Silberstein

BLACKWELL
Publishers

Copyright © Blackwell Publishers Ltd 2002

First published 2002

2 4 6 8 10 9 7 5 3 1

Blackwell Publishers Inc.
350 Main Street
Malden, Massachusetts 02148
USA

Blackwell Publishers Ltd
108 Cowley Road
Oxford OX4 1JF
UK

Library of Congress Cataloging-in-Publication Data has been applied for.

ISBN 0-631-22107-7 (hardback); 0-631-22108-5 (paperback)

British Library Cataloguing in Publication Data
A CIP catalogue record for this book is available from the British Library.

Typeset in 10 on 13 pt Galliard
by Best-set Typesetter Ltd., Hong Kong
Printed in Great Britain by T.J. International, Padstow, Cornwall

This book is printed on acid-free paper.

Contents

Contents

Notes on Contributors

Daniela M. Bailer-Jones studied Philosophy and Physics at the Universities of Freiburg, Oxford and Cambridge, receiving an M. Phil. in Physics (1993) and a Ph.D. in Philosophy of Science (1998) from the University of Cambridge. She taught at the University of Paderborn (1998–2000), and was at the University of Bonn until be coming a Fellow at the Center of Philosophy of Science of the University of Pittsburgh in the summer of 2001. Her main research interest is scientific models.

James Bogen, having retired after many years at Pitzer College, is now an adjunct professor in the University of Pittsburgh HPS Department. His publications include papers on topics in the theory of knowledge, including methodology in the neurosciences.

Craig Callender is an Assistant Professor of Philosophy at the University of California at San Diego. He was formerly a Senior Lecturer at the London School of Economics, where he also worked at times with the *British Journal for the Philosophy of Science* and *Mind*. With Nick Huggett, he recently edited a book entitled *Physics meets Philosophy at the Planck Scale* (2001). He has published and lectured extensively on topics in the philosophical foundations of modern physics.

Carl F. Craver was Assistant Professor of Philosophy, Florida International University, and moved last Fall, to Washington University, Saint Louis. He has a Ph.D. from The University of Pittsburgh, Department of History and Philosophy of Science, and an M.S. from University of Pittsburgh, Department of Neuroscience. His primary research areas are philosophy of neuroscience, with particular emphasis on mechanisms, mechanical explanation, and theory construction.

Paul Griffiths was educated at Cambridge and the Australian National University, and taught at Otago University in New Zealand and the University of Sydney,

Australia before moving to the Department of History and Philosophy of Science at the University of Pittsburgh. He is author, with Kim Sterelny, of *Sex and Death: An Introduction to Philosophy of Biology* (1997) and editor, with Susan Oyama and Russell Gray, of *Cycles of Contingency: Developmental Systems and Evolution* (2001).

Rick Grush received his joint doctorate in Cognitive Science and Philosophy from UC San Diego in 1995. From 1995 to 1998, he held positions at the PNP Program at Washington University on St. Louis, and the Center for Semiotic Research at the University of Aarhus, Denmark. From 1998 to 2000, he was at the University of Pittsburgh, where he also served as Associate Director of the Center for Philosophy of Science for the 1999–2000 academic year. He is currently in the Philosophy Department at UC San Diego. His work involves understanding the physical basis of the mind.

Alan Hájek is an Associate Professor of Philosophy at the California Institute of Technology, Pasadena, California. He works mainly in the foundations of probability and decision theory, epistemology, philosophy of science and philosophy of religion. His publications have dealt with such topics as: probabilities of conditionals; the interpretation of probability; the relationship between conditional and unconditional probability; Bayesian epistemology and philosophy of science; infinite decision theory and Pascal's Wager; Hume's miracles argument; and Moore's paradox.

Ned Hall, Associate Professor of philosophy at MIT, works mainly on metaphysics, philosophy of science, and, more specifically, philosophy of quantum physics. His current research on quantum physics focuses on the measurement problem, and on implications of and problems for the usual quantum mechanical treatment of identical particles. In metaphysics and philosophy of science, his work has included investigations into the connections between probability theory and the logic of conditionals, the epistemology and metaphysics of objective probability, and the analysis of causation.

Carl Hoefer is Lecturer in the Department of Philosophy, Logic and Scientific Method at the LSE, and Co-Director of the Centre for Philosophy of Natural and Social Science. He works in the areas of philosophy of space and time (especially general relativity) and metaphysics.

Harold Kincaid is Professor of Philosophy and Director, Center for Ethics and Values in the Sciences at the University of Alabama at Birmingham. He is the author of *Philosophical Foundations of the Social Sciences* (1996), *Individualism and the Unity of Science* (Rowman and Littlefield, 1997), and numerous articles on topics in the philosophy of social science.

Peter Machamer is Professor of History and Philosophy of Science at the University of Pittsburgh and Associate Director of The Center for Philosophy of Science. He edited the *Cambridge Companion to Galileo* (1998), and was co-editor of *Scientific Controversies* (2000) and *Theory and Method in Neuroscience* (2001). He is currently working on a book about interpretation in science and art and maybe publishing a collection of his essays on the seventeenth century.

Roberta L. Millstein is an Assistant Professor in the Department of Philosophy at California State University, Hayward. She received her A.B. from Dartmouth College with a double major in Computer Science and Philosophy. She earned her Ph.D. in Philosophy, with a minor in the History of Science and Technology, at the University of Minnesota. She teaches courses in the history and philosophy of science (including courses in science and ethics), and publishes articles in the philosophy of biology.

Lynn Hankinson Nelson is Professor of Philosophy at the University of Missouri-St. Louis. She is co-author with Jack Nelson of *On Quine* (2000), co-editor with Jack Nelson of *Feminism, Science, and the Philosophy of Science* (1996 and 1997), guest editor of a special issue of *Synthese* devoted to feminism and science (1995), and the author of *Who Knows: From Quine to Feminist Empiricism* (1990).

Laura Ruetsche is Assistant Professor of Philosophy at the University of Pittsburgh. Her interests include the foundations of physical theories, the epistemology of science (including feminist approaches), and Plato.

Michael Silberstein is Associate Professor of Philosophy at Elizabethtown College. He has published and delivered papers in both Philosophy of Science and Philosophy of Mind. His primary areas of research and interest are philosophy of physics and the philosophy of cognitive-neuroscience respectively.

Jim Woodward is Professor of Philosophy and Executive Officer for the Humanities at the California Institute of Technology. He is completing a book on explanation.

John Worrall is Professor of Philosophy of Science and Co-Director of the Centre for Philosophy of Natural and Social Science at the London School of Economics. He was editor of the *British Journal for the Philosophy of Science* from 1974 to 1982, and editor of the collected works of Imre Lakatos. He has published widely on topics in general philosophy of science and history and philosophy of nineteenth century physics. He is currently finishing a book on theory-change in science and developing an interest in issues in the methodology of medicine.

Preface

This volume was conceived by Michael Silberstein, who then contacted Blackwell about the idea. Peter Machamer joined the project before the final presentation was made to Blackwell. It has been a collaborative effort thereafter.

The conception for the chapters in this volume were drawn up along a number of parameters. First, we wished a good mix of authors, people established in the field as well as some younger scholars who would bring a fresh perspective to their chapters. Second, each author was charged with writing a three-part essay: the first part to review the problem; the second to assay the current state of the discipline with respect to the topic; and finally to prognosticate on the future and discuss where the field should be moving. All this was to be done within 8500 words! Most chapters stayed nearly within their limits and have accomplished the set task with great aplomb.

A third parameter was that the chapters should be written to be of use to those who are not specialists in the field or on the topic, but who wished a single source they could read that would bring them "up to speed." However, the chapters also were to be of interest to the specialists, and thus not merely introductory in nature. Obviously, different topics require different levels of expertise on the part of the reader, but we feel all of the chapters are accessible. This is compatible with the fact that some chapters are more technical and require a specialized knowledge on the part of the reader. For example, we felt no good use could come of having a chapter on quantum mechanics that eschewed the mathematics, a chapter on space–time that had to explain the basis of the general theory of relativity or a probability chapter that ignored the probability calculus. Such a book could have been put together but it would not be a guide, it would have been a popularizing introduction. Such was not our aim.

Finally, we sought to cover the basic topics where research in philosophy of science was, in our eyes, progressing. Due to space limitations, we have not covered everything we might have, nor that we would have liked. Something should have been said about the relation between sciences studies and philosophy of science and again about history of science and philosophy of science. We should have spent more time on the "continental" tradition and its relations to philosophy of science. Many of the special sciences are ignored. We had only so many chapters we could chose. Others might have chosen differently.

We think this book is good. Each chapter is written with care, and has substantive import. This is our judgment. The final evaluation will rest with you, the reader.

Acknowledgments and references are given in each chapter. In addition, Michael Silberstein would like to thank his love and best friend Elizabeth Newell for her kindness and patience and his assistant Michael Cifone for his invaluable help. He would also like to give a special thanks to Peter Machamer for his patience, thoughtful suggestions, hard work, long hours and without whom this book would not exist. He dedicates this volume to Elizabeth Newell and his son Christopher Robin Silberstein. Peter Machamer would like to thank Barbara Diven Machamer, and Michael and Tara Gainfort for their support and patience with many late dinners. His efforts are dedicated to Rachel, Courtney and Nico – grandchildren who make life special.

Chapter 1

A Brief Historical Introduction to the Philosophy of Science*

Peter Machamer

Philosophy of science is an old and practiced discipline. Both Plato and Aristotle wrote on the subject, and, arguably, some of the pre-Socratics did also. The Middle Ages, both in its Arabic and high Latin periods, made many commentaries and disputations touching on topics in philosophy of science. Of course, the new science of the seventeenth century brought along widespread ruminations and manifold treatises on the nature of science, scientific knowledge and method. The Enlightenment pushed this project further trying to make science and its hallmark method definitive of the rational life. With the industrial revolution, "science" became a synonym for progress. In many places in the Western world, science was venerated as being the peculiarly modern way of thinking. The nineteenth century saw another resurgence of interest when ideas of evolution melded with those of industrial progress and physics achieved a maturity that led some to believe that science was complete. By the end of the century, mathematics had found alternatives to Euclidean geometry and logic had become a newly re-admired discipline.

But just before the turn to the twentieth century, and in those decades that followed, it was physics that led the intellectual way. Freud was there too, he and Breuer having published *Studies in Hysteria* in 1895, but it was physics that garnered the attention of the philosophers. Mechanics became more and more unified in form with the work of Maxwell, Hertz and discussions by Poincaré. Plank derived the black body law in 1899, in 1902 Lorenz proved Maxwell's equations were invariant under transformation, and in 1905 Einstein published his paper on special relativity and the basis of the quantum. Concomitantly, Hilbert in 1899 published his foundations of geometry, and Bertrand Russell in 1903 gave forth his principles of mathematics. The development of unified classical mechanics and alternative geometries, now augmented and challenged by the new relativity and quantum theories made for period of unprecedented excitement in science.

What follows provides a brief historical overview of the problems and concepts that have characterized philosophy of science from the turn of the twentieth century until the present day. This is presented in the form of conceptual and problem-oriented history because I believe that the real interest in philosophy of science and the lessons to be learned from its history are found in the topics it addressed and the methods it used to address them. Further, the cast of characters, and the specific articles and books can be easily researched by anyone who is interested. There is, appended a selective chronological bibliography of "classical" sources.

A few caveats need to be stated from the start. First, I deal almost exclusively with certain aspects of one Austro-Germanic-Anglo-American tradition. This is not because there was not interesting and important work in philosophy of science going on in France and elsewhere. I do this, first, because this tradition is the one that is formative for and dominant in contemporary American philosophy (for good or ill), and, second, because it is the tradition in which I was raised and about which I know the most. Another caveat is that space limitations and igno-rance often require the omission of many interesting nuances, qualifications and even outright important facets of the history of philosophy of science. What I try to do is run a semi-coherent thread through the twentieth century, in such ways that a developmental narrative can be followed by those who have not lived within the confines of the discipline. Many scholars would have done things differently. *C'est la vie!*

To provide some structure for the exposition, I shall break this text into three important periods:

- 1918–50s: Logical Positivism to Logical Empiricism
- 1950s through 1970s: New Paradigms and Scientific Change
- Contemporary Foci: What's "hot" today

Logical Positivism to Logical Empiricism: 1918–55

As was noted above, the forming spirit of twentieth century philosophy of science were the grand syntheses and breakthroughs (or revolutions) in physics. Relativity and, later, quantum theory caused scientists and philosophers alike to reflect on the nature of the physical world, and especially on the nature of human knowledge of the physical world. In many ways, the project of this new philosophy of science was an epistemological one. *If* one took physics as the par-adigmatic science, and *if* science was the paradigmatic method by which one came to obtain reliable knowledge of the world, then the project for philosophy of science was to describe the structure of science such that its epistemological under pinnings were clear. The two antecedents, that physics was the paradigmatic science and that science was the best method for knowing the world, were taken to be obvious. Once the structure of science was made precise, one could then see how far these lessons from scientific epistemology could be applied to others areas of human endeavor.

Another important background tradition needs to be described. Propositional and predicate logic became the model for clear reasoning and explicit statement. First in the work of Frege (in the 1880s–90s), and later with Russell and White-head (in the 19-teens), logic came to be regarded as the way to understand and clarify the foundations of mathematics. It became the ideal language for model-ing any cognitive enterprise. Simultaneously, Hilbert re-introduced to the world the ideal of axiomatization. Again this was a clarifying move to ensure that there were no hidden assumptions, and everything in a system was made explicit. This logico-mathematical language became the preferred form, because of its precision, into which philosophy of science had to be cast.

The epistemological project of the positivists was to explicate how science was grounded in our observations and experiments. Simultaneously, the goal was to provide an alternative to the neo-Kantianism that was the contemporaneously concurrent form of philosophy. Taking from the tradition of British empiricism, empirical grounding, or being based on the facts, was seen as the major difference between science and the other theoretical and philosophical pretenders to knowl-edge. This insight led the positivists to attempt to formulate and solve the problem of the nature of meaning, or more specifically, empirical meaning. What was it, they asked, that made statements about the world meaningful? This attempt to explicate the theory of meaning had two important parts: First, claims about the world would have to be made clear, avoiding ambiguity and the other confusions inherent in natural language. To this end, the positivists tried to restrict them-selves to talking about the language of science as expressed in the sentences of sci-entific theories, and attempted to reformulate these sentences into the clear and unequivocal language of first-order predicate logic. Second, they tried to develop a criterion that would show how these sentences in a scientific theory related to the world, i.e. in their linguistic mode this became the problem of how theoreti-cal sentences related to observation sentences. For this one needed to develop a procedure for determining which sentences were true. This method came to be codified in the verification principle, which held that the meaning of an empirical sentence was given by the procedures that one would use to show whether the sentence was true or false. If there were no such procedures then the sentence was said to be empirically meaningless.

The class of empirically meaningless sentences were said to be non-cognitive, and they included the sentences comprising systems of metaphysics, ethical claims and, most importantly, those sentences that made up theories of the pseudo-sciences. This latter problem, distinguishing scientific sentences from those only purporting to be scientific, came to known (following Karl Popper's work) as the demarcation problem.

The verification principle was thought to be a way of making precise the empirical observational, or experimental component of science. Obviously, the positivists, following in the empiricist tradition, thought, the basis of science lay in observation and in experiment. These were the tests that made science reliable, the foundation that differentiated science from other types of knowledge claims.

So, formally, what was needed was a set of sentences that bridged the gap from scientific theory to scientific experiment and observation. These sentences that tied theory to the world were called bridge sentences or reduction sentences. The set of sentences that described the world to which theoretical sentences were reduced or related was called the observation language. Sentences in the observation language were taken to be easily verifiable or decidable as to their truth or falsity.

So that these bridge sentences might be made very explicit, theories were themselves idealized as sets of sentences that could be put into an axiomatic structure, in which all their logical relations and deductions from them could be made explicit. The most important sentences in a scientific theory were the laws of science. Laws came in two types: universal and statistical. Universal Laws were sentences of the theory that had unrestricted application in space and time (sometimes they were explicitly said to be causal, and, later, they were held to be able to support counterfactual claims.) Idealized universal laws had the logical form:

$$(x)(Fx \supset Gx)$$

Since such a form could be used to clearly establish their logical implications. Obviously, this was an idealized form, since most of the laws of interest were from physics and had a much more complex mathematical form. Statistical laws only made their conclusions more or less probable.

Scientific explanation was conceived as deducing a particular sentence (usually an observation or basic sentence) from a universal law (given some particular initial conditions about the state of the world at a time). The particular fact, expressed by the sentence, was said to be explained if it could be so deduced. This was called the deductive-nomological model of explanation. "Nomos" is the Greek word for law. If, a particular sentence was deduced before the fact was observed, it was a prediction, and then later if it was verified, the theory from which it was deduced was said to be confirmed. This was the hypothetico-deductive model because the law was considered an hypothesis to be tested by its deductive consequences.

The names of some of the major players in this period of philosophy of science were Moritz Schlick, Rudolf Carnap, Otto Neurath, Hans Reichenbach, and Carl Hempel. There were two main groups, one centered in Vienna (Schlick, Carnap and Neurath), called the Vienna Circle that was established late in the 1920s, and the other, coming a bit later, in Berlin (Reichenbach and Hempel). There was a important third group in Warsaw, doing mostly logic and consisting of Alfred Tarski, Stanislau Lesnewski and Tadeusz Kotarbinski.

This view of science, as an idealized logically precise language which could have all its major facets codified, never worked. Throughout the history of logical positivism there were debates and re-formulations among its practitioners about the idealized language of science, the relations of explanation and confirmation, the adequate formulation of the verification principle, the independent nature of observations, and the adequacy of the semantic truth predicate. The static, universalist nature of science that was idealized by positivism proved to be wrong. The attempt to fix procedures and claims in a logically simplified language proved to be impossible. The neat, clear attempts at explicating explanation, confirmation, theory and testability, all proved to have both internal difficulties with their logical structures and external problems in that they did not seem to fit science as it was actually practiced.

The positivists themselves were the first to see the problems with their program, and, as they attempted to work out the philosophical difficulties, the positions changed shifted into what became called logical empiricism. This happened in the mid-to late 1930s, the same time that many of the group left Germany and Austria because of World War II and the rise of Adolph Hitler. Reichenbach left Germany immediately after Hitler took power in 1933 and went first to Istanbul, Turkey, Richard von Mises went also. Reichenbach then in 1938 went to UCLA in the USA. Neurath and Popper both ended up in England. Carnap, from Prague, and Hempel, from Berlin, came to the USA.

Here is bit more sociology of the how philosophy of science developed. The first modern program in history and philosophy of science (HPS) was set up at University College, London. A. Wolf first offered a history of science course in collaboration with Sir William Bragg and others in 1919–20. Then a "Board of Studies in Principles, Methods and History of Science" was established in 1922, and an M.Sc. was first offered in 1924. Wolf was the first holder of the chair in "History and Method of Science." In 1946, the Chair became full time with the appointment of Herbert Dingle. The London School of Economics' Department evolved after the appointment of Karl Popper to the Readership in Logic and Scientific Method in 1945. The same Wolf who was associated with U.C., London also held the Chair in Logic and taught courses at LSE, prior to Popper. The University of Melbourne in 1946 began teaching courses in HPS.

Erkenntnis, the journal of the Vienna Circle, or rather the Max Plank Society, was first published in 1930. This followed on the first congress on the Epistemology of the Exact Sciences held in Prague in September of 1929. In 1934 the journal, *Philosophy of Science*, published its first issue. William M. Malisoff, a Russian biochemist, was its first editor. Malisoff died unexpectedly in 1947, and C. West Churchman became editor. The Philosophy of Science Association was in existence in 1934. In 1948 the PSA had 153 members, and Philipp Frank was its President. In the discipline of history of science, the American History of Science Society was founded in 1924. The HSS journal *Isis*, had been started earlier in 1912 by George Sarton when he was still in Belgium.

Logical empiricism never had the coherence as a school that logical positivism had. Various influences began to make themselves felt after the late 1930s. One most important conceptual addition came from American born pragmatism. Its specific influences can be seen clearly in the post-1940 work of Hempel, and even Carnap; also in the work of American born, Ernest Nagel and W. V. O Quine. But, until the late 1950s, philosophers of science, despite significant changes in the programs and allowable methods, philosophers of science were still trying to work out and change things to fit into the goals and aspirations left by the positivists. Moreover, it ought to be noted clearly that virtually all the major moves that were to come later and so change the character of philosophy of science were first initiated by the original positivists themselves. This continuity was not noted by those who became famous during the next decades; they saw themselves as revolutionary and stridently anti-positivistic. By the late 1950s, philosophy of science included ever-increasing complex models, much looser claims, many new philosophical methods and increasingly vague philosophical goals.

New Paradigms and Scientific Change:
Late 1950s through the 1970s

While the logical positivists, and later the logical empiricists, were attempting to explicate and clarify the structure of science, another group of scholars had begun to transform an old activity into the modern academic discipline of history of science. The goal of much history of science was to examine historically significant intellectual episodes in science and to articulate these analytically in a way that exhibited the character of science at that particular historical moment and also showed that moment fit into the development and progress of science. Questions for which answers were sought were, e.g. about the nature of Galileo's physics, and what made it both continuous with and yet different from his medieval predecessors. Was Galileo the last of the Medievals or the first of the moderns? What was the nature of Galileo's methodology, and how did he frame explanations? Was Galileo's use of mathematics in physics really revolutionary? Did Galileo really use experiments in some modern sense? Of course, it was not just Galileo who was of interest, historians of science studied all the heroes of modern science, and reached backwards into the Greek, Roman and Medieval periods. The attempt was to describe the actual practice of science of these thinkers and to discern what was peculiar to these historical periods. While history of science courses had been taught in a number of places, by the mid-1960s history of science was an established enterprise with programs and departments in Universities that trained graduate students in the discipline. Actually, the University of Wisconsin started its department in 1942, but World War II kept it from being staffed until 1947. Harvard offered degrees in History of Science, but their department was started only in 1966.

In the late 1950s, philosophers too began to pay more attention to actual episodes in science, and began to use actual historical and contemporary case studies as data for their philosophizing. Often, they used these cases to point to flaws in the idealized positivistic models. These models, they said, did not capture the real nature of science, in its ever-changing complexity. The observation language, they argued, could not be meaningfully independent of the theoretical language since the terms of the observation language were taken from the scientific theory they were used to test. All observation was theory-laden. Yet, again, trying to model all scientific theories as axiomatic systems was not a worthwhile goal. Obviously, scientific theories, even in physics, did their job of explaining long before these axiomatizations existed. In fact, classical mechanics was not axiomatized until 1949, but surely it was a viable theory for centuries before that. Further, it was not clear that explanation relied on deduction, or even on statistical inductive inferences. The various attempts to formulate the deductive-nomological model in terms of necessary and sufficient conditions failed not only because counter-examples were found, but also because explanation seemed to be more complex phenomena when one looked at examples from actual sciences. Even the principle of verification itself failed to find a precise, or even minimally adequate, formulation.

All the major theses of positivism came under critical attack. But the story was always the same – science was much more complex than the sketches drawn by the positivists, and so the concepts of science – explanation, confirmation, discovery – were equally complex and needed to be rethought in ways that did justice to real science, both historical and contemporary. Philosophers of science began to borrow much from, or to practice themselves, the history of science in order to gain an understanding of science and to try to show the different forms of explanation that occurred in different time periods and in different disciplines.

Debates began to spring up about the theory ladeness of observation, about the continuity of scientific change, about shifts in meaning of key scientific concepts, and about the changing nature of scientific method. These were both fed by and fed into philosophically new areas of interest, areas that had existed before but which had been little attended to by philosophers. The social sciences, especially sociology, became of considerable interest, as did evolutionary biology. These fields provided not only new sciences to study and to be contrasted with physics, but also new models and methods which were then borrowed to study science itself.

By the early 1960s, as the result of the work of Thomas Kuhn – and concurrently Norwood Russell Hanson and Paul Feyerabend – the big philosophical question had become: Were there revolutions in science? The problem of scientific change, as it was called, dealt with issues of continuity and change.

Kuhn had argued that science in one period is characterized by a set of ideas and practices that constitute a paradigm, and when problems or anomalies begin to accumulate in a given paradigm, there often was introduced a new paradigm which, in fact and in logic, repudiated the old and supplanted it. (This model was not unlike Gaston Bachelard's view about crises in science leading to *rupture*.) This concept of a revolutionary paradigm shift implied that scientific change was discontinuous, and that the very meaning of the same terms, e.g. "mass", changed from their use in one paradigm (Newtonian) to their use in the new paradigm (Einsteinian). This was called meaning variance. One methodological implication for philosophers of science, clearly, was that to study science, one had to confine oneself to a historically dominant paradigm and one could not look for more general, trans-paradigmatic models that covered all science, except maybe for the process of paradigm change itself.

Many philosophers made a job of criticizing Kuhn's paradigms and his program. They began to search for alternative, general models of scientific change that were more accurate in describing episodes in science, more sensitive in analyzing the parts of science that actually underwent change, and that avoided the ambiguities and unclarities of Kuhn. So, talk of paradigms gave way to research programmes (Lakatos) and then to research traditions (Laudan). Another group of philosophers began to look at explanations in different periods and disciplines to find out if there could be general principles that could be said to apply to all explanations, and thus undercut the meaning variance thesis. Yet, other thinkers, including some philosophers, began to take Kuhn's claims about practices seriously, argued, as had some historians of science earlier, that science could not be explained solely in terms of its concepts and internal structure. One needed, it was held, to understand the social and political settings in which such concepts were developed to understand how they became acceptable and why they were thought to be explanatory.

It should be noted also that many of the more purely philosophical moves (including those of Hanson, Kuhn and Feyerabend) had been influenced by the new dominance of the more central philosophical practices of ordinary language philosophy, inspired to a large extent by the work of the later Wittgenstein. This was still philosophy which dealt with analyzing language, but the language was no longer just the formal a language of logic, but the various language games the comprised the various disciplines of human endeavor. New directions in linguistics, spurred on by Chomsky and his followers, had also changed the way people, including philosophers, looked the problem of syntax, semantics, and meaning. Even basic epistemology itself began to be questioned. W. V. O. Quine (1969) announced to world that philosophy of science was philosophy enough, and epistemology had to be naturalized and was part of natural science.

By the mid 1960s, logical positivism and logical empiricism was quite out of fashion in Anglo-American philosophy. At this time, philosophical analysis was the key mode of operation, and the logicism that had provided the guiding model for the earlier philosophical work, was superseded by the study of real scientific language and by the complexities uncovered in studying the history of science. During this period Indiana University founded its Department of History and Philosophy of Science (1960), which was followed a decade later by the institution of HPS at the University of Pittsburgh (1971). Adolph Grünbaum was president of the Philosophy of Science Association in 1968. (The preceding President was Ernest Nagel.) The PSA seems to have waned somewhat during the post war years, but Grunbaum began the tradition of biennial meetings that continues to this day.

The result for philosophy of science was invigorating, exciting, and devastating. General characterizations of scientific change proved to be just as intractable as earlier general models of scientific explanation. The laudable tendency to explore the nature of sciences other than physics and to examine in detail cases from the history of many sciences left philosophers without a "paradigm." There was little consensus about the nature of explanation, confirmation, theory testing or, even, scientific change. Yet science itself, more than ever, was recognized by the populace at large, as a (if not the) major force in human life, and philosophy of science had become a discipline to stand along side of ethics, epistemology and metaphysics. But there was intellectual disarray over its nature in the philosophical community at large. In fact, some philosophers, following Paul Feyerabend took the intellectual confusion as evidence that science had no identifiable structure, and proffered the view that in science, as in art, "anything goes." All evidence and proof is just rhetorical, and those with the best rhetoric, or the most power (Foucault), become the winners, i.e. their theories became the ones accepted. Luckily, this epistemological relativism was not followed by many philosophers, though, as we shall see below in some contemporary communities this idea still flourishes.

A consensus did emerge among philosophers of science. It was not a consensus that dealt with the concepts of science, but rather a consensus about the "new" way in which philosophy of science must be done. Philosophers of science could no longer get along without knowing science and/or its history in considerable depth. They, hereafter, would have to work within science as actually practiced, and be able to discourse with practicing scientists about what was going on. This was a major shift in the nature of philosophy. It is true that most of the early positivists were trained in science, usually physics. But this scientific training had led them to try to make philosophy scientific after the image of their own philosophical–logical model of science. In contrast, from the 1950s on, more and more philosophers had been trained by the Oxbridge inspired analytic philosophers, who adhered to Wittgenstein's dictum that philosophy was a *sui generis* enterprise and so had nothing to do with, and nothing to learn from, science. It is no wonder that students of philosophy so trained found it hard to figure out what philosophers of science should be doing, and as a result turned either to science itself or to various forms of sociology of science, which was taken to be legitimate because it was a sub-discipline of an actual science (sociology). Ironically, despite this confusion about goals, there were more philosophers of science than ever before.

Contemporary Foci and Future Directions

The turn to science itself meant that philosophers not only had to learn science at a fairly high level, but actually had to be capable of thinking about (at least some) science in all its intricate detail. In some cases philosophers actually practiced science, usually theoretical or mathematical. This emphasis on the details of science led various practitioners into doing the philosophy of the special sciences. Currently, there are philosophers of space-time, who variously specialize in special or general relativity theory, and philosophers of quantum theory and quantum electro-dynamics. There do not seem to be any philosophers of plasma physics. Fairly recently, philosophy of chemistry has become somewhat of a "hot" research area. Philosophers of biology continue to work on problems in evolutionary theory, and finally some study molecular biology, which is the area in which almost all biologists work. Work on genetics has been around for some time, but usually connected to evolutionary biology. Work on biological development is just start-ing and is seen to be increasingly important.

With the explosion of health care, philosophy of medicine also became a newly emergent and important field of research. Philosophy of the social sciences still continues to be worked upon, but sociology as the paradigmatic social science has been replaced by anthropology, except for those people who work in science studies which still treats sociology with some respect. Philosophy of economics, especially game theoretic modeling, is a somewhat popular field today. This is inter-esting since the game theory model had been started in the 1940s (von Neumann and Morgenstern), and then mostly dropped in 1960s, only to be revived by biol-ogists using game theory to model evolution and by experimental economists trying to find an empirical model for studying economic behavior; these then influ-enced philosophers of economics who revived game theory as tool for economic analysis.

One of the most innovative and biggest changes has come in the area that used to be known as philosophy of psychology. Philosophy of psychology used to be tied to philosophical psychology, to philosophy of mind, and to behaviorism and cognitive psychology, especially to questions about the nature of the mental. In a way it still is, but the "cognitive revolution" hit philosophy quite hard. Cognitive studies now includes many of those working in experimental psychology, neuro-science, linguistics, artificial intelligence, and philosophers. There are many aspects to this re-defined field, including work on problems of representation, explana-tory reduction (usually to neuroscience), and even confirmation. Confirmation theory has used techniques from artificial intelligence to re-establish a modern form of older confirmation functions as developed originally by Carl Hempel. Cognitive problem solving has even been used by some to model the nature of science itself. A new direction to be explored are the relations of neuroscience to traditional philosophical problems, such a representation and knowledge.

Historically, it is of note that cognitive science began to emerge in the mid-1950s, close to the time that the shift away from logical positivism began. Many of the intellectual forces that caused the philosophical change were also the causes of the emerging new cognitive paradigm, but, even more importantly, one needs to note the impact of the computer and its related ways of acting and thinking. The computer was not only a tool for calculation, reasoning and processing, but also became also a model for thinking about human beings, and, even, for think-ing about science.

One interesting implication of this work in the specialized sciences is that many philosophers have clearly rejected any form of a science/philosophy dichotomy, and find it quite congenial to conceive of themselves as, at least in part of their work, "theoretical" scientists. Their goal is to actually make clarifying and, sometimes, substantive changes in the theories and practices of the sciences they study.

A very different current trend is exhibited by those philosophers of science who have become part of the science studies movement, which is dominated by historians and sociologists. This movement focuses on the social dimensions of science (as opposed to the "outmoded" intellectual aspects.) In one sense the social study of science grew out of the dispute between internalist and externalist historians of science, which was resolved in favor of the externalists when the discipline of history itself shifted to quantitative social history and away from intellectual history. From another direction the work of the epistemological relativists, whom I referred to earlier, fits nicely with the relativism thought to characterize historical periods and with cultural (and ethical) relativism that is rampant in much of cultural anthropology. Essentially the view here is that science is a human social activity not unlike any other and so is subject to historical and cultural contingencies. In order to study such activities we must look at the socio-cultural milieu in which scientists are raised, trained, and in which their work occurs. So, for example, we should study the laboratories in which scientists work and describe how these function to self-validate knowledge claims issued from the laboratory. Moreover, we should study the conventions of discourse that comprise the "rules" by which scientists' influence and exert power over one another. For example, in the seventeenth century there were codes of conduct that English gentleman "had" to adhere to, and these provided (somehow) the structure of the debates and experimental practices for the members of the Royal Society. A concomitant belief held by most of the science studies group, though it is not necessarily implied by their position, is the relativism of different or competing claims. That is, it is a historical, cultural and/or epistemic peculiarity that a given group of scientists holds the views that they do. From this, it is presumed to follow that no one view is any better than any other. You are what your time and culture have made you, and that's an end to it.

Such claims for relativism often lead people to worry about values and their status, for cultural relativism is closely tied with ethical relativism. But questions about the relations between values and science also arose from even more pressing sources. Perhaps the most important and influential questions about values arose from medicine. The practical problems of medical ethics began to make themselves felt due to changes in the practice of medicine and in medical technology. All of a sudden, there were urgent questions concerning life and death, physician-patient relations, and informed consent that had to be answered in pragmatically expeditious ways. This coincided with, and was in part responsible for, a shift in philosophical ethics away from the theoretical, from meta-ethics, towards the practical. Philosophers, of ethics and of science, became involved in consulting about the day to day decisions in hospitals and about the re-writing of health care policies. Philosophers of science are especially useful here because they actually know some of the science that is involved in making informed decisions, and they have often studied various aspects of decision making and the use of evidence.

This practical side of ethics in the sciences has other dimensions too. Codes of ethics for the various professions, e.g. engineers, have become "hot" topics for philosophical research. One of the more interesting and important new fields that philosophers of science dealing with values are involved in have to do with issues concerning how science is used to base regulatory decisions, e.g. concerning lead or dioxins or global warming. Also, there is work being done of the values that are implicitly or explicitly involved in the actual doing of scientific research. For example, what values are assumed in choosing a certain type of experimental paradigm, or, more generally, what values are assumed in giving more money to AIDS research rather than malaria (which is back with us in a big way.) The feminist movement of the late 1960s, also brought many value questions to the fore, and some excellent work has been done on how gender assumptions have influenced scientific practice.

This practical side of the "new" philosophy of science, I believe, derives from the same need for relevance that pushed other thinkers into dealing with the special sciences. There is an, often unacknowledged, awareness that philosophy must become important in ways that go beyond the hallowed halls of academe. The logical positivists, though some of them had studied physics, had little influence on the practice of physics, though their criteria for an ideal science and their models for explanations did have substantial influence on the social sciences as they tried to model themselves on physics, i.e. on "hard" science. The analytic philosophers of the mid-1950s onwards had little influence outside of the Universities in which they taught. They were content to defend their professional turf as being a thing unto itself and in some ways were quite proud to be "irrelevant" to the concerns of ordinary life, despite the ironic emphasis on ordinary language. By the 1980s, this intellectual isolationism had begun to break down, philosophers, and especially philosophers of science, had to get involved in the real world, the world of science.

I end this little essay by noting that the old questions and topics that had been raised by the logical positivists, and even in previous 2000 years, have not disappeared. Philosophers of science still puzzle over what makes a good explanation, what kind of evidence provides what kind of confirmation for theory, and what is the difference between science and pseudo-science. These are the perennial questions of philosophy of science. Today, we still try to answer them in specific ways that will have effects on science and the larger world. Philosophers of science have been instrumental in showing the non-scientific status of creationism and some versions of sociobiology and, now, evolutionary psychology. They have discussed fruitfully the role of scientific evidence in making decisions about nuclear energy plants or about levels of toxicity in our environment. They have asked hard questions about how to discover mechanisms such that finding them allows us to understand how systems of molecular biology or neuroscience work. And they have continued to elucidate and elaborate the unclarities and confusions in the special sciences.

Of course, there is much left to do. There are always more puzzles than people, more problems than solutions. The twentieth century saw many changes in what are taken to be the important puzzles and problems, but even more importantly, these same years have seen changes in how people need to be trained to approach problems and in what solutions to problems must look like. Maybe this past century has only taught us that there are no simple answers to truly complex questions. Yet, with this realization comes the awareness that there must be pragmatic answers provided in a timely and efficacious manner. Decisions must be made, and, hopefully, philosophy of science can help us to see how they may be made in better ways.

Note

*   Thanks to Adolph Grünbaum, Noretta Koertge, David Lindberg, Nick Maxwell, Wesley Salmon and John Worrall for information regarding the history of philosophy of science and founding of institutions and departments. Many thanks to Merrilee Salmon, Paolo Parrini, Ted McGuire and Aristides Baltas for their help and comments on an earlier draft of this essay. An even earlier draft was given as a lecture at The Catholic University of America, and I thank those present who gave me good feedback, especially Bill Wallace.

Appendix: Selected Relevant Philosophical and Scientific Publications (1895–1969), their dates, and a few events

1895	Josef Breuer and Sigmund Freud, *Studies in Hysteria*
1897	Leon Brunschvig, *La Modalité du Judgment*
1899	David Hilbert, *Die Grundlagen der Geometrie*
	Max Plank derives black body law
	Sigmund Freud, *The Interpretation of Dreams*
1901	Ernst Mach, *Die Mechanik in ihrer Entwicklung*, 4th edn.
1902	Lorentz proves Maxwell's equations were invariant under transformations
	Henri Poincaré, *La Science et l'Hypothèse*
1903	Bertrand Russell, *Principles of Mathematics*
1905	Ernst Mach, *Erkenntnis und Irrtum*,
	Bertrand Russell, "On Denoting" *Mind*
	Albert Einstein, "Zur Elektrodynamik bewegter Koeper" *Annalen der Physik*

	General strike and revolution in Russia
	Sigmund Freud, "Three essays on the Theory of Sexuality"
1906	Pierre Duhem, *La Theorie Physique. Son Objet. Sa Structure*
	Albert Einstein and Paul Ehrenfest, *hv* indivisible unit of energy
1907	Hans Hahn, Otto Neurath and Philipp Frank in Vienna
1908	Ernst Zermelo, "Untersuchungen uber die Grundlagen der Mengenlehre I" *Mathematische Annalen*
	Emile Meyerson, *Identite et Realite*
1910–13	Russell and A. N. Whitehead, *Principia Mathematica*
1911	Arthur Sommerfeld introduces phase-integral form of quantum law
	Einstein, "Uber den Einfluss der Schwerkraft auf die Ausbreitung des Lichtes" *Annalen der Physik*
	Solvay Congress, Brussels
1913	Edmund Husserl, *Ideen zu Einer reinen Phanomenologie und Phanomenologischen Philosophie, vol. 1*
	J. B. Watson, "Psychology as the Behaviorist sees it" *Psych. Rev.*
	Niels Bohr, publishes on the atom (*Phil. Mag.*)
1914	Russell, *Our Knowledge of the External World as a Field for Scientific Method in Philosophy*
	WWI (till 1918): Franz Ferdinand assassinated
	Easter Rising in Ireland
	Russian Revolution
1915	Sommerfeld explains fine structure of spectral lines
	Max Plank estimates value for *h* (*Phys. Rev.*)
1916	Einstein "Die Grundlage der allgemeinen Relativitatstheorie" *Annalen der Physik*
1917	Robert Millikan, *The Electron*
1918–19	Bertrand Russell, "Philosophy of Logical Atomism", *Monist* Moritz Schlick, *Allgemeine Erkenntnislehre*
	Arthur Eddington observes eclipse confirming general relativity
	Niels Bohr's "Principle of Correspondence"
1920	N. R. Campbell, *Physics, the Elements*
1921	Ludwig Wittgenstein, *Tractatus Logico-Philosophicus* [*Logische-Philosophische Abhandlung*] English version 1922
	J. M. Keynes, *A Treatise on Probability*
1922	Moritz Schlick to Vienna as professor of inductive sciences
	Leon Brunschvig, *L'Expérience Humaine e la Causalité Physique*
1923	David Hilbert, "Die Logische Grundlagen der Mathematik" *Mathematische Annalen*
	Helene Metzger. Les Doctrines Chimiques Début du XVIIème à la Fin du XVIIIème Siècle
1925	Erwin Schrödinger develops wave mechanics
1926	Rudolf Carnap to Vienna as instructor in philosophy
	Niels Bohr shows equivalence of matrix and wave mechanics

1927	Werner Heisenberg formulates indeterminacy principle
1928	Verein Ernst Mach (Ernst Mach society) founded
	Rudolf Carnap, *Der Logische Aufbau der Welt*
	David Hilbert, *Grundzuge der Theoretische Logik* (3rd edn. 1949 by Hilbert and Ackermann)
1927	P. W. Bridgman, *The Logic of Modern Physics*
	Charles Lindberg makes first solo transatlantic flight
1929	Carnap, Hahn and Neurath, *Wissenschaftliche Weltauffassung, Der Wiener Kris*
	Ernst Mach Society Congress held in Prague
	Wall Street Crash
1930	*Erkenntnis* founded (till 1940)
	Gödel's Completeness Theorem
1931	Carnap to Prague, Feigl to Iowa
	Gödel's Incompleteness Theorem
1932	E. A. Burtt, *The Metaphysical Foundations of Modern Science* (revised edn.)
1933	Hitler appointed Chancellor
1934	Carnap, *Logische Syntax der Sprache*
	M. R. Cohen and E. Nagel, *Introduction to Logic and Scientific Method*
	Gaston Bachelard, *Le Nouvel Esprit Scientifique*
	Philosophy of Science first published
	Hitler becomes *Führer* of Germany (till 1945)
1935	Karl Popper, *Logik der Forschung* (English, 1959)
	Kurt Koffka, *Principles of Gestalt Psychology*
1936	Carnap appointed at Chicago
	Alfred Tarski "Der Wahrheitsbegriff in den Formalisierten Sprachen" *Studia Philosophica*
	Carnap, "Testability and Meaning" *Philosophy of Science* (and 1937)
	A. J. Ayer, *Language, Truth and Logic*
	Spanish Civil War (to 1939)
1938	Ernst Mach Society formally dissolved (publications of the society forbidden in Germany)
	Waismann and Neurath to England
	Zilsel and Kaufmann to USA (Menger and Gödel already there too)
	Erkenntnis moved to The Hague, and renamed *Journal of Unified Science*
	Claude Shannon, "A Symbolic Analysis of Relay and Switching Circuits" *Trans. of Am. Inst. of Electrical Engineers*
	Alexandre Koyre, *Etudes Galileennes*
	B. F. Skinner, *The Behavior of Organisms*
	Hans Reichenbach, *Experience and Prediction*
	WWII (to 1945)

1940	*Journal of Unified Science* discontinued
	Carl G. Hempel "Studies in the Logic of Confirmation I & II", *Mind*
	Clark L. Hull, *The Principles of Behavior*
1947	Carnap, *Meaning and Necessity*
	J. von Neumann and O. Morgenstern, *Theory of Games and Economic Behavior*
1948	C. G. Hempel and Paul Oppenheim, "Studies in the Logic of Explanation", *Philosophy of Science*
	J. H. Woodger, *Biological Principles*
	Norbert Wiener, *Cybernetics*
1949	H. Feigl and W. Sellars (eds.), *Readings in Philosophical Analysis*
	Herbert Butterfield, *The Origins of Modern Science, 1300–1800*
	Anneliese Maier, *Die Vorlaufer Galileis im 14 Jahrhundert*
	Hans Reichenbach, *The Theory of Probability*
1951	Reichenbach, *The Rise of Scientific Philosophy*
1952	Carnap, *Logical Foundations of Probability*
	Georges Canguilhem, *La Connaissance de la Vie*
1953	Wittgenstein, *Philosophical Investigations (Philosophische Untersuchungen)*
	H. Feigl and M. Brodbeck (eds.), *Readings in Philosophy of Science*
	W. V. O. Quine, *From a Logical Point of View*
	Stephen Toulmin, *Philosophy of Science*
	R. B. Braithwaite, *Scientific Explanation*
1954	Gustav Bergmann, *The Metaphysics of Logical Positivism*
	A. R. Hall, *The Scientific Revolution, 1500–1800*
	Nelson Goodman, *Fact, Fiction, and Forecast*
	Leonard J. Savage, *The Foundations of Statistics*
1955	Canguilhem succeeds Gaston Bachelard as Professor of Philosophy at the Sorbonne and Directeur of Institut d'Histoire des Sciences et des Techniques
1956	Ernest Nagel, *Logic without Metaphysics*
	J. O. Urmson, *Philosophical Analysis*
	Herbert Feigl and Michael Scriven, *Minnesota Studies in the Philosophy of Science, Vol. 1*
1958	Norwood Russell Hanson, *Patterns of Discovery*
	Marshall Clagett, *The Science of Mechanics in the Middle Ages*
	E. H. Gombrich, *Art and Illusion: A Study in the Psychology of Pictorial Representation*
	M. Clagett (ed.), *Critical Problems in the History of Science*
	Paul Feyerabend, "An Attempt at a Realistic Interpretation of Experience" *Proc. Aristotelian Society*
1959	Morton Beckner, *The Biological Way of Thought*

1960   W. V. O. Quine, *Word and Object*
1961   Ernest Nagel, *The Structure of Science*
1962   Thomas Kuhn, *The Structure of Scientific Revolutions*
        Mary Hesse, *Models and Analogies in Science*
        Israel Scheffler, *The Anatomy of Scientific Inquiry*
        Robert G. Colodny, *Frontiers of Science and Philosophy* (first volume
            of the Pittsburgh series)
1965   Hempel, *Aspects of Scientific Explanation*
        Paul Feyerabend, "Problems of Empiricism" in R. G. Colodny (ed.),
            Beyond the Edge of Certainty
        Michel Foucault, *Les Mots et les Choses*
1968   Imre Lakatos, "Criticism and the Methodology of Scientific
            Research Programmes"
        W. V. O. Quine, "Epistemology Naturalized" lecture delivered
            (published 1969)
1969   Foucault, *L'Archeolgie du Savoir*

Further reading

Contemporary presentations of the basic issues in philosophy of science
Merrilee Salmon, et al., *Philosophy of Science*, (by the Department of History & Philosophy
    of Science, University of Pittsburgh), Prentice-Hall, 1991
A collection of readings which cover the field of philosophy of science
Baruch Brody and Richard Grandy (eds.), *Readings in the Philosophy of Science*, 2nd edn,
    Prentice Hall, 1989
Historical overviews of the history of positivism
J. Alberto Coffa, *The Semantic Tradition from Kant to Carnap*, Cambridge: CUP, 1991
Michael Friedman, *Reconsidering Logical Positivism*, Cambridge CUP, 1999
Frederick Suppe, Critical Introduction, to *The Structure of Scientific Theories*, 2nd edn,
    Urbane, Ill.: University of Illinois Press, 1977
*A systematic treatment of the main parts of the logical positivist/empiricist program: Quite
    difficult in parts*
Israel Scheffler, *The Anatomy of Inquiry*, New York: Borzoi Books, 1964
A review of the critics of positivist/empricist program.
Israel Scheffler, *Science and Subjectivity*, Indianapolis. Bobbs Merrill, 1967

Chapter 2

Philosophy of Science: Classic Debates, Standard Problems, Future Prospects

John Worrall

The Background

Immanuel Kant's celebrated investigation of human knowledge started from the assumption that we have achieved rock-solid, indubitable knowledge – in geometry through Euclid and in physics through Newton – and from the question of how this was possible (especially in view of Hume's demonstration of the invalidity of inductive inference). Contemporary philosophy of science is a rich and multi-faceted enterprise and so any one way of viewing it will inevitably leave out much of importance and interest. Nonetheless, many of the classic debates and areas of current concern can be introduced by investigating how Kant's questions require modification in the light of the development of science since his time and by investigating the attempts made to answer those modified questions.

Two radical – apparently "revolutionary" – changes of fundamental theory occurred in the early twentieth century, those associated with the theory of relativity and with quantum theory. The former had the more direct effect on Kant's presuppositions and questions. If, at any rate, we think of geometry as a synthetic description of the fundamental structure of space, then Einstein's revolution involved the rejection of Euclidean geometry in favor of the Riemannian version of non-Euclidean geometry. Instead, for example, of two straight lines that are parallel being extendable indefinitely without intersecting, the new geometry states that any two straight lines (geodesics) eventually intersect. Far from being certainly true, Euclidean geometry (at least as a "physical geometry") is – it seems – not even true. Similarly, although Newton's theory (of mechanics plus universal gravitation) continues to be empirically adequate over a wide range of phenomena (basically motions involving velocities small compared to that of light), its fundamental claims about the structure of the universe – that space is infinite, that gravitation acts at-a-distance, that time is absolute so that two events simultaneous in one reference frame are simultaneous in all – are entirely rejected by relativity theory. Again, far from being certainly true, Newtonian physics is, it seems, not even true. Indeed, given that relativity theory denies action at a distance, suggests that space is finite (though unbounded), and entails that two events that are simultaneous in one frame of reference will *not* be simultaneous in another frame that moves relatively to the first, it is difficult for many to see intuitively how Newton's theory could count as even "close to the truth" (supposing for sake of argument that Einstein's theory were the truth).

These developments transform Kant's question into a dilemma. Is there some way of interpreting (or reinterpreting?) scientific theories so that the apparently radical nature of the revolutionary shift from classical to relativistic physics becomes just that – merely apparent? If so, then it might still be possible to argue that science *when properly understood*, delivers, if not outright certainty, then some close approximation to it. If not, if we simply have to accept that scientific development has involved revolutionary change at the most fundamental theoretical level, then we presumably cannot reasonably rule out the possibility of still further revolutions in the light of which our current theories will seem just as false as Newtonian theory now seems to us. And in that case, the question becomes what makes science special at all from the epistemic point of view?

Why is Science Special from the Epistemic Point of View?

Let's begin on the second horn of this dilemma – conceding for the sake of argument that the apparently revolutionary shifts are real. In that case, there is no prospect of continuing to hold that scientific theories are *proved* or established by unquestioned empirical data. What is it, then, that makes science and the methods of science special from an epistemic point of view? (There are of course some thinkers – mostly sociologists of science – who would reject this question, and insist that the conclusion we ought to draw from the existence of scientific revolutions is that science is just one human system of beliefs amongst others (such as the Azande system of magic) with no justified claim to any special epistemic status. But the staggering predictive success of our theories in "mature" science is so strongly at odds with this view that it is difficult to take seriously.)

Demarcation and falsifiability

The question of what makes science special is often called "the demarcation problem." One celebrated answer – directly motivated by the Einsteinian revolution – is Karl Popper's falsifiability criterion: science is special because, even though its theories are not provable from evidential statements, they are *refutable* by such statements. The Einsteinian revolution is – Popper (1959) suggested – a direct vindication of this view (and indeed that revolution was a major motivation for the view): the "revolution" was a great step forward because it involved the refutation of a highly falsifiable, but hitherto unfalsified theory (Newton's), and its replacement by a still more falsifiable – but not yet falsified – theory (Einstein's). In contrast, non-scientific claims – metaphysical claims, such as that God exists, or claims that Popper categorized as *pseudo*scientific, such as the claims of astrology or of Freudian psychoanalysis, are (allegedly) entirely unfalsifiable: no possible evidential statement could contradict any such claim and hence establish its falsity. Science is special because at least we can know when we are wrong.

It is now (almost) universally accepted that Popper's account fails. One issue – raised right at the beginning by Reichenbach, for example – was whether the problem of induction, and, in particular, the so-called "pragmatic problem," can ever be solved in a purely falsificationist way. It seems positively irrational not to base our technological interventions – in building say bridges or aeroplanes – on the best available scientific theories. But would this judgment be underwritten simply by the report that those best available theories are so far unrefuted – that is, unrefuted in tests already performed? We surely also need some sort of reason to think that the past test-record of those theories reflects their overall truth-likeness and therefore at least their likely performance in *future* tests. (It is, after all, perfectly possible given simply deductive considerations that theories that have performed relatively badly in the past will, in the future, perform better than ones that have performed relatively well so far.) It seems that we need, then, some sort of link between past performance in tests and overall truth (or at least overall empirical adequacy). But this is exactly the sort of inductive assumption that was anathema to Popper.

Difficulties with falsifiability – the Duhem problem

Moreover, fundamental issues also arise about the assumed falsifiability of scientific theories. The most direct problem here had already in fact been explained in impressive detail some thirty years before Popper's work by Pierre Duhem (1906). Scientists often talk about testing scientific theories, such as Newton's theory (of mechanics and gravitation) by comparing that theory's predictions – about, say, planetary positions – with the "data." But Duhem pointed out that if the deductive structure of any such test is analysed carefully then further premises – often called "auxiliary assumptions" – always turn out to be necessary if the deduction of the observation statements at issue is really to be valid. Nothing that we are likely to characterize as a "single theory" in science – Newton's theory or Maxwell's theory of electromagnetism or quantum theory or whatever – has any empirical consequence when considered "in isolation," further auxiliary assumptions are always needed. For example, no consequences about planetary positions at some given time t follow from Newton's theory (of mechanics plus universal gravitation) and nor do they follow from Newton's theory plus "initial conditions" about the positions of those planets at some earlier time t'. What is needed, in addition, is a whole set of other assumptions that are clearly themselves theoretical rather than in any sense "directly given" by observation – this set includes assumptions, for example, about the mass of the planet concerned and the number and masses of the other bodies in the solar system, not to mention assumptions about how light travels between the planet concerned and our telescope. (So, in particular, a – clearly theoretical – assumption is needed about the extent to which light is refracted in passing from "empty space" into the earth's atmosphere.)

This apparently minor logical point has major consequences. Suppose we have some observation sentence O and are happy to say that we can decide the truth value of O on the basis of observation or experiment. If contrary to Duhem, we could invariably take any "single" scientific theory T and deduce a range of such results O from it, then, just as Popper emphasized, if some such O were established as false on the basis of observation, then it would follow that T must be false as well. (The so-called "principle of retransmission of falsity" says that if some premise, in this case the theory T, entails deductively some conclusion, in this case the observation sentence O, then if that conclusion is false, so also must be the premise.) In fact, however, as Duhem's analysis showed, the deductive structure of any real test of any real scientific theory always involves auxiliary assumptions – often quite a large set of them. But if we can infer O only from a conjunction of sentences $T\&A_1\&\ldots\&A_n$, then should we decide, on the basis of observation or experiment, that O is false, all that we can infer is that *at least one of* the set of theoretical claims $T, A_1, \ldots, A_n$ is also false. (The principle of retransmission of falsity when applied to deductive inferences with more than one premise does not, of course, say that if the validly deduced conclusion is false, then so are all the premises, but only that *not all* the premises can be true – at least one must be false.) In particular, we cannot infer that it is T itself that is false.

Duhem's analysis does *not* show that observation results never supply good grounds for holding that some "central" theory T is false; but it does show that these are never conclusive and that something more than falsification must be involved. There might, for example, be independent grounds for thinking that the auxiliaries $A_1, \ldots, A_n$ are more likely to be true than is T. If so, then the fact that the falsity of O shows that not all of $T, A_1, \ldots, A_n$ can be true would supply good grounds for rejecting T. Or, and this is what generally in fact happens in cases of scientific theory-change, while a theoretical system built around theory T can be made to yield O only by adjusting some of the auxiliaries A_i exactly with the requirement in mind that O be entailed, an alternative system built around some alternative theory T' involving non *ad hoc* auxiliaries is independently empirically confirmed (that is turns out to predict some further empirical result O' which is then confirmed). Either suggestion, however, brings in ideas of confirmation that are foreign to Popper's scheme.

Confirmation – the attempt at an "objective" account

Why not then go straight for confirmation as the solution to the problem of what makes science special? The Einsteinian revolution was a constructive proof of the fact – which in any event ought in retrospect to have been obvious – that we can never conclusively *prove* general explanatory scientific theories on the basis of observation or experiment; Duhem's analysis showed that we can never conclusively *falsify* them either. But perhaps we can nonetheless *confirm* scientific theories on the basis of empirical results. Perhaps what distinguishes a better scientific theory from a good one is that the former is better confirmed by the evidence; perhaps what explains "revolutionary" shifts in scientific theory – for example, that from Newton to Einstein – is exactly that, given the evidence that had accumulated, Einstein's theory was the better confirmed theory; and finally perhaps what distinguishes scientific theories from non-scientific ones (whether metaphysical or pseudoscientific) is that the latter are not even capable of empirical confirmation. The claim that "God exists" fails to be scientific, not because it cannot be proved from evidence, not because it can never be falsified by evidence, but because it can never be confirmed (and therefore can never be disconfirmed either) by any possible – intersubjectively agreed – evidence.

As a general framework suggestion, this answer still seems to me viable (indeed perhaps when considered in a very general way, it is the only viable answer). The problem has been that of giving a more precise account of the notion of "confirmation" – a more precise account that delivers all the above judgments and that seems both coherent and philosophically defensible.

A number of "non-standard" approaches have been tried (perhaps most notably Clark Glymour's (1980, 1987) "bootstrapping" approach), which have run into their own difficulties. But most attempts to put flesh onto the skeleton of the confirmation approach have, unsurprisingly, involved the notion of probability. What confirmation delivers, it is suggested, is greater probability of being true: the change from Newton to Einstein was the change from one reasonably probable theory (of course probable in the light of the evidence) to another that is still more probable in the light of the evidence; theory-change in science can be explained as rational because in the light of accumulating evidence the relative probabilities of rival theories naturally change; and finally, non-scientific theories are those whose probability cannot be affected one way or the other by the evidence.

Although there are intimations of the approach much earlier in the history of thought, recent discussions of this idea really stem from Carnap's (1950) groundbreaking work. His initial idea was to produce an entirely objective version of the account by developing a probabilistic "inductive logic" as a generalization of deductive logic. The crucial notion in all accounts is the probability of some theory given (or conditional on) some evidence. Carnap's original idea was that such conditional probabilities measure degrees of partial entailment – to claim that the probability that Einstein's theory is true, given the evidence is, say, 0.8 means that the evidence entails Einstein's theory to degree 0.8. (Here full – deductive – entailment would of course be degree 1, that is, the probability of A given B is 1 whenever B deductively entails A.) This idea might then be used to supply the rationale for scientific revolutions if it could be shown that the newer theory – say Einstein's theory – has higher probability in the light of the evidence available at the time of the "revolution" than had the earlier theory – in this case Newton's. Intuitively, although the evidence of course entails neither theory, it comes closer to entailing Einstein's theory than to entailing Newton's.

This idea, for all its simplicity and appeal, fails. The basic problem is essentially the same as the one that afflicts the so-called classical account of probability – which defines the probability of some event A as the ratio of the number of "equally possible" cases in which A holds to the number of all the equally possible cases. (Intuitively the probability that a fair dice when rolled will finish with "6" up is $\frac{1}{6}$ because there are six equally possible cases and just one in which the event "6 up" occurs; the probability that an even numbered face will be uppermost in the same situation is $\frac{3}{6}$, i.e. $\frac{1}{2}$ since there are again six equally possible outcomes and in three of them the event "even number uppermost" is instantiated.) The difficulty concerns the notion of partitioning the set of all the possible events in some experiment into the "equally possible" ones. In general, there are different ways of doing this and it seems impossible to argue that only one such way is "correct." And yet with a different partition of the events into equally possible cases we arrive at different probabilities.

Although this approach and this difficulty for it were originally developed in the context of probabilities of various *events*, an entirely analogous approach, and an entirely analogous difficulty, can be developed when thinking, as Carnap did, of the probability that a particular *sentence* is true. Suppose, for example, we are interested in hypotheses about the contents of an urn known to contain, say, 50 balls, each of which is either black or white but in an unknown proportion; suppose further that we are (for some reason) unable to break open the urn and our evidence is restricted to drawing some number of balls from the urn, with replacement, and noting their colours. What constitute the equally likely cases here? All possible *proportions* of black to white balls – all 50 black, 49 black 1 white, 48 black 2 white, etc.? Or are the equally likely cases specified by assuming that each individual ball has the same chance of being white as of being black? It seems difficult indeed to argue that one of these notions is the "correct" one. But it is no surprise that the two yield quite different probabilities for various hypotheses. Suppose we are interested in the hypothesis that exactly half of the balls are white and our evidence is that we have drawn 10 balls, 6 of which are white. The inductive support given to that hypothesis by that evidence, the degree to which the evidence partially entails the hypothesis, will be quite different depending on which of these two ways we slice up the "equal possibilities"; and this makes it very difficult to claim that there is one objective probability for the hypothesis in the light of the evidence.

Confirmation – the Bayesian account

This and a range of other problems led those pursuing the idea that "confirmation is probability" – including eventually Carnap himself – to abandon this "objectivist" partial entailment approach. The currently most popular version of this general idea takes the probabilities at issue in confirmation theory in fact to measure simply a person's degree of belief in the proposition at issue. An agent is considered to have degrees of belief in every proposition available to her and in every logical combination of such propositions. Such an agent is "rational" if

(i) *at any given time*, those degrees of belief can be represented as probabilities (that is satisfy the probability calculus) and

(ii) changes in her degrees of belief *from one time to the next* satisfy something called the "principle of conditionalization."

Although a thoroughly subject- (or agent-)based approach, this account does have clear objective elements. For example, condition (i) requires that if an agent's degree of belief in the theory that the initial escape velocity of matter from the big bang was v_1 is d_1, while her degree of belief in the theory that the initial escape velocity of matter from the big bang was v_2 is d_2, then (assuming that she – properly – believes that it is not possible for the escape velocity to have both values!), she must believe that the theory that the escape velocity was *either v_1 or v_2* to degree $d_1 + d_2$. Also, if an agent has degree of belief d in some proposition P then she must have a degree of belief d' at least as high as d in any proposition Q that is a logical consequence of P.

Defenders of this view have produced various arguments for why condition (i) should be considered an absolute requirement on rationality. The most often-cited argument proceeds by identifying an agent's degrees of belief with fair betting odds (the worst odds at which the agent would be ready to bet on the proposition's being true) and showing that if those degrees of belief were *not* probabilities, did not satisfy the probability calculus, then the agent would be committed to accepting as fair a system of bets such that she would be bound to make a net loss, whatever way the world turned out to be (that is, which ever sentences were eventually accepted as true). This is the so-called "Dutch Book Argument."

A crucial notion in this approach is the *conditional probability* $p(a|b)$ – the probability that a holds on the assumption that b does. These are, of course, interpreted as measuring what your degree of belief in a *would be* if you came to accept b. The most important such conditionals for a theory of confirmation will of course be of the form $p(T|e)$ where T is some theory and e some statement that can be checked on the basis of observation or experiment. Principle (ii) in this impressively austere approach then says something like the following. Suppose that all that happens of any epistemic relevance concerning some particular theory T between two successive stages in science t_1 and t_2 is that some empirical statement *e* that is simply *potential* evidence at t_1 has been checked and actually found to hold (that is, has become real evidence, an accepted part of "background knowledge") by time t_2. How should the agent's degrees of belief in *T* at times t_1 and t_2 be related? Given the understanding of p($T|e$) as measuring the degree of belief in *T* that you would have if you were to come to know *e*, advocates of this approach have suggested that it is obvious that the agent's "new" degree of belief in *T* at t_2 should be her "old" degree of belief in *T* conditional on *e*. That is, introducing subscripts on the probabilities for clarity;

$$p_{t2}(T) = p_{t1}(T|e)$$

And this is the "principle of conditionalization."

Conditional probabilities like p($T|e$) are calculated using Bayes' theorem, which, in its simplest form, says

$$p(T|e) = \frac{p(T)p(e|T)}{p(e)}$$

Because of the frequent use of Bayes' theorem, the approach we have been discussing is called the Bayesian approach to theory-confirmation, or – for reasons made clearer shortly – the personalist Bayesian approach.

Bayesianism has a number of pleasing features. First, as already mentioned, it is impressively austere, appearing at any rate to define "inductive rationality" via only two assumptions. Second, it gives a gratifyingly simple account of what it takes for a theory to be confirmed by evidence: *e* confirms *T* just in case *e* raises *T*'s probability, i.e. just in case p($T|e$) > p(T). And third, it is easy to see that this simple account captures a number of firmly entrenched intuitive judgments about confirmation. It is, for example, part of scientific folklore that if a theory passes a "severe test" (in Popper's terminology) then this confirms the theory more highly than would a less severe test – where a test is severe to the extent that its outcome is highly improbable in the light of background knowledge. One frequently cited example here is the prediction by Fresnel's wave theory of light that if the "shadow" of a small opaque disk held in the light emerging from a point source is carefully examined then the centre of the "shadow" will be seen to be illuminated, and illuminated indeed to precisely the same extent as it would have been had no opaque object been interposed. The usual story is that the idea that there should be such a "white spot" was so improbable in the light of background knowledge, that, once Poisson had shown that Fresnel's theory implied its existence, the scientific establishment was fully confident that Fresnel's goose had been cooked. The account of confirmation under consideration, using Bayes's theorem, straightforwardly captures this intuition. According to the Bayesian formula, the extent to which *e* confirms *T* (i.e. the difference between p(T) and p($T|e$)) is greater the smaller is p(e) – i.e. the less likely *e* is according to background knowledge. (Remember that any probability lies in the interval (0,1).)

Virtues like these, combined with major difficulties in alternative approaches, have convinced many contemporary commentators that Bayesianism is essentially "the only game in town" when it comes to providing a clear-cut, formal theory of confirmation (as opposed to simply some unsystematic list of intuitive judgments about theory-evidence relations). If so, then the only game in Confirmation Town leaves philosophers of science with a lot of work to do in adding to its rules.

Problems with Bayesianism

Of the difficulties facing the personalist Bayesian approach, I outline here one relatively specific "internal" problem and one issue that seems to me a major, general difficulty for the whole approach. The more specific difficulty has come to be known as the "problem of old evidence." There has been much discussion in philosophy of science going back to debates between John Stuart Mill and William Whewell (and beyond) about the relative confirmational value of a theory's predicting hitherto unknown "new" evidence and of its simply explaining already known "old" evidence. Certainly, many of the great confirmational successes for theories that are much heralded in the scientific folklore were predictions: the wave theory of light and the "white spot" at the centre of the "shadow" of an opaque disk (already mentioned) is one such example, and the prediction by the theory of general relativity of star shift (that stars would seem to be different distances apart during the day because of the gravitational effect of the sun) confirmed by Eddington's Eclipse Expedition is another. However, although there may well be some sort of special psychological effect of predictive success, it is difficult to see any principled reason why the time-order of theory and evidence should count *in itself*. Moreover, there are definitely cases where "old evidence" strikingly confirmed a theory – indeed confirmed it, in the eyes of the scientific cognoscenti, just as strongly as any piece of predicted "new" evidence could. Funnily enough, two such cases match the predictive successes just mentioned: Fresnel's explanation of straightedge diffraction (a phenomenon known for around 150 years when Fresnel proposed his theory) seems to have played just as strong a role as the "white spot" evidence in the acceptance of his theory; and, certainly, general relativity's success in accounting for the long-known "anomalous" precession of Mercury's perihelion counted for at least as much as its success with the "star shift." It seems clear that, whatever the truth about the "prediction versus accommodation" issue, it cannot be a blanket "old evidence always counts less." Yet, the Bayesian account of confirmation seems to yield the even stronger result that old evidence can *never count at all*.

This can be seen very easily from the Bayes formula and the fact that all probabilities in this approach are always implicitly relative to background knowledge – that is, to what we already take ourselves to know, at whatever stage of science we are considering. But if some piece of evidence *e* is "old" – already known, in background knowledge at some time t – then its probability at t, relative to that background knowledge, must of course be one. It follows, however, from Bayes formula and assuming that T deductively entails e so that $p(e|T) = 1$, that if $p(e) = 1$, then $p(T|e) = p(T)$. And that precisely means on the Bayesian account that e fails to confirm T.

There have been suggestions from its defenders for how this "old evidence problem" might be solved within the Bayesian framework, though none has won widespread assent. The more general problem seems to me, however, to have no possible solution within the purely personalist framework, but to require – at least – a major extension of it. The problem is that the Bayesian approach seems clearly too weak, to allow too wide a role to subjective opinion, to have any chance of capturing fully what is special about science.

Consult again the crucial Bayesian formula. The Bayesian agent is taken to be a perfect deductive logician, so that if T deductively entails e (usually *modulo* background knowledge) then she must assign a value of 1 to the term $p(e|T)$ – and similarly if T is a well-defined probabilistic hypothesis then she must assign whatever probability T – objectively – assigns to e. The other terms in the formula are however taken to be agent-relative. In particular, the so-called *prior probability* of T, $p(T)$, measuring the degree of belief that an agent has in the theory T ahead of whatever evidence we are now proposing to take into account is subjective – there is no truth of the matter as to what this prior probability is, the Bayesian simply takes it as a fact about a particular agent that she has a certain degree of belief.

It is true, of course, that, in applying this apparatus to some particular theory as it and the evidence for it develop over time, the Bayesian will usually tell a story of how the current prior for T is the end result of a series of applications of the principle of conditionalization on earlier pieces of evidence. But, even then, this series will, of course, have started with some initial prior which will then, by definition, be "purely subjective." Bayesians cite various interesting theorems about the "washing out" of priors which show that, in certain circumstances, two agents with radically different priors on some theory T will nonetheless converge to the same probability for T as evidence of certain kinds comes in. The fact however that in such circumstances (which may not in any event match real cases) any two agents will, in the – of course never actually attained – limit, agree hardly seems sufficient to capture what we generally think of as scientific rationality.

It will surely be generally agreed that, given all the evidence that we currently have from the fossil record, homologies, and various experiments, not to mention the results of various dating techniques, that the Darwinian theory of evolution together with its view of the earth as extremely ancient is altogether more rationally believable *now* than the "scientific" creationist view that the earth was created essentially as it now is, stocked with essentially the "kinds" that it currently has, in 4004 BC. If ever there was a non-defeasible desideratum on an adequate account of the relationship between scientific theories and evidence this is surely it. Yet, it is trivial to show that given any relative degrees of belief in Darwinism (D) and

Creationism (C) – say $p(D) = 0.000001$ and p (C) = 0.999999 – it is entirely possible for an "agent" to have arrived at those degrees in full accordance with Bayesian principles. She could have conditionalized away on all the evidence and still have arrived at degrees of belief that any satisfactory account ought surely to brand as absurd. Of course, this will require the supposition that the agent started the process – ahead of the consideration of any evidence – with even more extreme priors. But the personalist Bayesian explicitly eschews any restrictions on these priors. Any proof that such a "scientific" Creationist is bound to agree with us Darwinians in the indefinite long run is no consolation – it seems clear that the creationist holds a view *now* that is counter to good scientific reasoning, and the Bayesian just cannot deliver that judgment.

The way forward?

Here, then, is a problem that, in my view, remains very much open to future research. Personalist Bayesianism seems at best to capture only a part of scientific rationality. It needs to develop and to defend further requirements – placing at least restrictions on acceptable priors. It is by no means clear how this is to be done, however, within a genuinely Bayesian context. The alternative of course would be to develop another "game in town" – another different systematic attempt to capture good scientific confirmational practice in a precise, and philosophically defensible, way.

One – altogether more radical – suggestion that has been taken up by many recent philosophers is that the sort of approach embodied in Bayesianism and similar enterprises involves an entirely mistaken set of aims and priorities. According to the currently (and increasingly) strong movement towards a "naturalized" philosophy of science, philosophers have for too long been obsessed with traditional issues bequeathed to us to by the likes of Descartes and Hume. We should not be looking for anything like a *logic* of science or of scientific confirmation. Any such system would, in any event, itself rest on assumptions (assumptions which moreover must certainly go beyond deductive logic); and, as centuries of philosophy ought to have taught us, we should be powerless against the sceptic who then asks for justification of those principles themselves. We cannot ask for, and so should not seek, any firmer ground than science itself on which to build our epistemological claims. Philosophy of science should be pursued in a naturalized, scientific way, simply recording the methods of science.

The naturalizing movement with its greater emphasis on philosophers knowing about the details of science has undoubtedly led to many significant improvements. (Though it has to be said that it is easy to get the – of course, absurd – impression from recent treatments that earlier philosophers (the likes of Reichenbach, Hempel and Popper, not to mention still earlier figures like Poincaré and Duhem) knew nothing of the details of science!) Following Kuhn (1962), Lakatos (1970) and others, we now have a much more nuanced view of scientific theory-construction; we have a much richer set of descriptive tools for analysing science and its development involving models, idealizations and the like and a better understanding of the intricacies of scientific "observation." But, as for the general idea that a fully naturalized view can somehow establish the specialness of science, without any rate *vicious* circularity, by itself adopting a scientific approach – this seems to me a very difficult line to argue. The problem again remains an open one for future investigation.

Accumulation in Science, Despite "Revolutions"?

I explained at the beginning how the Einsteinian revolution turned Kant's problem into a dilemma. So far, we have been investigating the prospects for a program that admits the revolutionary nature of scientific change and tries, none the less, to rescue the epistemic specialness of science. Attempts to escape the *other* horn of the dilemma involve conceding that the idea of scientific rationality would indeed be in deep trouble if scientific development were as "revolutionary" as it might at first appear to be and therefore accepting the challenge of arguing that once science, and in particular scientific theories, are *properly* understood, the revolutionary nature of scientific theory-change disappears (or perhaps "largely" disappears). We should now investigate this second possibility.

Revolution in permanence? The "pessimistic meta-induction"

First, let's be clear about the extent of the apparent difficulty. As many commentators would see it, the relativistic and quantum revolutions are simply the tip of the iceberg and their chief effect ought to have been to take off the blinkers so that philosophers could see that "revolutions" (of varying degrees of magnitude) are, in fact, ubiquitous in science. Long before the turn of the century, and even allowing for the sake of argument that science only really started with "the" Scientific Revolution, there had been plenty of less well-publicized but none the less definite cases of seemingly radical theory-change in science. Consider, for example, the history of optics – even when restricted to the modern era. In the eighteenth century, the theory that light consists of material corpuscles had been widely accepted only to be replaced in the early nineteenth century by the theory that light sources do not emit matter but rather energy – matter within the light source vibrates and causes the neighboring particles of the all-pervading "luminiferous ether" to vibrate and hence these vibrations spread through the ether until absorbed by some receptor or other (such as the human eye). This theory, in turn, was replaced by what might be called the mature version of the Maxwell electromagnetic theory of light that denies the existence of the mechanical ether and attributes light instead to the "vibrations" of electric and magnetic field vectors.

Then, of course, as part and parcel of the quantum revolution came the photon theory with its probability waves. From particles to vibrations in an elastic solid, to changing strengths of a *sui generis* electromagnetic field, to photons governed by probability waves – these seem radical shifts indeed. And, of course, according specially to Kuhn (1962), similar revolutions took place in all other branches of science too.

Instances of revolutionary change supply the premises for the "pessimistic meta-induction" that has received a good deal of attention in philosophy of science in the past few decades. This argument is simply an elaboration of the problem from which I began. It is surely a characteristic of revolutionary theory-change that the new theory contradicts the old so that, if we assumed for the sake of argument that the new theory were true, we would be forced to the conclusion that the older theory was false. But what possible grounds could we have for thinking that scientific revolutions are now at an end – that we now have the final theories in all scientific fields? Newtonians in the eighteenth and nineteenth centuries believed – on the basis of very strong evidence – that the fundamental truth about the universe had been discovered; and they turned out to be wrong. No physicist in the nineteenth century, again on good evidential grounds, dreamed that the fundamental processes in nature might be inherently probabilistic and yet that, according to most views, is precisely what presently accepted theories are telling us is true. As we saw, theories of the basic constitution of light have undergone radical shifts. From the standpoint of the current photon theory, the theory that light consists of vibrations transmitted through an all-pervading elastic solid ether looks about as false as any theory could be – after all, for one crucial thing, the newer theory denies entirely the existence of such an all-pervading mechanical ether.

How, then, can we have any faith that that currently accepted photon theory will not, in its turn, eventually be replaced by a theory in whose light it will appear just as false as it itself makes the classical wave theory appear? And, if the findings of science are, at this fundamental level, as transient as this account makes them seem, how can we have any confidence in the process? Even if we could produce persuasive arguments for the methods of science as characterizing a rational process – that is, even if we could solve the problems sketched earlier with, say, some de luxe theory of confirmation – then even so, if that "rational" process produces conclusions that are subject to periodic radical, chalk-and-cheese change, it seems difficult to see why we should regard science as so special.

Notice that no one is asserting here that the "pessimistic meta-induction" is by any means a compelling argument – it is after all inductive and not deductive. It is perfectly *possible* that our scientific predecessors were unlucky (or misguided) and that we have now hit on the truth. And indeed the intuition underwriting the programs discussed earlier is exactly that science, and scientific theories, have improved over time. But it is difficult to see that improvement as in any sense qualitative – nineteenth century physicists had a good deal of evidence for their theories. We now have a good deal more evidence in the light of which very different theories seem true. But, then, since science will presumably continue to "improve" and evidence continue to accumulate, what grounds could be given for holding that our current theories will resist radical change in the light of that accumulating evidence? The pessimistic meta-induction does not need to establish that we have good grounds for thinking that our current theories will eventually be "radically" replaced; the weaker conclusion that we have no good grounds for thinking that they will not be so replaced is sufficient to pose the problem.

Resisting "pessimism" by restoring an essentially cumulative view

Instrumentalism   So how, then, could philosophers of science before 1962 have been blind to what ought to have stared them in the face? The answer is that many of them at least were not at all blind to this phenomenon. Although we tend to think of the "pessimistic meta-induction" as a new philosophical argument, starting with Hilary Putnam or Larry Laudan (1981), in fact it can be found fully formed in Poincaré's (1905) *Science and Hypothesis*:

> The ephemeral nature of scientific theories takes by surprise the man of the world. Their brief period of prosperity ended, he sees them abandoned one after the other; he sees ruins piled upon ruins; he predicts that the theories in fashion today will in a short time succumb in their turn, and he concludes that they are absolutely in vain. This is what he calls the *bankruptcy of science*.

As the way he introduces it suggests, Poincaré was not only aware of the problem he was confident that he had an answer to it:

> [The "man of the world's"] scepticism is superficial; he does not take account of the object of scientific theories and the part they play, or he would understand that the ruins may still be good for something. No theory seemed established on firmer ground than Fresnel's which attributed light to the movements of the ether. Then if Maxwell's theory is preferred today, does it mean that Fresnel's work was in vain? No, for Fresnel's object was not to know whether there really is an ether, if it is or is not formed of atoms, if these atoms really move this way or that; his object was to predict optical phenomena.

Underneath the apparently radical theory-changes (producing the seeming "ruins") there is, Poincaré suggests, a steady accumulation of "real" knowledge in science.

There are two importantly different versions of this claim – versions which Poincaré himself did not *always* clearly differentiate (though I think there is, in the end, no doubting his preferred position). As it stands, the last part of this quotation suggests an "instrumentalist" view of science. Scientific theories, like Fresnel's theory of light, may *seem* to make true-or-false assertions about the underlying structure of reality, about material ethereal atoms held in place by elastic forces and about the vibrations of those atoms, which we cannot of course directly observe, but which allegedly constitute light and hence explain the optical phenomena that we *can* observe. However, the real role of scientific theories is not even to attempt to describe a reality "underlying" the phenomena, but instead merely to codify those phenomena in a coherent, efficient and "simple" way, and hence to enable their prediction. And at the level of "phenomena" – the results of experiments, such as various interference, diffraction and polarization experiments – Maxwell's theory, while attributing those phenomena to a radically different process, none the less agrees (exactly) with Fresnel's theory. Maxwell's theory, of course, goes on to make further predictions – about, for example, radio waves; but where the two theories both make empirical predictions they always exactly agree.

There has been some discussion in the literature of so-called "Kuhn loss" of empirical content – (alleged) cases where some observational or experimental result correctly accounted for by the deposed theory in some "revolution" is not correctly accounted for by the newer theory. Kuhn's own examples of this alleged methodological phenomenon are entirely unconvincing. There are undoubtedly cases in the history of science where a new theory is accepted despite the fact that it cannot at that stage account for some already known phenomenon and where the older theory (which has, of course, at that stage the advantage of longevity) gave at least some sort of account of that same phenomenon. A good example from optics is prismatic dispersion – according to the simplest models of the elastic solid (or indeed elastic fluid) ether, all waves, no matter what their frequency, would travel through it at the same velocity and yet the phenomenon of prismatic dispersion (exhaustively studied, of course, long before Fresnel's wave theory by Newton and others) establishes that the different monochromatic components of solar light travel through the material of the prism (usually glass) at different velocities. The corpuscular theory of light, deposed in the early nineteenth century "wave revolution," gives hints of an explanation – for example a "fixed force of refraction" with the different monochromatic rays corresponding to particles with different masses. But this explanation was known to run into enormous difficulties. If there are any genuine cases of "Kuhn loss" in which some phenomenon was *satisfactorily* explained by the pre-revolutionary but not by the post-revolutionary theory, then they are few and far between. Moreover, it is of the nature of science that any "losses" would be high on the agenda for work aimed at making them good. This is true even where the older "explanation" is highly flawed – in the example just discussed, for instance, a central thrust of the wave optics research program after Fresnel was precisely to develop a detailed mechanical account of the ether that yielded dispersion.

It seems difficult to deny, I suggest, that the development of science has been, at least to a very good approximation, cumulative *at the observational or experimental level*. This need not mean that the "post-revolutionary" theory has exactly the same empirical consequences as the pre-revolutionary one (though in a restricted domain). That happens to be true in the Fresnel–Maxwell case cited by

Poincaré, but the more usual pattern is the one exemplified in the shift from Newtonian classical to Einsteinian relativistic physics. Every *precise* observational consequence of special relativity theory is strictly inconsistent with the corresponding observational consequence of classical theory. Those conflicting observational consequences, none the less, explain the same data across a wide range, because they are, within that range, *observationally indistinguishable*. It follows then that the apparently radical theory changes brought about by "scientific revolutions" pose no problem for the instrumentalist – as concerns what that account sees as the real purposes of science, there is essential continuity across scientific change. Science is special because it delivers more and more of the epistemic "goods" – it is just that those goods do not consist of ever deeper, ever "truer" explanatory theories but rather of ever wider codifications of ever more phenomena.

An interesting more recent slant on this old position is provided by Bas van Fraassen's (1980) "constructive empiricism." Van Fraassen countenances no positivist reduction of the theoretical claims of science – if a theory asserts that electrons exist, it asserts they exist: the claim cannot be regarded as merely shorthand for some complicated set of observational sentences or as some sort of non-assertive "inference licence"; and such a theory is either true or false (in the regular Tarski correspondence sense) depending on how the world really is. However, to explain the rationality of what goes on in science, there is no need to involve considerations of whether such a theoretical claim *is* true (indeed as we have been seeing such involvement poses major problems for ideas about rationality). Scientists should be seen as "accepting" theories, not as true, but only as *empirically adequate*. Although van Fraassen does not directly address the issue of (apparently) radical theory-change, his position provides the basis for a response identical to the one just considered – the progress of science through theory-change can be seen as the development of ever more empirically adequate theories, each new theory revealing that its predecessor was indeed highly empirically adequate but over a restricted range.

Although I shall not discuss them here, there are, of course, many problems with this instrumentalist view – all of them associated in one way or another with the fact that the view does not seem to give proper weight to the role of theory, especially in the *development* of science.

Resisting "pessimism" by restoring an essentially cumulative view

Positivism and structural realism   Instrumentalism, at least in the way I am interpreting it here, allows that successive theories in science contradict one another, and hence allows that theory-change leaves "ruins" (to use Poincaré's term) in its wake. The instrumentalist insists, however, that there is none the less accumulation at the level that science is really all about – the codification of phenomena. There are ruins but they are insignificant.

A different view – a version of which Poincaré himself in fact adopted – is that, when properly viewed, *there are no ruins.* Once the cognitive content of scientific theories is correctly analyzed, we see that the apparent ruins are just that – (merely) *apparent.* It may seem as though Fresnel's theory, for example, makes ontological claims about a medium with the constitution of an elastic solid pervading the whole of space and about the particles of that medium vibrating in certain ways. In fact, however, when we understand properly what the theory says we see that this is not really the case.

One extreme version of this general line is, of course, an outright empiricism or positivism. This sees the real cognitive content of a "theoretical" claim as somehow "reducing to" some (infinite) set of observation sentences. In the case of Fresnel's theory, for example, all the apparent theoretical talk about ether particles, in fact, "reduces to" assertions about interference and diffraction patterns and the like. The logical empiricists did not, in fact, pay much direct attention to theory-change, and developed their account of theories to solve different problems. But if their account could have been made to work, then clearly the phenomenon of theory-change would present it with no problem, assuming that, as I have claimed, the development of science is essentially cumulative at the empirical level.

It has for a long time now been very widely accepted that any such empiricist account is untenable. Certainly, various particular attempted reductive analyses did not work; and the general view, as in the case of instrumentalism, is that no such account can do real justice to the role of theory, particular its heuristic role in the development of science.

The account that Poincaré himself endorsed is different (at least pre-analytically) from both instrumentalism and empiricism or positivism. Having said the Fresnel's theory was not in vain despite its displacement by Maxwell's, because it still allows us to predict optical phenomena as before, he elaborates as follows:

> The differential equations [in Fresnel's theory] are always true [that is, they are carried over into Maxwell's theory], they may always be integrated by the same methods and the results of this integration still preserve their value.
>
> It cannot be said that this is reducing physical theories to practical recipes; these equations express relations, and if the equations remain true, it is because the relations preserve their reality. They teach us now, as they did then, that there is such and such a relation between this thing and that; only the something which we then called *motion*, we now call *electric [displacement] current.* But these are merely the names of the images we substituted for the real objects which Nature will hide for ever from our eyes. The true relations between these real objects are the only reality we can attain . . . (Poincaré, 1905).

Hence, Poincaré claims a continuity across theory-change in science that extends not merely to the observational, but also to the structural level – as is evinced, at any rate in the case he discusses, by the retention of the mathematical equations (and hence of the observational consequences). All that is "lost" are preferred "names of images." The real cognitive content is preserved entirely in tact.

Another problem that seems to me still very much an open one for current philosophy of science is whether some version of Poincaré's *structural realism* can be elaborated, extended to all cases of theory-change and be shown to avoid collapse into outright empiricism. If not, is there any serious hope for any form of scientific realism? The idea that one can retain the view that Newton's theory may be "approximately true" despite the Einsteinian revolution seems to me implicity to presuppose some such (apparently) reduced form of realism. Otherwise, at the "ontological" level, we do seem to have not approximation but outright rejection (of absolute space, absolute simultaneity, action-at-a-distance and so on).

Other Issues

I have tried to build my introductory account of some central issues in philosophy of science around a theme. But, just as I said it would from the outset, any such thematic treatment is bound to leave out much of value. I have not touched on some central issues – such as scientific explanation, the notion of causality and others. Many of these will be dealt with in what follows. I have also not been able to discuss those very important areas of philosophy of science which overlap with theoretical work in the sciences themselves. Analyses of conceptual issues in the theory of general relativity, quantum mechanics and statistical mechanics have all been at the forefront – and have, in turn, raised in especially sharp ways general philosophical issues about determinism, locality and the like. More recent work has seen an extension into the foundations of biology – particularly the structure of Darwinian theory and of genetics; and, especially via interest in causal models, into the foundations of the social sciences.

References

Carnap, R. (1950): *Logical Foundations of Probability*. Chicago: University of Chicago Press.

Duhem, P. (1906): *The Aim and Structure of Physical Theory*, English translation, Princeton: Princeton University Press, 1954.

Glymour, C. (1980): *Theory and Evidence*. Princeton: Princeton University Press.

Glymour, C., Scheines, R., Sprites, P. and Kelly, K. (1987): *Discovering Causal Structure: Artificial Intelligence, Philosophy of Science and Statistical Modeling*. New York: Academic Press.

Kuhn, T. S. (1962): *The Structure of Scientific Revolutions*. Chicago: University of Chicago Press.

Lakatos, I. (1970): "Falsification and the Methodology of Scientific Research Programmes," in I. Lakatos and A. Musgrave (eds.), *Criticism and the Growth of Knowledge*, Cambridge: Cambridge University Press, 91–196.

Laudan, L. (1981): "A Confutation of Convergent Realism," *Philosophy of Science*, 48, 19–49.

Poincaré, H. (1905): *Science and Hypothesis*, English translation, New York: Dover Books.

Popper, K. R. (1959): *The Logic of Scientific Discovery*. London: Hutchison.

Van Fraassen, B. (1980): *The Scientific Image*. Oxford: Oxford University Press.

Further reading

Boyd, R. (1973): "Realism, Underdetermination and the Causal Theory of Evidence," *Nous*, 7, 1–12.

Cartwright, N. (1989): *Nature's Capacities and their Measurement*. New York: Oxford University Press.

Earman, J. (1989): *World Enough and Space-Time*. Cambridge: MIT Press.

Earman, J. (1992): *Bayes or Bust? A Critical Examination of Bayesian Confirmation Theory*. Cambridge: MIT Press.

Earman, J. and Glymour, C. (1980): "Relativity and Eclipses: The British Eclipse Expeditions of 1919 and their Predecessors," *Historical Studies in the Physical Sciences*, 11, 49–85.

Howson, C. and Urbach, P. (1993): *Scientific Reasoning: The Bayesian Approach, second edition*. Chicago: Open Court.

Kitcher, P. (1985): *Vaulting Ambition: Sociobiology and the Quest for Human Nature*. Cambridge: MIT Press.

Mayo, D. G. (1996): *Error and the Growth of Experimental Knowledge*. Chicago: University of Chicago Press.

Popper, K. R. (1963): "Science: Conjectures and Refutations," in *Conjectures and Refutations*, London: Routledge and Kegan Paul, 33–57.

Redhead, M. L. (1987): *Incompleteness, Nonlocality and Realism: A Prolegomenon to the Philosophy of Quantum Mechanics*. Cambridge: Cambridge University Press.

Sklar, L. (1993): *Physics and Chance: Philosophical Issues in the Foundations of Statistical Mechanics*. Cambridge: Cambridge University Press.

Sober, E. (1984): *The Nature of Selection: Evolutionary Theory in Philosophical Focus*. Cambridge, MA: MIT Press.

Worrall, J. (1989): "Fresnel, Poisson and the 'White Spot': The Role of Successful Prediction in Theory-acceptance" in D. Gooding, T. Pinch and S. Schaffer (eds.) *The Uses of Experiment*, Cambridge: Cambridge University Press, 135–57.

Worrall, J. (1999): "Two Cheers for Naturalised Philosophy of Science," *Science and Education*, 8(4), July, Special Edition.

Chapter 3

Explanation

Jim Woodward

Although the subject of explanation has been a major concern of philosophy since Plato and Aristotle, modern philosophical discussion of this topic, at least as it pertains to science, begins with the so-called deductive-nomological (DN) model of explanation in the middle of the twentieth century. This model has many advocates but unquestionably the most detailed and influential statement is due to Carl Hempel (1965).

The DN Model

The basic idea of the DN model is that explanations have the structure of sound deductive arguments in which a law of nature occurs as an essential premise. One deduces the *explanandum*, which describes the phenomenon to be explained, from an *explanans*, consisting of one or more laws, typically supplemented by true sentences about initial conditions. The model is intended to apply both to the explanation of "general regularities" by other laws and the explanation of particular events, although subsequent developments have largely focused on the latter. The derivation of facts about planetary trajectories (e.g. Kepler's laws) from the laws of Newtonian mechanics, the gravitational inverse square law and appropriate information about initial conditions is a paradigmatic illustration of the pattern of explanation that the DN model attempts to capture.

The DN model is meant to capture explanation via deduction from deterministic laws and this raises the obvious question of the explanatory status of statistical laws. Hempel claims that there is a distinctive sort of statistical explanation, which he calls inductive-statistical or IS explanation, involving the subsumption of individual events (like the recovery of a particular person from streptococcus infection) under statistical laws (such as a law specifying the probability of recovery, given that penicillin has been taken). The details of Hempel's account are complex, but the underlying idea is roughly this: an IS explanation will be good to the extent that its explanans confers high probability on its explanandum. Although once a flourishing area of research, the structure of statistical explanation has received relatively little attention recently.[1] In what follows, I will largely ignore it.

Much of the appeal of the DN model lies in the undeniable fact that in some areas of science, such as physics, many explanations do seem to involve derivations from laws. However, the DN model (or at least the version of the model I will discuss) is committed to a good deal more than this commonplace observation. It claims that *all* explanations conform to the requirements of the model, and that everything conforming to those requirements is an explanation. We need to ask whether these claims are correct and whether the key components of the model such as the notion of a law, are sufficiently clear and well-understood to play the role the model assigns to them. I begin with this second issue and then turn to whether the DN requirements are necessary and sufficient for explanation.

Laws

There is general agreement among defenders of the DN approach that laws are (at least) regularities or uniformities – they tell us that if a system exhibits certain properties, it will always or with a certain probability exhibit others. However, not all regularities – even exceptionless regularities – are laws. To take a stock example, while "all spheres of uranium have a mass of less than 10^5 kg" is regarded as a law (since the critical mass for uranium is only a few kilograms), the syntactically similar generalization, "all spheres of gold have a mass of less than 10^5 kg," although presumably true is no law and hence cannot play the role of nomological premise in a DN explanation. The problem of distinguishing genuine laws from such "accidental regularities" is thus central to a defense of the DN model.

Most philosophers, including both defenders and critics of the DN model, have assumed that an adequate account of laws must satisfy certain "empiricist" strictures. These are rarely explained with any precision, but amount in practice to the requirement that the account be "reductive": notions like "law," "cause," and "explanation" are seen as belonging to a family of closely interrelated concepts that must, on pain of "circularity," be explicated in terms of concepts that lie outside of this family like "regularity." A number of criteria for lawfulness that are thought to meet these strictures have been proposed: laws are said

1 to be exceptionless generalizations
2 to contain only purely qualitative predicates and make no reference to particular objects or spatio-temporal locations
3 to support counterfactuals

4 to be confirmable by a limited number of instances in a way that accidental generalizations are not, and

5 to be integrated into some body of systematic theory and play a unifying role in inquiry in a way that accidental generalizations do not.

While each set of criteria has its defenders, I think that a fair summary of current discussion is that none, either singly or in combination, is generally accepted. Many, perhaps most, paradigmatic laws violate certain of the criteria such as (1). Others, such as (2) seem both unclear and overly restrictive and have been abandoned in most recent discussions. Criteria (3) and (5) are, as formulated, both vague and arguably satisfied by accidental as well as lawful generalizations.[2] Criterion (4) looks fundamentally confused from the perspective of any modern treatment of confirmation.[3]

Given the absence of a satisfactory account of lawhood, it is natural to wonder whether the contrast between laws and non-laws can play the central role it is assigned in the DN model. If we cannot say what laws are, why should we accept the DN claim that they are required for successful explanation? One possible response is that although there may be no generally accepted account of laws, there is at least general agreement about which generalizations count as laws and this is all the DN model requires. In fact, however, there seems to be no such agreement. The so-called special sciences – biology, psychology, economics and so on – are full of generalizations that appear to play an explanatory role and/or to describe causal relationships and yet fail to satisfy many of the standard criteria for lawfulness. For example, although Mendel's law of segregation (M) is widely used in evolutionary models, it has a number of exceptions, such as meiotic drive. Other widely used generalizations in the special sciences have very narrow scope in comparison with paradigmatic laws, hold only over restricted spatio-temporal regions, and lack explicit theoretical integration. There is considerable disagreement over whether such generalizations are laws. Some philosophers suggest that such generalizations satisfy too few of the standard criteria to count as laws but can nevertheless figure in explanations; hence we should abandon the DN requirement that all explanations must appeal to laws. Others – e.g. Mitchell (1997) – emphasizing different criteria for lawfulness, conclude instead that generalizations like (M) are laws and hence no threat to the requirement that explanations invoke laws. In the absence of an adequate account of laws, it is hard to evaluate these competing claims.

Motivation

Putting aside these unclarities surrounding the notion of law, why suppose that all (or even some) explanations have a DN or IS structure? Hempel appeals to two central motivating ideas. The first connects the information provided by a DN argument with a certain conception of what it is to achieve understanding:

a DN explanation answers the question "Why did the explanandum-phenomenon occur?" by showing that the phenomenon resulted from certain particular circumstances, specified in $C_1, C_2, \ldots, C_k$, in accordance with the laws $L_1, L_2, \ldots, L_r$. By pointing this out, the argument shows that, given the particular circumstances and the laws in question, the occurrence of the phenomenon was to be expected; and it is in this sense that the explanation enables us to understand why the phenomenon occurred (Hempel, 1965, p. 337).

IS explanation involves a natural generalization of this idea: it shows that the explanandum-phenomenon was to be expected, on the basis of a law, with high probability.

The second main motivation for the DN/IS (hereafter DN) model has to do with the role of causation in explanation. Whether or not all explanations are causal – itself a disputed question in the theory of explanation – there is general agreement among philosophers that many explanations cite information about causes. However, most philosophers, including advocates of the DN model like Hempel, have been unwilling to take the notion of causation as primitive in the theory of explanation. Instead, they have regarded the notion of causation as at least as much in need of explication as the notion of explanation and have sought an account of causation meeting the reductionist or empiricist requirements described above in connection with notion of law. While there are many forms that a theory of causation might take, advocates of the DN model have generally accepted a broadly Humean or regularity theory of causation, according to which (very roughly) all causal claims imply the existence of some corresponding law or regularity linking cause to effect. This is then taken to show that all causal explanations "imply," perhaps only "implicitly," the existence of some law and hence that laws are "involved" in all such explanations, just as the DN model claims.

To illustrate of this line of argument, consider

(Ex1) The impact of my knee on the desk caused the tipping over of the inkwell.

(Ex1) is a so-called singular causal explanation, advanced by Michael Scriven (1962) as a counterexample to the claim that the DN model describes necessary conditions for successful explanation. According to Scriven, (Ex1) explains the tipping over of the inkwell even though no law or generalization figures explicitly in (Ex1) and (Ex1) appears to consist of a single sentence, rather than a deductive argument. Hempel's response (1965, p. 360) was that (Ex1) should be understood claiming there is a "law" or regularity linking knee impacts to tipping over of inkwells. It is the claim that some such law holds that "distinguishes" (Ex1) from "a mere sequential narrative" in which the spilling is said to follow the impact but without any claim of causal connection. We should think of this law as the nomological premise in the DN argument that, according to Hempel, is "implicitly" asserted by (Ex1). Critics have in turn responded that the claim that (Ex1) implies, in virtue of its meaning, the existence of an underlying DN argument looks implausible, given the fact that people use and understand such explanations even if they lack the concepts like "deductively valid argument" and "law of nature."

Counterexamples

While (Ex1) is a potential counterexample to the claim that the DN model provides necessary conditions for explanation, several other examples challenge the claim that the DN model provides sufficient conditions.

Many explanations exhibit directional or asymmetric features to which the DN model appears to be insensitive. From information about the height (h) of a flag pole, the angle ϕ it makes with the sun, and laws describing the rectilinear propagation of light one can deduce the length (s) of its shadow – such a derivation is arguably an explanation (call it (Ex2)) of s. It is equally true that from s, these same laws, and ϕ, one can deduce h. Such a derivation (Ex3) although it apparently meets all of the criteria for an acceptable DN argument, is no explanation of why the flagpole has this height (Bromberger, 1966).

There are other kinds of explanatory irrelevancies besides those associated with the directional features of explanation. Consider a well-known example due to Wesley Salmon (1971).

(Ex4)   (L) All males who take birth control pills regularly fail to get pregnant.
John Jones is a male who has been taking birth control pills regularly.
John Jones fails to get pregnant.

(L) appears to meet the criteria for lawfulness accepted by Hempel and many other writers.[4] Despite this, (Ex4) is no explanation of why Jones fails to get pregnant.

Since both of these derivations show that their putative explananda were "nomically expectable," they seem to cast doubt on the whole idea that explaining an outcome is (just) a matter of showing that it was to be expected on the basis of a law.

One obvious diagnosis of both examples is that they neglect the role that causation plays in explanation. The height of the flagpole causes the length of its shadow and this is why we find a derivation of the former from the latter explanatory. By contrast, the length of the shadow is an effect, not a cause of the height of the flagpole and this is why we don't regard a derivation of h from s as explanatory. Similarly, taking birth control pills does not cause Jones' failure to get pregnant and this is why (Ex4) is not an acceptable explanation.

As explained above, advocates of the DN model would not regard this diagnosis as very illuminating, unless accompanied by some positive account of causation. We should note, however, that an apparent lesson of (Ex3) and (Ex4) is that the regularity account of causation favored by DN theorists is at best incomplete: the occurrence of c, e and the existence of some law linking them (or x's having property P and x's having property Q and some law linking these) is at best a necessary and not a sufficient condition for the truth of the claim that c caused e or x's having P is causally or explanatorily relevant to x's having Q. Contrary to what is often claimed – see, for example Kim (1999, p. 17) – we can not argue that explanations like (Ex1) have an implicit DN structure on the grounds that instaniations of such a structure "guarantee" that c is causally or explanatorily relevant to e.

The SR Model

To a significant extent, subsequent developments in the theory of explanation represent attempts to capture the features of causal or explanatory relevance that appear to be left out of examples like (Ex3) and (Ex4), usually within the empiricist constraints described above. Wesley Salmon's statistical relevance (or SR) model (Salmon, 1971) attempts to capture these features in terms of the notion of statistical relevance (conditional dependence relationships). On the SR model, a request for explanation will take the following canonical form: Why does this member x of the class characterized by attribute A have attribute B? Define a *homogenous partition* of A as a set of subclasses or cells C_i of A that are mutually exclusive and exhaustive, where $P(B|A.C_i) \neq P(B|A.C_j)$ for all $C_i \neq C_j$ and where no further statistically relevant partition of any of the cells $A.C_i$ can be made with respect to B – that is, there are no additional attributes D_k in A such that $P(B|A. C_i) \neq P(B|A. C_i. D_k)$. Then an SR explanation of why A is B consists of

(i)   the prior probability of B within $A: P(B|A) = p$
(ii) a homogeneous partition of A with respect to B, $(A. C_1, \ldots A.C_n)$, together with the probability of B within each cell of the partition: $P(B|A.Ci) = p_i$, and
(iii) The cell of the partition to which x belongs.

To employ one of Salmon's examples, suppose we want to construct an SR explanation of why x who is a teenager (= A) is delinquent (= B). Suppose further that there just two attributes and no others that are statistically relevant to B in A – gender (M or F) and whether residence is urban (U) or rural (R), with the probability of B conditional on A and each the four possible conjunctions of these attributes being different. Then {$A.M.U$, $A.M.R$, $A.F.U$, $A.F.R$} is a homogenous partition of A with respect to B and the SR explanation will consist of

(i)   a statement of the probability of being a delinquent within the class of teenagers

(ii) a statement of the probability of delinquency within this class as we condition on each of the four possible combinations of attributes, and

(iii) the cell to which x belongs.

Intuitively, the idea is that this information tells us about the relevance of each of these combinations of attributes to being delinquent among teenagers and has explanatory import for just this reason. As an additional illustration, suppose that in the birth control pills example (Ex4) the original population T includes both genders. Then

$$P(Pregnancy|T. Male. Takes\ birth\ control\ pills) = P(Pregnancy|T. Male)$$

while

$$P(Pregnancy|T. Male. Takes\ birth\ control\ pills)$$
$$\neq P(Pregnancy|T. Takes\ birth\ control\ pills)$$

assuming that birth control pills are not always effective for women. In this way, we can capture the idea that among males, taking birth control pills is explanatorily irrelevant to pregnancy, while being male *is* relevant.

The SR model has a number of features that have generated substantial discussion, but I want to focus on what I take to be the central motivating ideas of the model:

(i)   explanations cite causal relationships and

(ii) causal relationships are captured by statistical relevance relationships.

The fundamental problem with the SR model is that (ii) is false – as a substantial body of work[5] has made clear, casual relationships are greatly underdetermined by statistical relevance relationships. Consider Salmon's example of a system in which atmospheric pressure A is a common cause of the occurrence of a storm S and the reading of a barometer B with no causal relationship between B and S. Salmon claims that B is statistically irrelevant to S given A – i.e. $P(S|A.B) = P(S|A)$ but A remains relevant to S given B – i.e. $P(S|A.B) \neq P(S|B)$ and thus that A is explanatorily (causally) relevant to S while B is not. However, many other causal structures are compatible with these statistical relevance relationships. Structures in which B causes A which in turn causes S will, if we make assumptions like Salmon's connecting causation and probability, lead to exactly the same statistical relevance relationships. In these structures, unlike Salmon's example, B is causally (and presumably explanatorily) relevant to S. Similarly, the statistical relevance relationships among A, B and S, will not tell us whether we are dealing with a system in which, say, A causes B which causes S and in which A also directly causes S, independently of B, or one in which the direction of the causal arrow from A to B is reversed, so that B causes A. A mere list of statistical relevance relationships, which is what the SR model provides, does not tells us which causal or explanatory relationships are operative.

The Causal Mechanical Model

In more recent work, Salmon (1984) acknowledges this and abandons the attempt to characterize explanation or causal relationships in purely statistical terms. His new account, which he calls the Causal Mechanical (CM), attempts to capture the "something more" involved in causal/explanatory relationships over and above facts about statistical relevance. The CM model employs several central ideas. A *causal process* is a physical process, like the movement of a particle through space, that is characterized by the ability to transmit its own structure in a continuous way. A distinguishing feature of causal processes is their ability to transmit marks. Intuitively a mark is some local modification to the structure of a process, as when one scuffs the surface of a baseball. A baseball is a causal process and one expects the scuff mark to persist as the baseball moves from one spatio-temporal location to another, even in the absence of further interventions or interactions. Causal processes contrast with *pseudo-processes* which lack the ability to transmit marks. An example is the shadow of a moving physical object. Intuitively, Salmon's idea is that, if we try to mark the shadow by modifying its shape at one point (for example, by altering a light source or introducing a second occluding object), this modification will not persist unless we continually intervene to maintain it as the shadow occupies successive spatio-temporal positions. *Causal interactions* occur when one causal process spatio-temporally intersects another and produces a modification of it structure. An example would be a collision between two particles which alters the direction and kinetic energy of both.

According to the CM model, an explanation of some event E will trace the causal processes and interactions leading up to E (Salmon calls this the *etiological* aspect of the explanation), or at least some portion of these, as well as describing the processes and interactions that make up the event itself (the *constitutive* aspect of explanation). In this way, the explanation shows how E "fit[s] into a causal nexus" (1984, p. 9).

The suggestion that explanation involves "fitting" an explanandum into a causal nexus does not of course give us any very precise characterization of just what the relationship between E and other causal processes and interactions must be if information about the latter is to explain E. But rather than belaboring this point, I will focus on the intuitive idea behind this suggestion and examine what implies for some specific examples.

Suppose that a cue ball, set in motion by the impact of a cue stick, strikes a stationary eight ball with the result that the eight ball is put in motion and the cue ball changes direction. The impact of the stick also transmits some blue chalk to the cue ball which is then transferred to the eight ball on impact. The cue stick, the cue ball and the eight ball are causal processes and the collision of the cue stick with the cue ball and the collision of the cue and eight balls are causal interactions. Salmon's intuitive idea is that citing such facts about processes and interactions explains the motion of the balls after the collision; by contrast, if one of these balls casts a shadow that moves across the other, this will be causally and explanatorily irrelevant to its subsequent motion since the shadow is a pseudo-process.

However, as Christopher Hitchcock shows in an illuminating paper (Hitchcock, 1995) the information about causal processes and interactions just described leaves out something important. The usual elementary textbook "scientific explanation" of the motion of the balls following collision proceeds by deriving that motion from information about their masses and velocity before the collision, the assumption that the collision is perfectly elastic, and the law of the conservation of linear momentum. We think of the information conveyed by this derivation as showing that it is the mass and velocity of the balls, rather than, say, their color or the presence of the blue chalk mark, that is explanatorily relevant to their subsequent motion. However, it is hard to see what in the CM model allows us to pick out the linear momentum of the balls, as opposed to various other features, as explanatorily relevant. Part of the difficulty is that to express such relatively fine-grained judgments of explanatory relevance (that it is linear momentum rather than chalk marks that matter) we need to talk about relationships between properties or magnitudes and it is not clear how express such judgments in terms of facts about causal processes and interactions. Both the linear momentum and the blue chalk mark communicated to the cue ball by the cue stick are marks that are transmitted by the spatio-temporally continuous causal process consisting of the motion of the cue ball, and which then are transmitted via an interaction to the eight ball.

Ironically, as Hitchcock goes on to note, a similar observation may be made about (Ex4). Spatiotemporally continuous causal processes that transmit marks as well as causal interactions are at work when male Mr. Jones ingests birth control pills – the pills dissolve, components enter his bloodstream, are metabolized or processed in some way and so on. Similarly, causal processes (albeit different processes) and spatio-temporally continuous paths are at work when female Ms. Jones takes birth control pills. Intuitively, it looks as though the relevance or irrelevance of the birth control pills does not just have to do with whether the actual processes that lead up to Mr. Jones non-pregnancy are capable of mark transmission but rather (roughly) with the contrast between what happens in actual situation in which Jones takes the pills and an alternative situation in which Jones does not take the pills. It is because the outcome (non-pregnancy) would be the same in both cases if Jones is male that the pills are explanatorily irrelevant. This links explanatory relevance to counterfactuals – a point to which I will return.

A second, not unrelated set of worries has to do with how we are to apply the CM model to more complex systems which involve a large number of interactions among what from a fine grained level of analysis are distinct causal processes. Suppose that we have a mole of gas, confined to a container, with volume V_1, at pressure P_1, and temperature T_1. The gas is then allowed to expand isothermally into a larger container of volume V_2. One standard way of explaining the behavior of the gas – its rate of diffusion and its subsequent equilibrium pressure P_2 –

appeals to the generalizations of phenomenological thermodynamics – e.g., the ideal gas law, Graham's law of diffusion, etc. Salmon appears to regard putative explanations based on at least the first of these generalizations as not really explanatory because they do not trace continuous causal processes – the individual molecules are causal processes but not the gas as a whole. However, it is obviously impossible to trace the causal processes and interactions represented by each of the 6×10^{23} molecules making up the gas and the successive interactions (collisions) it undergoes with every other molecule. The usual statistical mechanical treatment, which Salmon presumably would regard as explanatory, does not attempt to do this. Instead, it makes certain general assumptions about the distribution of molecular velocities and the forces involved in molecular collisions and then uses these, in conjunction with the laws of mechanics, to derive and solve a differential equation (the Boltzmann transport equation) describing the overall behavior of the gas. This treatment abstracts radically from the details of the causal processes involving particular individual molecules and instead focuses on identifying higher level variables that aggregate over many individual causal processes and that figure in general patterns that govern the behavior of the gas. A plausible version of the causal mechanical model will need to avoid the conclusion that an explanation of the behavior of the gas must trace the trajectories of individual molecules and provide an alternative account of what tracing causal processes and interactions means for such a system. Such an extension of the CM model has not yet been developed. A similar point holds for other complex systems.[6]

There is another aspect of this example that is worthy of comment. Even if, per impossible, an account that traced individual molecular trajectories were to be produced, there are important respects in which it would not provide the explanation of the macroscopic behavior of the gas that we are looking for. This is because there are a very large number of different possible trajectories of the individual molecules in addition to the trajectories actually taken that would produce the macroscopic outcome that we want to explain. Very roughly, given the laws governing molecular collisions one can show that almost all (i.e., all except a set of measure zero) of the possible initial positions and momenta consistent with the initial macroscopic state of the gas, as characterized by P_1, T_1, and V_1, will lead to molecular trajectories such that the gas will evolve to the macroscopic outcome in which the gas diffuses to an equilibrium state of uniform density through the chamber at pressure P_2. Similarly, there is a large range of different microstates of the gas compatible with each of the various other possible values for the temperature of the gas and each of these states will lead to a different final pressure P_2*. It is an important limitation of the strategy of tracing actual individual molecular trajectories that it does not, at least as it stands, capture or represent this information. Explaining the final pressure P_2 of the gas seems to require identifying both the full range of (counterfactual and not just actual obtaining) conditions under which P_2 would have occurred and the (counterfactual) conditions under which it would have been different. Just tracing the causal processes (in the form of actual molecular trajectories) that lead to P_2, as the CM model requires, omits this information about what would happen under these counter-factual conditions.

Unificationist Models

The final account of explanation that we will examine is the *unificationist* account. The basic idea was introduced by Michael Friedman (1974) but its subsequent development has been most associated closely with Philip Kitcher (1989). One possible assessment of the DN model is that it (or something broadly like it) is correct as far as it goes – it states plausible necessary conditions on explanation – but that it needs to be supplemented by some additional condition X which avoids the counterexamples to the sufficiency of the model described above. This is roughly Kitcher's view. Explanations are derivations from premises that include generalizations of considerable scope (whether or not we regard these as laws) but such derivations must also meet an additional condition $= X$ having to do with unification. The underlying idea is that explanatory theories are those that unify a range of different phenomena. Such unifications clearly have played an important role in science; paradigmatic examples include Newton's unification of terrestrial and celestial theories of motion and Maxwell's unification of electricity and magnetism.

Kitcher attempts to make this idea more precise by suggesting that explanation is a matter of deriving as many descriptions as possible of different phenomena by using the same "argument patterns" over and over again – the fewer the patterns used, the more "stringent" they are in the sense of imposing restrictions on the derivations that instantiate them, and the greater the range of different conclusions derived, the more unified our explanations. Kitcher does not propose a completely general theory of how these considerations – number of conclusions, number of patterns, and stringency of patterns – are to be traded off against one another, but he does suggest that, in many specific cases, it will be clear enough what these considerations imply about the evaluation of particular candidate explanations. His basic strategy is to argue that the derivations we regard as good explanations are instances of patterns that taken together score better according to the criteria just described than the patterns instantiated by the derivations we regard as defective explanations. Following Kitcher, let us define the *explanatory store E(K)* as the set of argument patterns that maximally unifies K, the set of beliefs accepted at a particular time in science. Showing that a particular derivation is an acceptable explanation is then a matter of showing that it belongs to the explanatory store.

As an illustration, consider Kitcher's treatment of the problem of explanatory asymmetries. Our present explanatory practices – call these P – are committed to the idea that derivations of a flagpole's height from the length of its shadow are not explanatory. Kitcher contrasts P with an alternative systemization in which such derivations are regarded as explanatory. According to Kitcher, P includes the use of a single origin and development (OD) pattern of explanation, according to which the dimensions of objects – artifacts, mountains, stars, organisms etc. – are traced to "the conditions under which the object originated and the modifications it has subsequently undergone" (1989, p. 485). Now consider the consequences of adding to P, an additional pattern S (the shadow pattern) which permits the derivation of the dimensions of objects from facts about their shadows. Since the OD pattern already permits the derivation of all facts about the dimensions of objects, the addition of S to P will increase the number of argument patterns in P and will not allow us to derive any new conclusions. On the other hand, if we were to drop OD from P and replace it with the shadow pattern, we would have no net change in the number of patterns in P but would be able to derive far fewer conclusions than we would with OD, since many objects do not have shadows from which to derive their dimensions. Thus OD belongs to the explanatory store, and the shadow pattern does not. Kitcher's treatment of other problem cases in the theory of explanation is similar – for example, derivations like (Ex4) above are claimed to instantiate patterns that belong to a totality of patterns that are less unifying than the totality to which the pattern instantiated by a derivation that just appeals to a generalization about all males failing to become pregnant.

What is the role of causation on this account? Kitcher claims that "the 'because' of causation is always derivative from the 'because' of explanation" (1989, p. 477). That is, our causal judgments simply reflect the explanatory relationships that fall out of our (or our intellectual ancestors') attempts to construct unified theories of nature. There is no independent causal order over and above this which our explanations must capture.

Although the idea that explanation has something to do with unification is intuitively appealing, Kitcher's particular way of cashing out the idea seems problematic. His treatment of the flagpole example obviously depends heavily on the contingent truth that some objects do not cast shadows. But wouldn't it still be inappropriate to appeal to facts about the shadows cast by objects to explain their dimensions in a world in which all objects cast enough shadows (they are illuminated from a variety of different directions etc.) so that all of their dimensions can be recovered?[7]

The matter becomes clearer if we turn our attention to a variant example in which, unlike the shadow example, there are clearly just as many backwards derivations from effects to causes as there are derivations from causes to effects. Consider, following Barnes (1992), a time-symmetric theory like Newtonian mechanics, as applied to a closed system like the solar system. Call derivations of the state of motion of the particles at some future time t from information about their present positions (at time t_0), masses, and velocities, the forces incident on them between t_0, and the laws of mechanics *predictive*. Now contrast such derivations with *retrodictive* derivations in which the present motions of the particles are derived from information about their future velocities and positions at t, the forces operative between t_0 and t and so on. It looks as though there will be just as many retrodictive derivations as predictive derivations and each will require premises of exactly the same general sort – information about positions, velocities, masses etc. and the same laws. Thus, the pattern or patterns instantiated by the retrodictive derivations looks exactly as unified as the pattern or patterns associated with the predictive derivations. However, we think of the predictive derivations and not the retrodictive derivations as explanatory and the present state of the particles as the cause of their future state and not vice-versa. It is far from obvious how considerations having to do with unification could generate such an explanatory asymmetry.

Examples of this sort cast doubt on Kitcher's claim that one can begin with the notion of explanatory unification, understood in a way that does not presuppose causal notions, and use it to derive the content of causal judgments. This conclusion is reinforced by a more general consideration: The conception of unification underlying Kitcher's account is, at bottom, one of descriptive economy or information compression – deriving as much from as few assumptions or via as few patterns of inference as possible. However, there are many schemes and procedures in science that involve information compression and unified description but don't seem to provide information about causal relationships. This is true of many classificatory schemes including schemes for biological classification, and schemes for the classification of geological and astronomical objects like rocks and stars. If I know that individuals belong to a certain classificatory category (e.g. Xs are mammals), I can use this information to derive a great many of their other properties (Xs have backbones, hearts, their young are born alive, etc.) and this is a pattern of inference that can be used repeatedly for many different sorts of Xs. Nonetheless, and despite the willingness of some philosophers to regard such derivations as explanatory (X is white because X is a polar bear and all polar bears are white), most scientists think of such schemes as "merely descriptive" and as telling us little or nothing about the causes or mechanisms that explain why Xs have hearts or are white. Similarly, there are numerous statistical procedures (factor analysis, cluster analysis, multi-dimensional scaling techniques) that allow one to summarize or represent large bodies of statistical information in an economical, unified way and to derive more specific statistical facts from a much smaller set of assumptions by repeated use of the same pattern of argument. For example, knowing the "loading" of each of n intelligence tests on a single common factor g, one can derive $n(n-1)/2$ conclusions about pairwise correlations among these tests. Again, however, it is doubtful that this "unification" tells us anything about causal relationships.

Conclusion and Directions for Future Work

What conclusions/morals may we draw from this historical sketch? What are the most promising directions for future work? Any proposals about these matters will be tendentious, but with this caveat in mind, I suggest the following. First, many of the limitations of the theories reviewed above may be traced to their failure to satisfactorily capture causal notions. A more adequate account of causation is thus one of the most important items on the agenda for future work on explanation. The approach I regard as most promising differs from those described above – it takes counterfactual dependence to be the key to understanding causation and hence explanation. To motivate this approach, note that an obvious diagnosis of the difference between the acceptable and defective explanations described above is that the former but not the latter exhibit a pattern of counterfactual dependence between explanans and explanandum in the following sense: in the good explanations but not the bad ones, changing the explanans variables will be associated with a corresponding change in the explanandum. Thus, the birth control pills are causally and explanatorily irrelevant to Mr. Jones' pregnancy because whether he becomes pregnant does not depend counterfactually on whether he takes pills. We might establish this absence of counterfactual dependence by doing an experiment in which we observe that manipulating whether males take birth control pills is associated with no change in whether they become pregnant. Similarly, if we change the length of a flagpole while leaving other causally relevant factors undisturbed, the length of its shadow will change, but changing the shadow's length by changing the elevation of a light source or the angle the pole makes with the ground or in any other way that does not involve directly changing the flagpole's length will not result in a change in the pole's length. In this sense, the length of the shadow is counterfactually dependent on (and is explained by) the length of the pole and not vice versa. Again, changing whether there is a blue spot on the cue ball will change not change the subsequent motion of the balls but changing their linear momentum will. In this sense, the subsequent motion counterfactually depends on (and is explained by) the momentum but not the spot.

This view of the connection between explanation and counterfactual dependence allows us to deal with a puzzle that will have occurred to the alert reader. On the one hand, derivations from laws or other general principles seem to play an explanatory role in many areas of science. On the other hand, (Ex3) and (Ex4) seem to show that not all such derivations are explanatory and (Ex1) seems to show that not all explanations take the form of derivations. We may resolve this puzzle by rethinking the role of derivational structure in explanation. According to the DN model, the role of derivation from a law is to show that the explanandum phenomenon was to be expected. I suggest instead that explanations explain in virtue of conveying information about patterns of counterfactual dependence. Derivation from a law is sometimes a very effective way of conveying such information, as when a derivation of the subsequent motion of the cue balls from the conservation of linear momentum and their prior momenta shows us in a very detailed and fine grained way exactly how the subsequent motion of the balls would have been different in various ways if their prior momentum had been different in various ways. However, not all derivations from laws convey such information about counterfactual dependence and when they do not, as in the case of (Ex3), there is no explanation. Moreover, there are other ways of conveying such counterfactual information besides explicit derivation and as long as information is conveyed, one has an explanation. Thus, (Ex1) tells us about the counterfactual dependence of the ink tipping on the knee impact and is explanatory for just this reason – we need not see it as explanatory in virtue of instaniating an implicit DN structure, which in any event is not sufficient for explanatoriness in the absence of counterfactual dependence. Other representational devices such as diagrams and graphs similarly convey information about counterfactual dependence without consisting of explicit derivations.

There are many counterfactual theories of causation in the philosophical literature – David Lewis' theory (1973) is probably the best known.[8] For the most part, however, philosophers of science have been unwilling to make extensive use of counterfactual notions in developing theories of explanation. This attitude is partly due to suspicion that counterfactuals fail to meet the empiricist strictures described at the start of this chapter, but it has been exacerbated by features of the very influential semantics for counterfactuals developed by Lewis. Although the semantics is a wonderful achievement, its appeal to trade-offs along different dimensions of "similarity" across "possible worlds" and to "miracles" that violate laws of nature leaves it opaque how counterfactual claims can be tested by ordinary empirical evidence and seems to have little contact with scientific practice. The result has been to make counterfactuals look scientifically disreputable. Recently, however, this situation has changed. Judea Pearl and others – see especially Pearl (2000) – drawing on a substantial preexisting traditions in disciplines like statistics, experimental design, and econometrics have provided rigorous formal frameworks for exploring the connection between causation and counterfactuals. They have also emphasized the very close connection (gestured at above) between counterfactuals and experimentation, and have explored the ways in which even when experimentation is not possible, statistical evidence may be brought to bear on causal claims; in the latter connection, see especially, Spirtes et al. (1993). Although I lack the space to defend this judgment, I think this work goes a long way toward making counterfactuals and accounts of explanation and causation based on counterfactuals scientifically respectable. The task then becomes one of working out in detail how various causal and explanatory notions can be captured within this counterfactual/experimentalist framework – work of this sort is already underway[9] and, in my judgment, represents one of the most promising future directions in the theory of explanation. I will also add the prediction that the best work in this area will make use of formal machinery like systems of equations and directed graphs – machinery that is both richer than representational devices standardly employed by philosophers (logic, probability theory unsupplemented by anything else) and closer to the machinery employed by science itself. Neither logic nor probability theory by themselves can capture the modal and counterfactual elements that are central to explanation.

"Laws of nature" is also a topic on which much work remains to be done. There are many questions that need to be answered. Which if any of the traditional criteria for lawfulness can be reformulated in a defensible way? Is it possible to draw a relatively sharp distinction between laws and non-laws at all and, if so, does this distinction coincide with the distinction between those generalizations that can figure in explanations and those that cannot, as DN theorists claim? If there is no clear distinction, what follows for the theory of explanation? What are the advantages and disadvantages of thinking of the generalizations of the special sciences as laws even though they lack many of the features traditionally assigned to laws? My suspicion is that progress on these issues will require abandoning the "all As are Bs" framework for representing laws traditionally favored by philosophers in favor of a focus on examples of real laws, which are represented by equations of various sorts which have a much richer structure.

The issue of reductionism also merits rethinking. A great deal of work on explanation, including the accounts described above, seems animated by the assumption that, without a full reduction, no interesting progress has been made. This attitude is not self-evidently correct. Some non-reductionist theories of causation/explanation (e.g., c explains e if c produces e, with no further account of "production") do seem completely unilluminating. But not all non-reductive theories are trivial in the way just illustrated. Non-reductive theories can be interesting and controversial in virtue of conflicting with other reductive or non-reductive theories and suggesting different assessments of particular explanations. For example, even if the CM model fails to fully meet empiricist strictures, it will still disagree with counterfactual theories (including non-reductive versions of such theories) in its assessment of explanations that appeal to action at a distance or otherwise fail to trace continuous causal processes, since counterfactual theories presumably will regard such explanations as legitimate. Relatedly, even if we opt for a non-reductive account of some notion within the circle of concepts that includes "cause," "counterfactual," etc., we still face many non-trivial choices about exactly how this notion should be connected up with or used to elucidate other notions of interest – choices that can be made in more or less defensible ways. Finally, even in the absence of a fully reductive account of explanation, it may be possible to show how particular explanatory/causal claims can be tested by making use of other particular causal claims and correlational information. My own view is that, in their enthusiasm for reductive accounts, philosophers have often misdescribed the structure of the explanatory claims they have hoped to reduce. I also think that many of the empiricist constraints imposed on accounts of explanation have been abandoned elsewhere in philosophy and have little justification. Regardless of whether this is correct, the entire subject would benefit from a more explicit discussion of the rationale for the constraints that are standardly imposed.

Notes

1   Woodward (1989) argues it is a misconception that statistical theories explain individual outcomes. Instead, they explain features of probability distributions such as expectation values.

2   For example, the paradigmatically accidental generalization "All the balls in this urn are red" arguably supports the counterfactual "If a ball were drawn from this urn, it would be red." If we want to use support for counterfactuals to distinguish laws, we need to be more precise about which counterfactuals are supported by laws but not by accidental generalizations. Criterion (5) is arguably satisfied by accidental cosmological uniformities such as the generalization that at a sufficiently large scale the mass distribution of the universe is uniform, since these play a unifying role in cosmological investigation. Several of the objections to unificationist theories of explanation discussed below also appear to tell against this criterion.

3   Virtually all recent treatments of confirmation, whether Bayesian or non-Bayesian, agree that "positive instances" by themselves never confirm generalizations, whether lawful or accidental. Instead, it is only in conjunction with background assumptions that positive instances or any other form of evidence can be confirming. Once this is recognized, it becomes clear that in conjunction with the right background assumptions, accidental generalizations are just as confirmable by a limited number of instances as lawful generalizations. For example, in conjunction with the information that an appropriate small sample has been drawn randomly from the US population, the sample can accidental generalizations about political attitudes in that population.

4   Some readers may respond that (L) is not a bona-fide law but this just illustrates again that defense of the DN model requires a more adequate account of laws.

5   See especially Cartwright (1979) and Spirtes et al. (1993).

6   For more on this theme, see Woodward (1989).

7   Kitcher's implausible assumption that there is a single OD pattern of explanation also invites further comment. While the assumption may make little difference to the particular example under discussion, for reasons described in Barnes (1992), it raises the important issue of whether there are non-arbitrary criteria for counting or individuating patterns of argument.

8   My own defense of a counterfactual theory of explanation can be found in Woodward (1984) and Woodward (2000).

9   In addition to Pearl (2000) see, for example, Hitchcock (2001).

References

Barnes, E. (1992): "Explanatory Unification and the Problem of Asymmetry," *Philosophy of Science*, 59, 558–71.

Bromberger, S. (1966): "Why Questions," in R. Colodny (ed.), *Mind and Cosmos: Essays in Contemporary Science and Philosophy*, Pittsburgh: University of Pittsburgh Press, 86–111.

Cartwright, N. (1979): "Causal Laws and Effective Strategies," *Nous*, 13, 419–37.

Friedman, M. (1974): "Explanation and Scientific Understanding," *Journal of Philosophy*, 71, 5–19.

Hempel, C. (1965): *Aspects of Scientific Explanation and Other Essays in Philosophy of Science*. New York: Free Press.

Hitchcock, C. (1995): "Discussion: Salmon on Explanatory Relevance," *Philosophy of Science*, 62, 304–20.

Hitchcock, C. (2001): "The Intransitivity of Causation Revealed in Equations and Graphs," *The Journal of Philosophy*, xcviii (6), 273–99.

Kim, J. (1999): "Hempel, Explanation, Metaphysics," *Philosophical Studies*, 94, 1–20.

Kitcher, P. (1989): "Explanatory Unification and the Causal Structure of the World," in W. Salmon and P. Kitcher (eds.), 410–505.

Lewis, D. (1973): "Causation," *Journal of Philosophy*, 70, 556–67.

Mitchell, S. (1997): "Pragmatic Laws," *PSA 96*, Supplement to *Philosophy of Science* 64(4), S468–S479.

Pearl, J. (2000): *Causality: Models, Reasoning and Inference*. Cambridge: Cambridge University.

Salmon, W. (1971): "Statistical Explanation and Statistical Relevance," in W. Salmon (ed.), *Statistical Explanation and Statistical Relevance*, Pittsburgh: University of Pittsburgh Press, 29–87.

Salmon, W. (1984): *Scientific Explanation and the Causal Structure of the World*. Princeton: Princeton University Press.

Salmon, W. and Kitcher, P. (eds.) (1989): *Minnesota Studies in the Philosophy of Science, Vol 13: Scientific Explanation*. Minneapolis: University of Minnesota Press.

Scriven, M. (1962): "Explanations, Predictions and Laws," in H. Feigl and G. Maxwell (eds.), *Minnesota Studies in the Philosophy of Science, volume III*, Minneapolis: University of Minnesota Press, 170–230.

Spirtes, P. Glymour, C. and Scheines, R. (1993): *Causation, Prediction and Search*. New York: Springer-Verlag.

Woodward, J. (1984): "A Theory of Singular Causal Explanation," *Erkenntnis*, 21, 231–62.

Woodward, J. (1989): "The Causal Mechanical Model of Explanation," in W. Salmon and P. Kitcher (eds.), 357–83.

Woodward, J. (2000): "Explanation and Invariance in the Special Sciences," *British Journal for the Philosophy of Science*, 51, 197–254.

Chapter 4

Structures of Scientific Theories[1]

Carl F. Craver

Introduction

A central aim of science is to develop theories that exhibit patterns in a domain of phenomena.[2] Scientists use theories to control, describe, design, explain, explore, organize, and predict the items in that domain. Mastering a field of science requires understanding its theories, and many contributions to science are evaluated by their implications for constructing, testing, and revising theories. Understanding scientific theories is prerequisite for understanding science.

The two dominant philosophical analyses of theories have sought an abstract formal structure common to all scientific theories. While these analyses have advanced our understanding of some formal aspects of theories and their uses, they have neglected or obscured those aspects dependent upon *nonformal* patterns in theories. Progress can be made in understanding scientific theories by attending to their diverse nonformal patterns and by identifying the axes along which such patterns might differ from one another. After critically reviewing the two dominant approaches (pp. 55–64), I use *mechanistic theories* to illustrate the importance of nonformal patterns for understanding scientific theories and their uses (p. 67).

The Once Received View (ORV)

Central to logical positivist philosophy of science is an analysis of theories as empirically interpreted deductive axiomatic systems.[3] This formal approach, the ORV,[4] emphasizes inferential patterns in theories. The primary virtue of the ORV (and some of its vice) lies in its association and fit with argument-centered analyses of, for example, explanation, prediction, reduction, and testing. The main commitments of the ORV are as follows.

Logical and extralogical vocabulary

According to the ORV, theories are linguistic structures composed of a logical and an extralogical vocabulary. The logical vocabulary contains the operators of first-order predicate calculus with quantifiers, variously supplemented with relations of identity, modality, and probability.[5] The extralogical vocabulary (V) contains the predicates that constitute the theory's descriptive terms. Theories systematize phenomena by exhibiting deductive and inductive inferential relations among their descriptive terms; this systematization provides a "logical skeleton" for the theory and "implicitly defines" the predicates in V (Nagel, 1961, p. 90).

Correspondence rules and the theory/observation distinction

The predicates of V, on the ORV, can be sorted into an observational vocabulary (V_O) and a theoretical vocabulary (V_T). Predicates in V_O are defined directly in terms of the observable entities and attributes to which they refer. The predicates in V_T refer to entities and attributes that cannot directly be observed; these predicates are defined indirectly via *correspondence rules* tethering them to predicates in V_O.

Correspondence rules give theories their empirical content and their explanatory and predictive power. Correspondence rules have been characterized as explicit definitions (including operational definitions), as reduction sentences (partially or conditionally defining the term within the context of a given experimental arrangement), or in terms of a more holistic requirement that the theory form an interpretive system with no part failing to make a difference to the observable consequences of the theory (Hempel, 1965, chs 4 and 8).

Laws of nature

On the ORV, the explanatory power of theories springs ultimately from the laws that are their axioms. Explaining an event or regularity (the explanandum), on the "covering law" account, is a matter of inductively and/or deductively systematizing (fitting) the explanandum into the axiomatic structure of the theory and thereby demonstrating that the explanandum was to be expected given the laws of nature and the relevant conditions.

Within the ORV, law statements (descriptions of laws) are canonically represented as universally quantified material conditionals (e.g., "For all x, if x is F then x is G"). Minimally, law statements are

(i) logically contingent
(ii) true (without exception)
(iii) universal generalizations, that are
(iv) unlimited in scope.

Requirement (iv) is generally understood to preclude the law's restriction to particular times and places. Many recommend the additional requirement that the regularity described by the law statement (v) hold by physical necessity. This requirement might be used to distinguish statements of law from merely accidental generalizations (Hempel, 1966, ch. 5), or to pick out those generalizations that support counterfactuals from those that do not (Goodman, 1983).

Theory construction, theory change, and derivational reduction

The ORV is commonly associated with a generalization/abstraction account of theory construction, a successional account of theory change, and a derivational account of intertheoretic reduction. The strictures of the ORV restrict its flexibility for analyzing theory construction and theory change.

The *generalization/abstraction account* depicts theory construction as a "layer cake" inference *first* from particular observations (via inductive generalization) to empirical generalizations constructed from V_O, and *then* from these empirical generalizations (via e.g., hypothetico-deductive inference) to laws of nature (constructed from V_T). This account is not mandated by the ORV, but its logical framing of the theory construction process (with its dichotomies of type and token, general and particular, observable and theoretical) naturally suggests such a picture; see, for example, Nagel (1961, ch. 5).

The ORV's analysis of meaning enforces a *successional account* of theory change. First, the ORV individuates theories too finely to illuminate the gradual and extended process of theory building. The weakening of correspondence rules to an "interpretive systems" requirement in effect ties the meaning of any term in V to its inferential relationships to all of the others. Even relatively insignificant changes, such as the development of a new experimental technique, produce an entirely different theory (Suppe, 1977). Understanding gradual theory construction requires a diachronic notion of theory that persists through such changes (Schaffner, 1993a, chs 3 and 9).

The ORV analyzes successional theory change as *intertheoretic reduction or replacement*. On the most sophisticated account – Schaffner's generalized reduction/replacement (GRR) model (1993a, ch. 9) – reduction is the deductive subsumption of one (corrected) theory by another (restricted) theory. The reduced theory often has to be corrected because it is literally false, and the reducing theory often has to be restricted because the reduced theory is a special case of the reducing theory. As more revision and restriction are required, it becomes more appropriate to describe the successor theory as replacing, rather than reducing, its predecessor.

Some reductions are interlevel; theories about one intuitive ontic level are deductively subsumed by theories at another intuitive ontic level (as in the putative reduction of the ideal gas laws to statistical mechanics). This derivational view of interlevel relations tends to enforce a stratigraphic picture of science and of the world – a picture in which ontological levels map onto levels of theory which in turn map onto fields of science (Oppenheim and Putnam, 1968). On this caricature, theories at each level develop in relative isolation until it is possible to derive the higher level theory from the lower. Schaffner's inclusion of correction and revision in the GRR model accommodates the fact that theories at different levels may co-evolve under mutual correction and revision (Churchland, 1986; Bechtel, 1988).

Criticisms of the ORV

Virtually every aspect of the ORV has been attacked and rejected, but there is no consensus as to where it went wrong. There are as many different diagnoses as there are perspectives on science and its philosophy.[6] Here, I focus on the limitations of the ORV for describing theories "in the wild" (i.e., as they are constructed, conveyed, learned, remembered, presented, taught and tested by scientists). The charges are that

- the ORV misdescribes theory structure(s) in the wild (p. 58)
- the ORV distorts theory dynamics in the wild (p. 60), and that
- the ORV's emphasis on laws of nature makes it inapplicable to many accepted theories (p. 62).

Theory structure in the wild

The ORV is not typically defended as an accurate description of theories in the wild; rather, it is a regimented reconstruction of their shared inferential structure. A descriptive gulf between the ORV and theories in the wild can nonetheless suggest

(i)   that there are important structures of scientific theories that are neglected, de-emphasized, or at best awkwardly accommodated by the ORV, and
(ii) that there are significant aspects of the ORV that are peripheral to the uses of theories in the wild.

Attention to inferential structure pays dividends for regimenting arguments, but inferential patterns do not exhaust the useful patterns in scientific theories.

Multiple, partial, and incomplete theory formulations are neglected or homogenized Theories in the wild are sometimes written in a natural language; they are also charted, graphed, diagrammed, expressed in equations, explicated by exemplars, and (increasingly) animated in the streaming images of web pages. Only rarely are theories represented in first-order predicate calculus. Even the theories most amenable to tidy treatment on the ORV can be given different equivalent logical formulations and can be scripted with different formalisms, and these differences often significantly influence how the theories are used and how they represent the patterns in a domain. Regimenting theories into the ORV structure obscures the diverse representational tactics used by scientists when they deploy, express, and teach their theories; see, for example, Nersessian (1992).

Representations of theories in the wild are also often *partial or incomplete*. Trumpler's (1997) historical study of the development and refinement of different visual representations of the Na^{2+} channel is an excellent example. The theory, in this case, is partially represented by a host of representations (e.g., images of primary, secondary, and tertiary protein structure, circuit diagrams, current-to-voltage graphs, cartoons of possible mechanisms like that shown in Figure 4.1), none which represents the theory of how the Na^+ channel works in its entirety. Learning this theory involves internalizing these representations and mastering the reticulate connections among them. Theories in the wild are also frequently incomplete as they are cobbled together over time. Such incompleteness blocks derivational arguments, but is treated as an innocuous fact of life in science as practiced.

Nomological patterns emphasized over causal/mechanical patterns   Many criticisms of the "covering law" account of explanation turn on the importance of *causal/mechanical* rather than merely *nomological* patterns in our examples of intuitively good explanations. There are many now familiar examples – propagated in part by W. Salmon (1984; 1989): the elevation of the sun and the height of the flagpole explain the length of the pole's shadow and not vice versa; falling barometric pressure, and not the falling mercury in the barometer, explains the ensuing storm; and the current positions of the planets can be explained on the basis of their positions yesterday but not on the basis of their future wandering. Examples of this sort (and similar counter-examples to inductive explanations) can be used to argue for the explanatory importance of explicitly causal/mechanical patterns rather than merely inferential or nomological patterns; see Salmon (1989) but also see Kitcher (1989). Such criticisms apply equally to the descriptive adequacy of the ORV for accommodating and highlighting causal/mechanical patterns in theories; see page 67.

Mathematical structures are awkwardly accommodated   Finally, the restriction of the ORV to the first-order predicate calculus awkwardly accommodates the mathematics, statistics, and probabilities required for expressing the theories of, for example, quantum mechanics, relativity, and population genetics. As proponents of a model-based view of theories have emphasized (p. 64), set-theoretic (Suppes, 1967) and state-space approaches (Suppe, 1989) to representing theories naturally accommodate these mathematical relations and, in many cases, are, in fact,

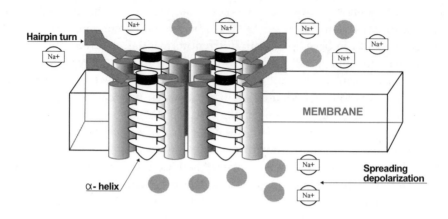

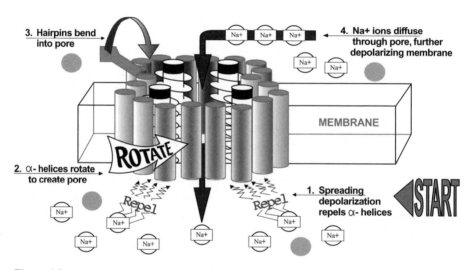

Figure 4.1

the representational conventions favored by the scientists (McKinsey and Suppes, 1953a, 1953b; Suppe, 1989).

Theory dynamics in the wild

A second major criticism of the ORV's descriptive adequacy is that it neglects or distorts the *dynamics* of scientific theories – the protracted process of generating, evaluating, revising, and replacing theories over time. For example, Darden (1991, ch. 2) argues that discovery has been neglected by traditional ORV-based approaches; Lloyd (1988) develops her alternative account of scientific theories to highlight aspects of theory testing that are neglected on the ORV; and Schaffner (1993a) emphasizes the importance of developing a diachronic account of theories. Close attention to science and its history have revealed aspects of theory dynamics that are neglected, or awkwardly accommodated within the ORV's strictures.

The generalization/abstraction account of theory building treats theory building as the joint application of inductive generalization and hypothetico-deduction. These strategies are incomplete, and leave unanswered questions about which inductive generalizations to draw (Goodman, 1983) and about how scientists generate the hypotheses from which to deduce predictions.

Successional accounts of theory change neglect or distort the gradual and piecemeal character of theory building. In the wild, grand clashes between rival hypotheses are infrequent and isolated compared to the more common process of articulating, refining, and elaborating a single theory over time. However, making sense of this gradual and piecemeal process of cobbling a theory together requires a diachronic notion of theories with criteria of individuation that accommodate such gradual changes. Arguments for the theory-ladenness of observation statements gloss successional theory change as a paradoxical choice among incommensurable theories (Kuhn, 1962; Feyerabend, 1965), obfuscating the reasoning involved in theory change over time. Furthermore, the ORV obscures the targeted nature of theory construction because the theory's ramified meaning structure makes it difficult to target praise or blame at parts of the theory. For these reasons, the ORV diverts attention from gradual and piecemeal construction, evaluation, and revision of theories over time; see Darden (1991, ch. 2) and compare with Wimsatt (1976).

Finally, the ORV's derivational account of reduction has been the subject of a variety of attacks discussed in Chapter 5 of this volume. One criticism worth emphasizing here is that derivational reductions are largely peripheral to many cases of reduction and theory succession in the wild (Schaffner, 1974; 1993a) and are accomplished, if ever, long after the interesting science is completed (P. S. Churchland, 1986, ch. 9). The derivational account of intertheoretic reduction is also unforgiving of gaps in the deductive argument, although, in the wild (there are many good examples in molecular and evolutionary biology, neuroscience, and medicine), both the predecessor and the successor theory are partial and incomplete to the point that derivation is out of the question. Additionally, the relationship between levels, scientific fields and theories has proved significantly more complicated than the Oppenheim–Putnam stratigraphy would suggest; both theories and fields in the biological sciences, for example, are characteristically multilevel.

The rigid strictures of the ORV leave it ill-suited for dealing with gradual and piecemeal theory change and also for highlighting the nonformal patterns that scientists use to construct, evaluate, and revise their theories.

Theories and laws

A third objection to the ORV is that there are legitimate theories in the wild (in e.g., molecular and evolutionary biology, neuroscience, and medicine) that lack ORV-style laws. Many have denied the importance of laws in physics as well (Cartwright, 1983; Giere, 1999). It would be dogmatic and unmotivated to insist that these scientific products are not theories. It is more plausible either

(a)   to insist that these theories do contain ORV-style laws, or
(b)   to give up the law requirement altogether.

Most have chosen some variant of (b). Opponents of (a) argue that the central generalizations in such theories are nonuniversal or restricted in scope (see next subsection), that they are physically contingent (p. 62), or that law statements in the wild are typically either false or vacuous (p. 63). Most advocates of (b) have chosen either to replace (or redefine) the notion of a law with something less stringent (p. 63) or to sidestep the issue entirely (p. 64).

Laws, universality, and scope   ORV-style laws are universal, unrestricted and exceptionless. Rosenberg (1985), Schaffner (1993a), and Smart (1963) have each suggested that (most) biological theories fail to satisfy these requirements. Theories in these domains hold only on earth (they are, at best, "terrestrially universal"), they often hold only for particular species, and they have exceptions even within species. Even the best candidates for universal biological laws, such as the theories of the genetic code and protein synthesis, are unlikely to hold for exotic life forms (e.g., in distant solar systems), and are known to have earthbound exceptions. Viruses use RNA as their genetic material, and proteins can be synthesized without a DNA template (Darden, 1996, p. 410); see also Beatty (1981, 1995). Thus biological laws are often restricted to particular species, strains, and individuals. This feature is not unique to the laws of biology; see Lange (1995) and Giere (1988, ch. 3; 1999, ch. 6).[7]

Laws and necessity   A second difficulty for ORV-style laws in biological theories is that many of the generalizations in such theories hold only by the grace of evolution by natural selection, and so are evolutionarily contingent (Beatty, 1995). Such generalizations might not have come to hold and may, some day, no longer hold. But laws are supposed to express what must necessarily be the case rather than what is accidentally (or contingently) the case. Beatty thus raises a rather more specific form of quite general worries about the kind of necessity by virtue of which statements of law can sort accidental generalizations from nonaccidental laws or generalizations that support counterfactuals from those that do not.

   One important challenge, if one is to maintain these distinctions, to do so without running afoul of what Earman (1986) calls an "empiricist loyalty test" and Lewis (1986) calls the doctrine of "Humean Supervenience" (HS). HS is the requirement there be no difference in the laws of nature without there being a difference in past, present, or future occurrent facts (i.e., particulars, their manifest properties, and their spatiotemporal relations). As Roberts (1999) argues, denying HS

(i)   amounts to a commitment that knowledge of the laws of nature is in principle forever beyond our grasp (the "epistemological problem") and

(ii) leaves one unable to specify which set of true propositions is the extension of the term "law of nature" (the "semantic problem").

The importance and tenability of HS have been challenged by Carroll (1994). Yet reconciling nomological necessity with HS remains a major challenge for the philosophy of science. Some are driven by these empiricist intuitions in HS into denying that there is any form of physical or natural necessity; this cement or glue is to be found only in models and not in the phenomena in their domains. Still others have sought this natural necessity in causal relations among objects, processes, or events. These suggestions are well beyond the scope of the present discussion.

Laws in the wild are typically inaccurate or vacuous   A third challenge for the ORV's emphasis on laws is that the best examples of laws hold only under a range of conditions that typically do not obtain, that cannot obtain or that cannot exhaustively be described (and so are glossed by so-called "*ceteris paribus*" clauses). Many laws hold only under extreme conditions (e.g., in the absence of air resistance, or assuming all other gravitational effects are negligible), and many specify what will happen under idealized conditions (e.g., assuming frictionless planes and point masses). In an effort to spell out the law's *ceteris paribus* conditions, one risks turning laws into meaningless truisms, i.e., the theory holds unless it does not hold (Hempel, 1965, pp. 166–7). On the other hand, unless all possibly confounding conditions are included in the law statement, the law is inaccurate. Criticisms of this sort have been most rigorously pursued by Cartwright (1983) and Giere (1999); for counter-arguments, see Earman and Roberts (1999).

Weakening the law requirement   One response to criticisms of ORV-style laws is to replace them with a weaker alternative. However, there is no foreseeable consensus as to what that alternative should be. Schaffner (1993a) distinguishes universal generalizations$_1$ and universal generalizations$_2$, the former applying to "all (terrestrial) organisms" (p. 121), and the latter "referring to the property illustrated by the phrase 'same cause (or same initial conditions and mechanisms), same effect'" (p. 121). Generalizations may have a restricted scope or known exceptions, but this does not detract from the fact that these generalization have the kind of necessity associated with the support for counterfactuals. Also, focusing on the importance of counterfactual support, Woodward (1997) has suggested that the required physical necessity can be supplied by "invariant" generalizations, those that hold under a range of interventions and so can be used to control or manipulate (and hence understand) some effect under conditions within that range (which may be rather limited). Still more pluralistically, Mitchell (2000) has suggested that ORV-style emphasis on universality, nonaccidentality, and unrestrictedness produces, "an impoverished conceptual framework that obscures much interesting variation in both the types of causal structures studied by the sciences and the types of representations used by scientists" (p. 243). In a similar spirit, Lange (1995) argues that laws of nature, as identified in scientific practice, need be neither exceptionless nor unrestricted to particular times and places. Instead, he suggests that statements of laws be identified by their functions in the practice of science and be characterized as warrants for reliable inferences (in the service of relevant purposes).

It is not necessary to abandon the ORV to accommodate theories without ORV-style laws of nature; one need only amend it by removing the law requirement or replacing it with something else. Some characterize laws as the axioms of the best system for describing the world, thus effectively removing the need to provide a conceptual analysis of law talk in terms of a checklist of properties they all share (Lewis, 1986). Others have sought to divorce the discussion of laws from discussion of theory structure by making claims about scope, necessity and universality extrinsic to the theory (p. 64).

Conclusion

Although the ORV neglects or distorts a wide range of interesting questions about science, an understanding of the logical patterns in scientific argument is indispensable for any account of the epistemology of science, and so the ORV is really the once *and future* received view, at least for some central questions in philosophy of science. Yet, the ORV is awkward at best in its treatment of theory building, laws, and the nonformal patterns exhibited by theories in the wild.

The "Model Model" of Scientific Theories

Some critics of the ORV have found its failings so systematic as to warrant an alternative formal approach to theory structure.[8] This alternative (or cluster of alternatives) has been dubbed the "semantic conception," the "nonstatement view," and the "models approach" to scientific theories. I will refer to it as the model model (MM).[9] MM was developed in part in response to criticisms of the sort discussed on page 58. MM offers a less restrictive framework for representing the nonformal patterns exhibited by theories but ultimately provides little guidance in characterizing and understanding these nonformal patterns.

Theories and models

The different versions of MM share a core commitment to viewing theories as an abstract specifications of a class of models.[10] The term "model" is notoriously ambiguous; meaning a representation or simulation (a scale model, map, or computer program), an abstraction (as in some mathematical models), an analogue (Bohr's planetary model of the atom), an experimental organism (as in the adult male Sprague–Dawley rat) or an experimental preparation (such as the amphetamine model of schizophrenia).

According to MM, a model is a structure that satisfies (i.e., renders true) a theory. The relationship between theories, models, and the real systems in the world can be understood as follows:

(i) *Theories* specify or define abstract or idealized systems.
(ii) *Models* are the structures that satisfy (or instantiate) these specifications or definitions (the abstract and idealized system is itself a model of the theory).
(iii) These models are more or less similar to, or homomorphic, with *real systems*, and so could be used to control and predict real systems if the real systems were sufficiently similar to the model.[11]

Theories as extralinguistic structures

Central to MM is the idea that theories are abstract extralinguistic structures quite removed from the phenomena in their domains. Theories are not identified with any particular representation. In this way, MM accommodates the diverse conventions for communicating theories in the wild (p. 58) as well as the mathematical structures that often compose theories (p. 59). Models may be partial, as are the diverse representations of the Na^+ channel, and they may very well be incomplete, giving MM a flexibility not available within the inferential strictures of the ORV. MM is motivated in part by its ability to accommodate the varied structures and states of completion of theories in the wild (Beatty, 1981; Beth, 1949; Lloyd, 1988; Suppe, 1977; van Fraassen, 1980, pp. 64–5).

Abstraction and idealization

According to MM, theories typically are not isomorphic to any real system; instead, they are more naturally thought of as homomorphic with, as replicas of (Suppe, 1989), or as similar to (Giere, 1999), real systems. Theories (and their models) are typically abstract and/or idealized. Theories are abstract to the extent that they describe real systems in terms of only a few of their relevant parameters, assuming that all others impact negligibly on the behavior of the system (Suppe, 1989, pp. 94–5). Theories are idealized if it is physically impossible for the real system to take on the allowable values of the parameters (e.g., point masses or friction-less planes).

On Suppe's counterfactual account of the relationship between theories/models and real systems, theories and models are *replicas* of real systems (Suppe says "phenomenal systems"). Replicas describe what a real system R would be like if it were isolated from the disturbing influence of parameters not included in the model M (Suppe, 1989, p. 95). Abstract models satisfy this requirement, since it is physically possible that R satisfy the conditions specified in M (perhaps under extreme experimental conditions). Idealized models satisfy this counterfactual requirement since the antecedent is physically impossible.[12] This counterfactual formulation is one means by which advocates of MM hope to sidestep the ORV's problems concerning laws of nature (p. 62).

MM, theories, and laws of nature

Suppe (1989) describes three varieties of laws appearing in scientific theories: laws of coexistence, laws of succession and laws of interaction.[13] Each of these may be deterministic or statistical. Laws of coexistence, such as the Boyle–Charles gas law, specify possible positions in the state space by describing equations fixing possible overall states of the system. Laws of succession, such as Newton's laws of motion, specify possible trajectories through the state space and so specify how the system, left to itself, will change over time. Finally, laws of interaction, specify the results of interaction between two or more systems, such as the interaction of a particle with a measuring device. These laws together define the class of models of the theory.

Advocates of MM split on the empirical status of both scientific theories of laws. Suppe's counterfactual account treats theories as empirical commitments as to how some real system would work if the abstracted variables were the only determinants of its behavior or if the idealizing conditions were met. Others – Beatty (1981), Giere (1999) and van Fraassen (1980) – see theories as definitions; theories define a class of models, and the empirical claims of science, as Beatty puts it, "are made *on behalf of* theories" (1981, p. 400, emphasis in original), asserting that some (type of) real system is an instance(s) of the kind of system defined by the theory (Giere, 1999, ch. 5).

These accounts are each motivated by difficulties with ORV-style laws (concerning scope, abstraction and idealization). The accounts differ as to whether theories express empirical commitments. On each account, questions about scope and universality are seen as external questions about the relation between a theory and the phenomena in its domain, questions to be answered by experiment and auxiliary hypotheses. This is a useful suggestion, since preoccupation with universality and unrestricted scope distracts attention from the fact that theories often have limited domains. Because theories are abstract and idealized, they typically do not apply universally. Abstract theories apply only to real systems for which the influence of extraneous variables is negligible; idealized theories literally have a scope of zero.

Each of these MM approaches to laws provides tools to grapple with issues of universality, scope, abstraction, and idealization. Suppes' approach is *prima faciea* more appealing because it sustains the reasonable claim that theories express empirical commitments. Neither approach clarifies the necessity of laws. Giere (1999, p. 96) suggests that the necessity of laws statements should, like issues of scope, be considered external to theories. This suggestion is unattractive primarily because many uses of theories (including explanation, control, and experimental design) depend crucially upon notions of necessity; an account of theories cannot cavalierly dismiss problems with laws precisely because laws (or something else filling their role) are so crucial to the functions of theories in science.

MM and the nonformal structures of scientific theories

In a recent elaboration of MM, Schaffner argues that most theories in the biomedical sciences (e.g., the clonal selection theory of immunology) are typically "overlapping interlevel temporal models" of less than universal scope (1993a, ch. 3). In doing so Schaffner is the first to clearly recognize and explore this prevalent nonformal structure of theories in the biological sciences. He terms these theories "theories of the middle range" (1993a) and shows how they can be accommodated within MM; see also Suppe (1989, ch. 8). Schaffner's "models" are essentially the same as those described above. These models have nonuniversal domains, and they are typically constructed around different "standard cases" or "experimental models" that serve as prototypes and are all more or less similar to one another (hence "overlapping"). Schaffner also recognizes a temporal component to the organization of these theories; they depict temporal pathways of sequential events related by generalizations. Finally, these theories are "interlevel" in that they include entities at different Oppenheim and Putnam-style levels. Schaffner (1993a) is a hair's breadth away from recognizing that many theories are multilevel descriptions of mechanisms; he then toys with this idea (Schaffner, 1993b).

MM avoids some of the criticisms of the ORV, especially those problems relating to representational flexibility, the abstraction and idealization of theories, and perhaps problems with laws of nature. Yet, the added abstraction of MM renders it even less informative than the ORV about nonformal patterns in theories in the wild.

Mechanisms: Investigating Nonformal Patterns in Scientific Theories

While MM accommodates nonformal patterns better than the ORV, it does little to highlight or motivate the search for them. Attention to nonformal patterns provides important resources for understanding how theories are built and the diverse kinds of explanations that scientific theories provide. Consider one kind of theory, theories about mechanisms, and notice how the nonformal patterns of such theories are used in the construction, evaluation, and revision of theories over time.

Mechanisms and their organization

Mechanisms are entities and activities organized such that they realize of regular changes from start or setup conditions to finish or termination conditions (Wimsatt, 1976; Bechtel and Richardson, 1993; Glennan, 1996; Machamer et al., 2000, p. 2). Entities are the objects in mechanisms; they are typically described with nouns in linguistic representations. Activities are what these entities do; they are typically described with verbs or depicted with arrows. Together, these component entities and activities are organized to do something – to produce the *behavior of the mechanism* as a whole, to use the term suggested by Glennan (1996); behaviors are the "regular changes" that mechanisms realize.[14]

Types of mechanisms can be individuated on the basis of their overall behavior, their component entities and activities, or the way the components are organized. First, mechanisms can differ behaviorally – by the phenomena that they realize. In specifying the behavior of a mechanism, one immediately constrains the entities, activities, and organizational structures that are relevant to that behavior, and so places a global constraint on the search for the mechanism. Mechanisms can also be individuated by the (kinds of) entities and activities that constitute their components. Finally, mechanisms can be individuated by their active, spatial, and temporal organization. A mechanism's *active organization* includes activities and interactions (excitatory and inhibitory) of the mechanism's component entities (Wimsatt, 1974; Craver, 2001). *Spatial organization* includes the relative locations, shapes, sizes, orientations, connections, and boundaries of the mechanism's entities. Finally, a mechanism's *temporal* organization includes the orders, rates, durations, and frequencies of its activities (Craver and Darden, 2001).

Consider this example. The voltage sensitive Na^+ channels in Trumpler's (1997) discussion are crucial components in the mechanism for producing action potentials, the electrical waves propagated as signals through neurons (this is the behavior of the mechanism as a whole). Neurons are electrically polarized at their resting membrane potential (approximately $-70\,mV$). The intracellular fluid is negatively charged with respect to the extracellular fluid because of differences between intracellular and extrecellular ion concentrations. Depolarization is a positive change in the membrane potential. Neurons depolarize during an action potential when voltage-sensitive Na^+ channels open, selectively allowing Na^+ ions to flood the cell, thereby spiking the membrane potential (peaking at roughly $+50\,mV$). One plausible mechanism for the activation of the Na^+ channel is represented in Figure 4.1 (drawn from Hall's (1992) verbal description).

Here is how the mechanism works (shown in the bottom panel). First, a small initial depolarization of the membrane (resulting from chemical transmission at synapses or spreading from elsewhere in the cell) repels the evenly spaced positive charges composing the α-helix. Second, the alpha helix rotates in each of the four protein subunits composing the channel. The rotation of the helix changes the conformation of the channel, creating a pore through the membrane. Third, the pore is lined with a "hairpin turn" structure containing charges that select specifically Na^+ ions to flow into the cell by diffusion. This panel depicts the mechanism's active and temporal organization; it shows an orderly sequence of steps (repelling, rotating, opening, and diffusing), each systematically dependent on, and productively continuous with, its predecessor.

The top pannel depicts the set-up conditions for this mechanism, including the relevant entities (Na^+ ions, α-helices, hairpin turns), their relative sizes, shapes, positions, locations (e.g., the channel spans the membrane, and Na^+ ions fit through the pore), and the connections, compartments, and boundaries between them. Not represented in the diagram are such factors as temperature, pH, and the relevant ionic concentrations. Such factors are the background or standing conditions upon which the behavior of the mechanism crucially depends. Figure 4.1 thus nicely illustrates the active, spatial, and temporal organization of the components in the mechanism of Na^+ channel activation, but it nonetheless abstracts from several crucial parameters for the working of the mechanism.

Mechanism schemata

Mechanistic theories are mechanism schemata. Like MM-theories, mechanism schemata are abstract and idealized descriptions of a type of mechanism. They describe the behavior of the mechanism, its component entities and activities, their active, spatial, and temporal organization, and the relevant background conditions affecting the application of the theory. The scope of mechanism schemata can vary considerably, from no instances (for idealized descriptions) to universality, and any point between.

Levels

Mechanism schemata often describe hierarchically organized networks of mechanisms nested within mechanisms. In such schemata, higher-level activities (ψ) of mechanisms as a whole (S) are realized by the organized activities (ϕ) of lower-level components (Xs), and these are, in turn, realized by the activities (σ) of still lower-level components (Ps). The gating (σ) of the Na^+ channel (P) is part of the mechanism (X) for generating action potentials (ϕ), which is part of almost every brain mechanism involving electrical signals. The relationship between lower and higher mechanistic levels is a part-whole relationship with the additional restriction that the lower-level parts are components of (and hence organized within) the higher-level mechanism. Lower level entities (e.g. Xs) are proper parts of higher-level entities (S), and so the Xs are no larger, and typically smaller, than S; they are within S's spatial boundaries. Likewise, the activities of the lower-level parts are steps or stages in the higher-level activities. Exactly how many levels there are, and how they are to be individuated, are empirical questions that are answered differently for different phenomena (Craver, 2001).

Mechanistic hierarchies should not be confused with intuitive ontic hierarchies, which map out a monolithic stratigraphy of levels across theories, entities, and scientific fields. Mechanistic hierarchies are domain specific, framed with respect to some highest system S and its ψ-ing. The parts in mechanistic hierarchies are components organized (actively, spatially, temporally, and hierarchically) to realize the behavior of the mechanism as a whole. This distinguishes mechanistic wholes from mere aggregates (such as piles of sand), mere collections of improper parts (such as the set of 1-inch cubes that compose my dog, Spike), and mere inclusive sets (such as the albums in the Clash discography). There are no doubt many senses of "level" that are not sufficiently distinct in the philosophical literature. Sorting them out is an important and unresolved project in the philosophy of science (Simon, 1969; Wimsatt, 1974; Haugeland, 1998).

Varieties of mechanisms

Both the ORV and MM are pitched too abstractly to capture recurrent non-formal patterns exhibited by mechanism schemata: patterns in the organization of mechanisms that are crucial for understanding how these theories explain and how they are constructed over time. Consider one branch in a possible (nonexclusive) taxonomy of mechanisms.

Begin with *etiological mechanisms* and *constitutive mechanisms* (Shapere, 1977; Salmon, 1984, ch. 9). Etiological mechanisms (such as natural selection) include the organized entities and activities antecedent to and productive of the phenomenon to be explained (e.g., the mechanism by which a trait comes to be fixed in a population). Constitutive mechanisms (like the mechanism of Na^+ channel gating) realize (rather than produce) higher-level phenomena; these higher-level phenomena are contemporaneous with (rather than subsequent to) and composed of (rather than produced or effected by) the organized activities of lower-level components.

Etiological mechanisms include both *structuring mechanisms* and *triggering mechanisms*. Dretske (1995) has distinguished "structuring causes" from "triggering causes," on the grounds that the triggering cause T completes a set of otherwise insufficient preexisting conditions C thus making ($T + C$) a sufficient cause of the explanandum event or phenomenon E. For example, spreading depolarization (T), given the Na^+ channel setup (C), triggers the opening of the channel (E). A structuring cause U, in contrast, prepares the conditions C within which

T can be a triggering cause and so produces the mechanism linking *T* and *C* (1995, p. 124). For example, one may perhaps look to evolutionary theory to explain how the sodium channel came to activate under conditions of slight depolarization. In triggering mechanisms, *T* in *C* is sufficient for *E*; in structuring mechanisms, *U* produces the mechanism by which *T* is sufficient for *E*.

Two etiological varieties of structuring mechanisms are *selective* and *instructive mechanisms*. In selective mechanisms, a population of variants is produced (relatively) independently of environmental influences and then, by virtue of some critical environmental factor, the set of variants is changed such that certain traits are increasingly represented in the population. Examples of selective mechanisms include evolution by natural selection, clonal selection for antibodies in immunology, and perhaps neural Darwinism; each is discussed in Darden and Cain (1989). Instructive mechanisms (such as inheritance of adaptive acquired characteristics or pedagogy) are different in the first stage, since the production of adaptive variants is directly influenced by features of the population's environment.

Different types of mechanisms can be distinguished on the basis of recurrent patterns in their organization. Mechanisms may be organized in series, in parallel, or in cycles. They may contain branches and joins, and they often include feedback and feedforward subcomponents. Some mechanisms are redundantly organized, and some have considerable capacity for reorganization or plasticity in the face of damage. These recurrent patterns in mechanistic organization have been investigated by Wimsatt (1986), but there remains considerable work to be done in sorting out the axes along which mechanisms and schemata might differ.

Scientific theories exhibit a variety of patterns in domains of empirical phenomena, patterns that are invisible if one abstracts too far away from the details of scientific theories in the wild. Attention to these details pays dividends for understanding mechanistic explanation (next section) and the process of building multilevel mechanism schemata (p. 72).

Mechanistic explanation

Mechanism schemata explain not by fitting a phenomenon into a web of inferential relationships but by characterizing the mechanism by which the phenomenon is produced or realized. This suggestion is consistent with the MM-related account of explanation as pattern completion, or prototype activation (Giere, 1999, ch. 6; Churchland, 1989), but insists, in addition, on an explanatory role for the nonformal patterns in these theories. Not all patterns are explanatory; one goal is to distinguish those that are from those that are not. Salmon (1984) has suggested that at least one important kind of kind of explanation involves tracing pathways in a causal nexus; a phenomenon is explained by showing how that phenomenon fits into a pattern of causal processes and their interactions. Mechanistic patterns are further distinguished by their active, spatial, temporal, and hierarchical organization; and these features of mechanism schemata draw our attention to salient features relevant to the intelligibility provided by a description of a mechanism.

Scriven (1962) emphasizes the narrative structure of many explanations. There are no good stories without verbs. The verbs the Na^+ channel schema include "repelling," "rotating," "opening," and "diffusing." Verbs provide the productive continuity in the mechanism, intelligibly linking earlier stages to later stages. Substantivalists in the philosophy of science have emphasized static structures, occurrent events, entities and relations over dynamic activities, extended processes, changes and forces. Substantivalists nominalize or neglect active features of scientific ontology, the diverse kinds of changing that underlie regularities; they leave out the verbs. This neglect can be redressed with attention to types of activities, criteria for their individuation, and the differences between the scientific investigation of activities and entities (Machamer et al., 2000).

Emphasizing the importance of activities in mechanisms cannot sidestep the problems with laws of nature discussed on pages 62 and 67. An adequate account of mechanism schemata must await an account of how activities are different from mere regularities. Some progress on these problems will be gained by exploring the connections between the mechanistic perspectives on theory structure sketched here and recent work on laws (Lange, 1995; Roberts, 1999), invariant generalizations (Woodward, 1997), physical causality (Dowe, 1992), capacities (Cartwright, 1989; Glennan, 1997), and the pragmatics of laws (Mitchell, 2000). A fresh perspective might be provided by investigating the practices of scientists as they introduce, individuate, characterize, and describe the activities picked out by the verbs in mechanism schemata.

Constructing mechanism schemata

Attention to the nonformal patterns exhibited by theories has already yielded dividends in thinking about theory construction. For example, Bechtel and Richardson (1993) discuss decomposition and localization as research strategies in the construction of mechanistic theories. Craver and Darden (2001) have extended this work, showing that the construction of mechanism schemata typically proceeds gradually and piecemeal by revealing constraints on the mechanism, constraints from the behavior of the mechanism, the available entities and activities for the mechanism, and features of their active, spatial, temporal, and hierarchical organization. Finding such empirical constraints prunes the space of plausible mechanisms and often suggests potentially fruitful avenues for further research.

One goal in constructing a description of a mechanism is to establish a seamless productive continuity of the mechanism, without gaps, from beginning to end. In pursuit of this goal, researchers frequently forward chain, using known stages early in the mechanism to conjecture or predict stages that are likely to follow, and backtrack, using known stages late in the mechanism to conjecture or predict the entities, activities, or organizational features earlier in the mechanism. Non-formal aspects of theory structure are used by scientists to generate new hypotheses and to target the praise and blame from empirical tests at specific portions of the theory (Darden and Craver, 2001).

A second goal in constructing specifically multilevel mechanism schemata is to integrate the different levels together into a description of one coherent mechanism. *Interlevel integration* involves elaborating and aligning the levels in a hierarchy to show, for some X's ϕ-ing

(i) how it fits into the organization of a higher level mechanism for S's ψ-ing, and
(ii) how it can be explained in terms of the constitutive mechanism (the organized σ-ing of ps).

These levels are linked together through research strategies that exhibit the constitutive causal relevance of lower level organized entities and activities to higher level entities and activities. In this way, upward looking and downward looking research strategies combine to provide an integrated description of the pattern exhibited by a multilevel mechanism (Craver, 2001).

Conclusion

Scientific theories have many different structures, structures that exhibit patterns in diverse domains of phenomena. Inferential patterns are crucial to understanding some aspects of science and the way that it changes over time. But there is a great deal more to be said about these patterns than can be said by assimilating them to an inferential pattern. Nonformal patterns (such as mechanistic patterns) are also important for understanding how theories are used and constructed. Closer scrutiny of the diverse structures of scientific theories, especially mechanistic patterns, is likely to pay serious dividends for understanding science and scientific practice.

Notes

1 Thanks to Lindley Darden, Peter Machamer, and Ken Schaffner for their time and help.
2 Patterns can be understood, following Dennett (1991), either in terms of their ability to be recognized or in terms of their susceptibility to expression in something less than a "bit map"; see also Haugeland (1998); Toulmin's (1953) discussion of maps is in many ways similar to this notion of a pattern). A "domain" following Shapere (1977) is some body of items of "information" variously interrelated in a way that helps one to solve an important problem that science is ready to tackle at a given time (Shapere, 1977, p. 525).

3   Classic statements of the ORV can be found in Braithwaite (1953), Carnap ([1939] 1989), Duhem (1954), Hempel (1965, chs 4 and 8; 1966, ch. 6) and Nagel (1961, chs 5 and 6). Valuable critical expositions include Suppe (1977, 50–1; 1979; 1989) and Thompson (1989, chs 2 and 3). The ORV was developed primarily for the expression of physical theories, but it has been applied with debatable success to evolutionary biology and/or population genetics (Braithwaite, 1953; Hull, 1974; Ruse, 1973; Williams, 1970), and psychology (Skinner, 1945).

4   This inferential approach to scientific theories has been dubbed the "received" (Putnam, 1962) or "orthodox" (Feigl, 1970) view, the "statement view," the "syntactic conception" (Thompson, 1989), the "hypothetico-deductive" account (Lloyd, 1988), "the Euclidean ideal" (Schaffner, 1993a,b), and the "sentential" or "propositional" account (Churchland, 1989). I call it the ORV to flag its waning hold on the philosophy of science and to avoid enshrining in a name a single interpretation of either the ORV or of its shortcomings.

5   This image of theory structure was inspired at least in part by Russell and Whitehead's efforts to reduce mathematics to logic.

6   Some object to theory-centered approaches to the philosophy of science generally. Among these, "Globalists" focus on more inclusive units of analysis than theories, recommending such alternatives as disciplinary matrices or paradigms (Kuhn, 1962), fields (Darden and Maull, 1977; Darden, 1991), practices (Kitcher, 1993, p. 74), research programs (Lakatos, 1970), and traditions (Laudan, 1977). These global units of science include, in addition to theories, also experimental techniques, institutional practices, consensual standards and norms, organizations, and worldviews. "New Experimentalists," on the other hand, decenter theories in the analysis of science and center experimentation instead (Hacking, 1983; Galison, 1987; Rheinberger, 1997). Still others, with primarily epistemological concerns, have criticized correspondence rules, the theory/observation distinction, and the tenability of scientific realism (Achinstein, 1968; Putnam, 1962; Schaffner, 1969; van Fraassen, 1980). Suppe (1977) is the definitive history of this line of criticism.

7   One response to this line of criticism, one pursued by Waters (1998), is to argue that philosophers have mistakenly confused universal causal regularities with distributions (claims about how a trait or property is distributed across a population of organisms). One way of putting this is that the law $(x)(Fx \supset Gx)$ is true of everything (a universal causal generalization), although only some things satisfy the antecedent (a distribution).

8   Important statements and elaborations of the model model include Beth (1949), Giere (1979; 1988), Schaffner (1993a), Suppe (1977; 1989), Suppes (1967), and van Fraassen (1980). Beth (1949) applied this approach to Newtonian and quantum mechanics, and it has been worked out for theories in classical mechanics (McKinsey and Suppes, 1953), quantum mechanics (van Fraassen, 1991), evolutionary theory and population genetics (Beatty, 1980; 1981; Lloyd, 1988, ch. 2; Thompson 1989, ch. 5), sociobiology (Thompson, 1989), biological taxonomy (Suppe, 1989, ch. 7) and most recently, declarative memory and synaptic mechanisms in neuroscience (Bickle, 1998).

9   There is no consensus on how to draw the contrast between the ORV and MM. The most common approach relies on the distinction between syntax and semantics, a distinction that hardly clear in its own right and one that has been difficult to apply neatly

to ORV and MM. Another contrast is between ORV as a "statement view" of theories and MM as a "nonstatement view," but statements can be models and the components of the ORV might be reasonably interpreted as propositions rather than statements. Some have argued that anything representable in the ORV can be represented in MM and vice versa, minimizing the motivation to spell out the differences in detail. Little of significance has turned on getting this distinction right.

10 There are two classic formulations of MM: a *set theoretic formulation*, recommended by Sneed (1971), Stegmüller (1976), and Suppes (1967), according to which theories are structures represented by set theoretic predicates that define a class of models; and a *state-space approach*, favored by Beth (1949), Suppe (1989) and van Fraassen (1980), according to which theories are constraints on multidimensional state-spaces or configurations of sets of such spaces which define a class of models. Debates over the relative merits of these approaches can be safely neglected for present purposes (van Fraassen, 1972; Suppe, 1979).

11 I neglect a fourth element, a "phenomenal system" (Suppe, 1989) or an "empirical model" (Lloyd, 1988) that is constructed on the basis of data and intermediate between models and real systems.

12 This suggestion, if I understand it correctly, has the strongly counterintuitive consequence of rendering all idealized stems models of any given real system.

13 Suppe (1989) also includes laws of quasi-succession.

14 On one reasonable interpretation of this realizing relationship – modified from Kim (1995), discussing Lepore and Loewer (1987) – a mechanism M composed of the actively, spatially, and temporally organized ϕ-ing of Xs realizes S's ψ-ing just in case

(i)   it is physically impossible of S's ψ-ing to differ without there being some difference in M, and

(ii) S's ψ-ing is exhaustively explained by M (in an ontic and not necessarily epistemic sense).

This way of spelling out the realization relationship differs in that it specifies more precisely the character of the organizing relationships involved in realizing a higher-level phenomenon.

References

Achinstein, P. (1968): *Concepts of Science: A Philosophical Analysis*. Baltimore, MD: Johns Hopkins University Press.

Beatty, J. (1980): "Optimal-Design Models and the Strategy of Model Building in Evolutionary Biology," *Philosophy of Science*, 47, 532–61.

Beatty, J. (1981): "What's Wrong With the Received View of Evolutionary Theory?" in P. D. Asquith and R. N. Giere (eds.), *PSA 1980, vol. 2*, East Lansing, MI.: Philosophy of Science Association, 397–426.

Beatty, J. (1995): "The Evolutionary Contingency Thesis," in J. G. Lennox and G. Wolters (eds.), *Concepts, Theories and Rationality in the Biological Sciences*, Konstanz, Germany: University of Konstanz Press and Pittsburgh, PA: University of Pittsburgh Press, 45–81.

Bechtel, W. (1988): *Philosophy of Science: An Overview for Cognitive Science*. Hillsdale, N.J.: Erlbaum.

Bechtel, W. and Richardson, R. (1993): *Discovering Complexity: Decomposition and Localization as Strategies in Scientific Research*. Princeton: Princeton University Press.

Beth, E. (1949): "Towards an Up-to-Date Philosophy of the Natural Sciences," *Methodos*, 1, 178–85.

Bickle, J. (1998): *Psychoneuronal Reduction: The New Wave*. Cambridge, MA: MIT Press.

Braithwaite, R. (1953): *Scientific Explanation*. Cambridge: Cambridge University Press.

Carnap, R. ([1939] 1989): "Theories as Partially Interpreted Formal Systems," in B. A. Brody and R. E. Grandy, (eds.), *Readings in the Philosophy of Science*, Englewood Cliffs, NJ: Prentice Hall, 5–11. Reprinted from Carnap, *Foundations of Logic and Mathematics*, Chicago: University of Chicago Press.

Caroll, J. (1994): *Laws of Nature*. Cambridge: Cambridge University Press.

Cartwright, N. (1983): *How the Laws of Physics Lie*. Oxford: Clarendon Press.

Cartwright, N. (1989): *Nature's Capacities and their Measurement*. Oxford: Oxford University Press.

Churchland, P. M. (1989): *A Neurocomputational Perspective*. Cambridge, MA: MIT Press.

Churchland, P. S. (1986): *Neurophilosophy*. Cambridge, MA: MIT Press.

Craver, C. F. (2001): "Role Functions, Mechanisms, and Hierarchy," *Philosophy of Science*, 68(1), 53–74.

Craver, C. F. and Darden, L. (2001): "Discovering Mechanisms in Neurobiology: The Case of Spatial Memory," in P. K. Machamer, R. Grush and P. McLaughlin (eds.), *Theory and Method in the Neurosciences*, Pittsburgh, PA: University of Pittsburgh Press, 112–37.

Darden, L. (1991): *Theory Change in Science: Strategies from Mendelian Genetics*. New York: Oxford University Press.

Darden, L. (1996): "Generalizations in Biology," *Studies in the History and Philosophy of Science*, 27, 409–19.

Darden, L. and Cain, J. A. (1989): "Selection Type Theories," *Philosophy of Science*, 56, 106–29.

Darden, L. and Craver, C. F. (2001): "Interfield Strategies in the Discovery of the Mechanism of Protein Synthesis," *Studies in the History and Philosophy of Biology and Biomedical Sciences*, forthcoming.

Darden, L. and Maull, N. (1977): "Interfield Theories," *Philosophy of Science*, 44, 43–64.

Dennett, D. (1991): "Real Patterns," *Journal of Philosophy*, 88, 27–51.

Dowe, P. (1992): "Wesley Salmon's Process Theory of Causality and the Conserved Quantity Theory," *Philosophy of Science*, 59, 195–216.

Dretske, F. (1995): "Mental Events as Structuring Causes," in J. Heil and A. Mele (eds.), *Mental Causation*, Oxford: Clarendon Press, 121–36.

Duhem, P. (1954): *Aim and Structure of Physical Theory*. New York: Atheneum.

Earman, J. (1986): *A Primer on Determinism*. Dordrecht: D. Reidel.

Earman, J. and Roberts, J. (1999): "Ceteris Paribus, There's No Problem of Provisos," *Synthese*, 118, 439–78.

Feigl, H. (1970): "The 'Orthodox View' View of Theories: Remarks in Defense as Well as Critique," in M. Radner and S. Winkour (eds.), *Minnesota Studies in the Philosophy of Science, vol. 4*, Minneapolis: University of Minnesota Press, 3–16.

Feyerabend, P. (1965): "On the Meaning of Scientific Terms," *Journal of Philosophy*, 62, 266–74.

Galison, P. (1987): *How Experiments End*. Chicago: University of Chicago Press.

Giere, R. N. (1979): *Understanding Scientific Reasoning*. New York: Holt, Rinhart and Winston.

Giere, R. N. (1988): *Explaining Science: A Cognitive Approach*. Chicago: University of Chicago Press.

Giere, R. N. (1999): *Science Without Laws*. Chicago, IL: University of Chicago Press.

Glennan, S. S. (1996): "Mechanisms and the Nature of Causation," *Erkenntnis*, 44, 49–71.

Glennan, S. S. (1997): "Capacities, Universality, and Singularity," *Philosophy of Science*, 64, 605–26.

Goodman, N. (1983): *Fact, Fiction, and Forecast, 4th edn*. Cambridge, Ma: Harvard University Press.

Hacking, I. (1983): *Representing and Intervening*. Cambridge: Cambridge University Press.

Hall, Z. W. (1992): *An Introduction to Molecular Neurobiology*. Sunderland, Ma: Sinauer Associates, Inc.

Haugeland, J. (1998): *Having Thought: Essays in the Metaphysics of Mind*. Cambridge, Massachusetts: Harvard University Press.

Hempel, C. G. (1965): *Aspects of Scientific Explanation*. New York: Free Press.

Hempel, C. G. (1966): *Philosophy of Natural Science*. Englewood Cliffs, N.J.: Prentice-Hall, Inc.

Hull, D. L. (1974): *Philosophy of Biological Science*. Englewood Cliffs: Prentice Hall.

Kim, J. (1995): "The Non-Reductivist's Troubles with Mental Causation," in J. Heil and A. Mele (eds.), *Mental Causation*, Oxford: Clarendon Press, 189–210.

Kitcher, P. (1989): "Explanatory Unification and the Causal Structure of the World," in P. Kitcher and W. C. Salmon (eds.), *Scientific Explanation, Minnesota Studies in the Philosophy of Science XVIII*, Minneapolis: University of Minnesota Press, 410–505.

Kitcher, P. (1993): *The Advancement of Science*. Oxford: Oxford University Press.

Kitcher, P. and Salmon, W. C. (eds.) (1989): *Scientific Explanation, Minnesota Studies in the Philosophy of Science XVIII*. Minneapolis: University of Minnesota Press.

Kuhn, T. (1962): *The Structure of Scientific Revolutions*. Chicago: University of Chicago Press.

Lakatos, I. (1970): "Falsification and the Methodology of Scientific Research Programmes," in I. Lakatos and A. Musgrave (eds.), *Criticism and the Growth of Knowledge*, Cambridge: Cambridge University Press, 91–196.

Lange, M. (1995): "Are there Natural Laws Concerning Particular Species?" *Journal of Philosophy*, 92(8), 430–51.

Laudan, L. (1977): *Progress and Its Problems*. Berkeley: University of California Press.

Lepore, E. and Loewer, B. (1987): "Mind Matters," *Journal of Philosophy*, 84, 630–42.

Lewis, D. (1986): *Philosophical Papers, vol. 2*. Oxford: Oxford University Press.

Lloyd, E. A. (1988): *The Structure and Confirmation of Evolutionary Theory*. New York, NY: Greenwood Press.

Machamer, P. K., Darden, L. and Craver, C. F. (2000): "Thinking About Mechanisms," *Philosophy of Science*, 67, 1–25.

McKinsey, J. C. C. and Suppes, P. (1953a): "Axiomatic Foundations of Classical Particle Mechanics," *Journal of Rational Mechanics and Analysis*, 2, 253–72.

McKinsey, J. C. C. and Suppes, P. (1953b): "Transformations of Systems of Classical Particle Mechanics," *Journal of Rational Mechanics and Analysis*, 2, 273–89.

Mitchell, S. (2000): "Dimensions of Scientific Law," *Philosophy of Science*, 67, 242–65.

Nagel, E. (1961): *The Structure of Science: Problems in the Logic of Scientific Explanation.* New York, NY: Harcourt, Brace & World, Inc.

Nersessian, N. J. (1992): "How Do Scientists Think? Capturing the Dynamics of Conceptual Change," in R. N. Giere (ed.) *Cognitive Models of Science. Minnesota Studies in the Philosophy of Science, XV*, Minneapolis: University of Minnesota Press, 3–44.

Oppenheim, P. and Putnam, H. (1968): "Unity of Science as a Working Hypothesis," in H. Feigl, M. Scriven and G. Maxwell (eds.), *Concepts, Theories, and the Mind-Body Problem, Minnesota Studies in the Philosophy of Science II*, Minneapolis: University of Minnesota Press, 3–36.

Putnam, H. (1962): "What Theories Are Not," in E. Nagel, P. Suppes and A. Tarski (eds.), *Logic, Methodology and Philosophy of Science: Proceedings of the 1960 International Congress*, Stanford, CA: Stanford University Press, 240–51.

Rheinberger, H. (1997): *Towards a History of Epistemic Things.* Stanford: Stanford University Press.

Roberts, J. (1999): *"Laws of Nature: Meeting the Empiricist Challenge,"* doctoral dissertation, Department of Philosophy, University of Pittsburgh, Pittsburgh, PA.

Rosenberg, A. (1985): *The Structure of Biological Science.* Cambridge: University of Cambridge Press.

Ruse, M. (1973): *The Philosophy of Biology.* London: Hutchinson & Co. Ltd.

Salmon, W. C. (1984): *Scientific Explanation and the Causal Structure of the World.* Princeton: Princeton University Press.

Salmon, W. C. (1989): "Four Decades of Scientific Explanation," in P. Kitcher and W. C. Salmon (eds.), 3–219.

Schaffner, K. F. (1969): "Correspondence rules," *Philosophy of Science*, 36, 280–90.

Schaffner, K. F. (1974): "The Peripherality of Reductionism in the Development of Molecular Biology," *Journal of the History of Biology*, 7, 111–39.

Schaffner, K. F. (1993a): *Discovery and Explanation in Biology and Medicine.* Chicago: University of Chicago Press.

Schaffner, K. F. (1993b): "Theory Structure, Reduction, and Disciplinary Integration in Biology," *Biology and Philosophy*, 8, 319–47.

Scriven, M. (1962): "Explanations, Predictions, and Laws" in H. Feigl and G. Maxwell (eds.), *Scientific Explanations, Space, and Time, Minnesota Studies in the Philosophy of Science III*, Minneapolis: University of Minnesota Press, 170–230.

Shapere, D. (1977): "Scientific Theories and their Domains," in F. Suppe (ed.), *The Structure of Scientific Theories*, Urbana, IL: University of Illinois Press, 518–65.

Simon, H. A. (1969): *The Sciences of the Artificial.* Cambridge, MA.: MIT University Press.

Skinner, B. F. (1945): "The Operationalist Analysis of Psychological Terms," *Psychological Review*, 52, 270–7.

Smart, J. J. C. (1963): *Philosophy and Scientific Realism.* London: Routledge and Kegan Paul.

Sneed, J. (1971): *The Logical Structure of Mathematical Physics.* Dordrecht: Reidel.

Stegmüller, W. (1976): *The Structure and Dynamics of Theories.* Amsterdam: North-Holland.

Suppe, F. (1977): "Introduction," in F. Suppe (ed.), *The Structure of Scientific Theories*, Urbana, IL: University of Illinois Press. 3–241.

Suppe, F. (1979): "Theory Structure," in P. Asquith and H. Kyburg (eds.), *Current*

Research in Philosophy of Science, East Lansing, MI: Philosophy of Science Association, 317–38.

Suppe, F. (1989): *The Semantic Conception of Theories and Scientific Realism*. Urbana, IL: University of Illinois Press.

Suppes, P. (1967): "What is a Scientific Theory?" in S. Morgenbesser, (ed.), *Philosophy of Science Today*, New York: Basic Books, 55–67.

Thompson, P. (1989): *The Structure of Biological Theories*. Albany, NY: State University of New York Press.

Toulmin, S. (1953): *The Philosophy of Science*. London: Hutchinson and Co. Ltd.

Trumpler, M. (1997): "Converging Images: Techniques of Intervention and Forms of Representation of Sodium-Channel Proteins in Nerve Cell Membranes," *Journal of the History of Biology*, 30, 55–89.

van Fraassen, B. (1972): "A Formal Approach to Philosophy of Science," in R. Colodny (ed.), *Paradigms and Paradoxes*, Pittsburgh: University of Pittsburgh Press. 303–66.

van Fraassen, B. (1980): *The Scientific Image*. Oxford: Clarendon Press.

van Fraassen, B. (1991): *Quantum Mechanics: An Empiricists View*. New York: Oxford University Press.

Waters, K. (1998): "Causal Regularities in the Biological World of Contingent Distributions," *Biology and Philosophy*, 13, 5–36.

Williams, M. (1970): "Deducing the Consequences of Evolution," *Journal of Theoretical Biology*, 29, 343–85.

Wimsatt, W. (1974): "Complexity and Organization," in K. Schaffner and R. Cohen (eds.), *Proceedings of the 1972 Biennial Meeting, Philosophy of Science Association, Boston Studies in the Philosophy of Science XX*, Dordrecht, Holland: Reidel Publishing Company, 67–86.

Wimsatt, W. (1976): "Reductive Explanation: A Functional Account," in R. S. Cohen, G. Globus, G. Maxwell and I. Savodnik (eds.), *PSA-1974: Proceedings of the 1974 Biennial Meeting Philosophy of Science Association*, Dordrecht: D. Reidel, 205–67.

Wimsatt, W. (1986): "Forms of Aggregativity," in A. Donagan, A. N. Perovich, Jr. and M. V. Wedin (eds.), *Human Nature and Natural Knowledge*, Dordrecht: Reidel, 259–91.

Woodward, J. (1997): "Explanation, Invariance, and Intervention," *Philosophy of Science*, 64, Proceedings, S26–S41.

Chapter 5

Reduction, Emergence and Explanation

Michael Silberstein

Introduction: The Problem of Emergence and Reduction

Can everything be reduced to the fundamental constituents of the world? Or can there be, and are there, non-reducible, or emergent entities, properties and laws? What exactly do we mean by "reduction" and "emergent" when we ask such questions? For example, if everything can be reduced to the fundamental constituents of the world, does that preclude the existence of emergent entities, properties or laws? Obviously, the answers to many of these questions depend on what is meant by the terms "reduction" and "emergence." These terms are used in a variety of ways in the literature, none of which is uniquely privileged or uniform. Therefore, clarity is crucial to avoid confusion and equivocation. The first task of this chapter is to sort out and schematize the main versions of reduction and emergence, and then to turn to the current debates. The current state of the reductionism vs. emergentism debate is examined and the Final section looks toward future debate.

Historically, there are two main construals of the problem of reduction and emergence: ontological and epistemological; see Stephan (1992), McLaughlin (1992) and Kim (1999) for historical background.

- The ontological construal: is there some robust sense in which everything in the world can be said to be *nothing but* the fundamental constituents of reality (such as super-strings) or at the very least, *determined by* those constituents?

- The epistemological construal: is there some robust sense in which our scientific theories/schemas (and our common-sense experiential conceptions) about the macroscopic features of the world can be *reduced to* or *identified with* our scientific theories about the most fundamental features of the world?

Yet, these two construals are inextricably related. For example, it seems impossible to justify ontological claims (such as the cross-theoretic identity of conscious mental processes with neurochemical processes) without appealing to epistemological claims (such as the attempted intertheoretic reduction of folk psychology to neuroscientific theories of mind) and vice versa. We would like to believe that the unity of the world will be described in our scientific theories and, in turn, the success of those theories will provide evidence for the ultimate unity and simplicity of the world; things are rarely so straightforward.

Historically "reductionism" is the "ism" that stands for the widely held belief that both ontological and epistemological reductionism are more or less true. Reductionism is the view that the best understanding of a complex system should be sought at the level of the structure, behavior and laws of its component parts plus their relations. However, according to mereological reductionism, *the relations* between basic parts are themselves reducible to the intrinsic properties of the relata (see below). The ontological assumption implicit is that the most fundamental physical level, whatever that turns out to be, is ultimately the "real" ontology of the world, and anything else that is to keep the status of real must somehow be able to be 'mapped onto' or 'built out of' those elements of the fundamental ontology. Relatedly, fundamental theory, *in principle*, is deeper and more inclusive in its truths, has greater predictive and explanatory power, and so provides a deeper understanding of the world.

"Emergentism", historically opposed to reductionism, is the "ism" according to which both ontological and epistemological emergentism are more or less true, where ontological and epistemological emergence are just the negation of their reductive counterparts. Emergentism claims that a whole is "something more than the sum of its parts", or has properties that cannot be understood in terms of the properties of the parts. Thus, emergentism rejects the idea that there is any fundamental level of ontology. It holds that the best understanding of complex systems must be sought at the level of the structure, behavior and laws of the whole system and that science may require a plurality of theories (different theories for different domains) to acquire the greatest predictive/explanatory power and the deepest understanding.

The problem of reduction and emergence is (and has been) of great interest and importance in philosophy and scientific disciplines from physics to psychology; see *Philosophical Studies*, Vol. 95, 1999 and Beckermann et al., (1992), Blazer et al., (1984) and Sarkar (1998). It is always possible to divide claims about reductionism and emergentism. One may accept ontological reductionism but reject epistemological reductionism, and vice-versa, likewise for ontological emergentism and epistemological emergentism. Further, one may restrict the question of reductionism and emergentism to particular domains of discourse. For example, one might accept reductionism (epistemic and/or ontic) for the case of classical mechanics and quantum mechanics, but reject it (epistemic and/or ontic) for the case of folk psychology and theories from neuroscience.

The Varieties of Reductionism:
Ontological and Epistemological

The basic idea of reduction is conveyed by the "nothing more than . . ." cliché. If *X*s reduce to *Y*s, then we would seem to be justified in saying or believing things such as "*X*s are nothing other (or more) than *Y*s," or "*X*s are just special sorts, combinations or complexes of *Y*s." However, once beyond cliches, the notion of reduction is ambiguous along two principal dimensions: the types of items that are reductively linked and the nature of the link involved. To define a specific notion of reduction, we need to answer two questions:

- Question of the relata: Reduction is a relation, but *what types of things* may be related?
- Question of the link: *In what way(s)* must the items be linked to count as a reduction?

Let us first consider the question of the relata. The things that may be related have been viewed either as:

- real world items – entities, events, properties, etc. – which is the *Ontological* form of *Reduction*, or
- representational items – theories, concepts, models, frameworks, schemas, regularities, etc. – which is the *Epistemological* form of *Reduction*.

Thus, the first step in our taxonomy subdivides into two types of reduction. Each type further subdivides based on the specific kinds of relata in question. Ontological subdivisions include: parts and wholes; properties; events/processes; and causal capacities. Epistemological subdivisions include: concepts; laws (epistemically construed); theories; and models. (These lists are not intended to be exhaustive, but merely representative.)

The second question about the link was *in what way(s)* must the items be *linked* to count as a case of reduction? Again, there are a variety of answers on both the ontological and the epistemological side.

Question of the *ontological link*: How must things be related for one to ontologically reduce to the other? At least four major answers have been championed:

- Elimination
- Identity
- Mereological supervenience (includes "composition", "realization" and other related weaker versions of this kind of determination relation)
- Nomological supervenience/determination

The relative merits of competing claims have been extensively debated, but for present purposes it suffices to say a brief bit about each and give a general sense of the range of options.

Elimination

One of the three forms of reduction listed by Kemeny and Oppenheim in their classic paper on reduction (1956) was replacement, i.e., cases in which we come to recognize that what we thought were Xs are really just Ys. Xs are eliminated from our ontology, e.g., claims of demonic possession (Rorty, 1970; Churchland, 1981; Dennett, 1988; Wilkes, 1988, 1995).

Identity

Identity involves cases in which we continue to accept the existence of Xs but come to see that they are identical with Ys (or with special sorts of Ys). Xs reduce to Ys in the strictest sense of being the same thing as Ys. This may happen when a later Y-theory reveals the true nature of X to us. For example, we have come to see that heat is just kinetic molecular energy and that genes are just functionally active DNA sequences. However, the identity does not require elimination or deny the existence of the prior items, rather we see that two distinct theories have described or referred to the same entities/properties.

Mereological supervenience

Reductionism pertaining to parts and wholes goes by several names: "mereological supervenience," "Humean supervenience" and "part/whole reductionism." Mereological supervenience says that the properties of a whole are determined by the properties of its parts (Lewis, 1986, p. 320).

More specifically, mereological supervenience holds that all the properties of the whole are determined by the qualitative intrinsic properties of the most fundamental parts. Intrinsic properties being non-relational properties had by the parts which these bear in and of themselves, without regard to relationships with any other objects or relationships with the whole. Sometimes, philosophers say that intrinsic properties are properties that an object would have even in a possible world in which it alone exists. Paradigmatic examples include mass, charge, and spin. Further, intrinsic properties are much like the older primary qualities. It is notoriously difficult to define the notion of an intrinsic property or a relational property in a non-circular and non-question begging manner; nonetheless, philosophers and physicists rely heavily on this distinction (Lewis, 1986).

Nomological supervenience/determination

Fundamental physical laws (*ontologically construed*), governing the most basic level of reality, determine or necessitate all the higher-level laws in the universe. Mereological supervenience, on the one hand, says that the intrinsic properties of the most basic parts *determine* all the properties of the whole – this is a claim about part-whole determination (purely physical necessity). Nomological supervenience is about *nomic necessity*, the most fundamental laws of physics ultimately necessitate all the special science laws, and therefore these fundamental laws determine everything that happens (in conjunction with initial or boundary conditions). Thus, if two worlds are wholly alike in terms of their most fundamental laws and in terms of initial/boundary conditions, then we should expect them to be the same in all other respects.

In *epistemological reduction* one set of representational items is reduced to another. These representational items are all human constructions and often taken to be linguistic or linguistic surrogates, though this need not be the case. It was noted above that reduction relations might hold among at least four different kinds of representational items.

Concerning the *epistemological links (or relations) that do the reducing*, a diversity of claims have been made. Some relations, such as derivability, make sense as a relation between theories seen as sets of propositions but not among models or concepts. However, certain commonalities run through the family of epistemological-reductive relations. Most of the specific variants of epistemological reduction fall into one of four general categories:

- Replacement
- Theoretical-derivational (logical empiricist)
- Semantic/model-theoretic/structuralist analysis
- Pragmatic

Replacement

The analogue of elimination on the epistemological side would be replacement. Our prior ways of describing and conceptualizing the world might drop out of use and be superseded by newer more adequate ways of representing reality. For example, many of our folk psychological concepts might turn out not to do a good job of characterizing the aspects of the world at which they were directed, as happened with such concepts as demonic possession (Feyerabend, 1962).

Theoretical-derivational

The classic notion of intertheoretic reduction in terms of theoretical derivation, as found in Kemeny and Oppenheim (1956) or in Ernest Nagel's classic treatment (1961), descends from the logical empiricist view of theories as interpreted formal calculi statable as sets of propositions of symbolic logic. Intertheoretic reduction is the derivation of one theory from another; and so constitutes an *explanation* of the reduced theory by the reducing theory. This model treats intertheoretic reduction as deductive, and as a special case of deductive-nomological explanation. Thus if one such theory T_1 could be logically derived from another T_2, then everything T_1 says about the world would be captured by T_2. Because the theory to be reduced T_1 normally contains terms and predicates that do not occur in the reducing theory T_2, the derivation also requires some bridge laws or bridge principles to connect the vocabularies of the two theories. These may take the form of strict biconditionals linking terms in the two theories, and when they do such biconditionals may underwrite an ontological identity claim. However, the relevant bridge principles need not be strict biconditionals. All that is required is enough of a link between the vocabularies of the two theories to support the necessary derivation.

One caveat is in order. Strictly speaking, in most cases what is derived is not the original reduced theory but an image of that theory within the reducing theory, and that image is typically only a close approximation of the original rather than a precise analogue (Feyerabend, 1977; Churchland, 1985).

Nagel's account (1961) of intertheoretic reduction has become a standard for this type, and all alternative accounts are in one way or another amendments to it or reactions against it. So, let us look at it a little more closely, and see how problems for this account have arisen. Nagel distinguishes two types of reductions on the basis of whether or not the vocabulary of the reduced theory is a subset of the reducing theory. If it is – that is, if the reduced theory T_1 contains no descriptive terms not contained in the reducing theory T_2, and the terms of T_1 are understood to have approximately the same meanings that they have in T_2, then Nagel calls the reduction of T_1 by T_2 "homogeneous" (Nagel, 1961, p. 339).

From a historical perspective, this attitude is somewhat naïve (Sklar, 1967, pp. 110–11). The number of actual cases in the history of science where a genuine homogeneous reduction takes place are few and far between. One escape for the proponent of Nagel-type reductions is to distinguish explaining a theory (or explaining the laws of a given theory) from explaining it away (Sklar, 1967, pp. 112–13). Thus, we may still speak of reduction if the derivation of the approximations to the reduced theory's laws serves to account for why the reduced theory works as well as it does in its (perhaps more limited) domain of applicability.

The task of characterizing reduction is more involved when the reduction is heterogeneous, that is, when the reduced theory contains terms or concepts that do not appear in the reducing theory. Nagel takes as a paradigm example the (apparent) reduction of thermodynamics, or at least some parts of thermodynamics, to statistical mechanics. For instance, thermodynamics contains the concept of temperature (among others) that is lacking in the reducing theory of statistical mechanics. Nagel notes that "if the laws of the secondary science (the reduced theory) contain terms that do not occur in the theoretical assumptions of the primary discipline (the reducing theory) the logical derivation of the former from the latter is *prima facie* impossible" (Nagel, 1961, pp. 352–4). As a consequence, Nagel introduces two "necessary formal conditions" required for reduction to take place known as *connectability* and *derivability*. Connectability has to do with the bridge laws that relate the sets of terms from the theories in question. The consideration of certain examples lends plausibility to the idea that the bridge laws should be considered to express some kind of identity relation. For instance, Sklar notes that the reduction of the "theory" of physical optics to the theory of electromagnetic radiation proceeds by *identifying* one class of entities – light waves – with (part of) another class – electromagnetic radiation (Sklar, 1967, p. 120). In fact, if something like Nagelian reduction is going to work, it is generally accepted that the bridge laws should reflect the existence of some kind of synthetic identity.

One problem facing the theoretical-derivational account of intertheoretic reduction was forcefully presented by Feyerabend in "Explanation, Reduction, and Empiricism" (Feyerabend, 1962). Consider the term "temperature" as it functions in classical thermodynamics. This term is defined in terms of Carnot cycles and is related to the strict, nonstatistical zeroth law as it appears in that theory. The so-called reduction of classical thermodynamics to statistical mechanics, however, fails to identify or associate *nonstatistical* features in the reducing theory, statistical mechanics, with the nonstatistical concept of temperature as it appears in the reduced theory. How can one have a genuine reduction, if terms with their meanings fixed by the role they play in the reduced theory are identified with terms having entirely different meanings? Classical thermodynamics is not a statistical theory. The very possibility of finding a reduction function or bridge law that captures the concept of temperature and the strict, nonstatistical role it plays in the thermodynamics seems impossible (Takesaki, 1970; Primas, 1998).

Many physicists, now, would accept the idea that our concept of temperature and our conception of other exact terms that appear in classical thermodynamics such as "entropy," need to be reformulated in light of the alleged reduction to statistical mechanics. Textbooks, in fact, typically speak of the theory of "statistical thermodynamics."

Because of the problem mentioned above, as well as others, many philosophers of science felt that the theoretical-derivational model (Nagel, 1961) did not realistically capture the actual process of intertheoretic reduction. As Primas puts it, "there exists not a single physically well-founded and nontrivial example for theory reduction in the sense of Nagel (1961). The link between fundamental and higher-level theories is far more complex than presumed by most philosophers" (1998, p. 83). Therefore, alternative models of intertheoretic reduction abandon one or more *ontological assumptions* made by the theoretical-derivational account (i.e., the logical empiricist account):

1 Property/kind cross-theoretic (ontological) identities are to be determined solely by formal criteria such as successful intertheoretic reduction, e.g., smooth intertheoretic reduction is both necessary and sufficient for cross-theoretic identity.
2 Realism, scientific theories are more than mere "computational devices."

and/or one or more *epistemological assumptions*:

1 Philosophy of science is prescriptive rather than descriptive, e.g., philosophy of science should seek a grand, universal account of intertheoretic reduction.
2 Scientific theories are axiomatic systems.
3 Reduction = logical deduction, or at least deduction of a structure specified within the vocabulary and framework of the reduced theory or some corrected version of it.
4 Necessity of bridge laws or some other equally strong cross-theoretic connecting principles to establish synthetic identities.
5 Symbolic logic is the appropriate formalism for constructing scientific theories.
6 Scientific theories are linguistic entities.
7 Hardcore explanatory unification. Reduction is proof of displacement (in principle) showing that the more comprehensive reducing theory contains explanatory and predictive resources equaling or exceeding those of the reduced theory.
8 Intertheoretic reductions are an all or nothing *synchronic* affair as in the case of "microreductions" (Oppenheim and Putnam, 1958; Causey, 1977): the lower-level theory and its ontology reduce the higher-level theory and its ontology. Ontological levels are mapped one-to-one onto levels of theory which are mapped one-to-one onto fields of science.
9 The architecture of science is a layered edifice of analytical levels (Wimsatt, 1976).

Alternatives to the Nagel (1961) model are deemed more or less radical (by comparison) depending on which of the preceding tenets are abandoned. On the more conservative side, many alternative accounts of intertheoretic reduction merely modify (3) by moving to logico-mathematical deduction, but reject (4). For example, the requirement of bridge laws gets replaced by notions such as: "analog relation" – an ordered pair of terms from each theory (Hooker, 1981; Bickle, 1998), "complex mimicry" (Paul Churchland, 1989) or "equipotent image" (Patricia Churchland, 1986), to name a few. Many of these comparatively conservative accounts also reject (8), preferring to talk about a range of reductions, from replacement on one end of the continuum to identity on the other. More radical alternatives to the Nagel models are as follows.

Semantic/model-theoretic/structuralist analysis

This approach (the "semantic" approach for short), is regarded by some as comparatively radical because it rejects the conception of scientific theories as formal calculi formalizable in first-order logic and (partially) interpretable by connecting principles such as bridge laws. The semantic approach makes the following assumptions:

(i) Scientific theories are not essentially linguistic entities (sets of sentences), but are terms or families of their *mathematical models or mathematical structures*.

(ii) The formal explication of the structure of scientific theories is not properly carried out with first-order logic and metamathematics, but with *mathematics*, though the choice of mathematical formalisms will differ depending on who you read (Giere, 1988; Bickle, 1998; Batterman, 2000).

The semantic approach minimally rejects epistemological assumptions (2)–(6) and (8), i.e., rejects the derivation of laws and abandons truth preservation (everything the reduced theory asserts is also asserted by the reducing theory). On the semantic approach, the reduction relation might be conceived of as some kind of "isomorphism" or "expressive equivalence" between models (Bickle, 1998). However, as we shall shortly see, more radical versions of the semantic approach reject all the preceding epistemological assumptions held by the logical empiricist account of intertheoretic reduction.

Pragmatic

Success in real world representation is, in large part, a practical matter of whether and how fully one's attempted representation provides *practical causal* and *epistemic access* to the intended representational target. A good theory or model succeeds as a representation if it affords reliable avenues for *predicting, manipulating* and *causally* interacting with the items it aims to represent. It is the practical access that the model affords in its context of application that justifies viewing it as having the representational content that it does (Van Fraassen, 1989; Kitcher, 1989). If a lower-level theory about a specific domain provides superior *real-world explanatory* and *predictive* value compared to a higher-level theory representing the same domain, then the lower-level theory has met the ultimate test of successful intertheoretic reduction. Note that this contextual, pragmatic account of intertheoretic reduction is also highly *particularist*; it advocates adjudicating on a case-by-case basis; no universal theory of reduction is sought. This account rejects at least assumptions (1)–(6) in the epistemological category, and assumption (1) in the ontological category (Patricia Churchland, 1986). More radical versions reject all nine of the preceding epistemological assumptions.

Whereas the theoretical-derivational account (i.e., the logical empiricist account) of intertheoretic reduction (and its variants) only makes sense if you presuppose nomological and mereological supervenience; *in principle*, both the semantic and the pragmatic accounts of intertheoretic reduction are compatible with the failure of mereological supervenience and perhaps even nomological supervenience. We shall encounter specific versions of such accounts of reduction shortly.

While there are certainly mutually exclusive and competing accounts of intertheoretic reduction that represent each of our four *types*, there is no principled reason why the four types could not be synthesized into a single account. Schaffner's "generalized replacement-reduction" (GRR) model of intertheoretic reduction is one such attempt (Schaffner, 1992, 1998, 2000).

Though much more could be said about the many varieties of ontological and epistemological reduction and their respective faults and merits, the main versions may be graphically summarized (figure 5.1):

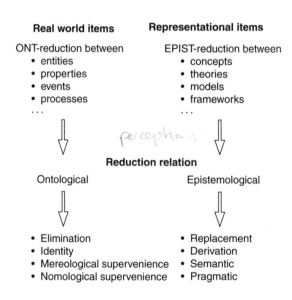

Real world items　　**Representational items**

ONT-reduction between　　EPIST-reduction between
- entities
- properties
- events
- processes
. . .

- concepts
- theories
- models
- frameworks
. . .

Reduction relation

Ontological　　Epistemological

- Elimination
- Identity
- Mereological supervenience
- Nomological supervenience

- Replacement
- Derivation
- Semantic
- Pragmatic

Figure 5.1

The Varieties of Emergence: Ontological and Epistemological

Emergence, like reduction, is interpreted in diverse ways (Silberstein and McGeever, 1999). Again, my aim is to survey the main variants.

The basic idea of emergence is roughly the converse of reduction. Though the emergent features of a whole or complex are not completely independent of those of its parts since they "emerge from" those parts, the notion of emergence nonetheless implies that, in some significant way, they *go beyond* the features of those parts. There are many senses in which a system's features might be said to emerge, some of which are relatively modest (Rueger, 2000a,b; Batterman, 2000; Bedau, 1997) and others which are more controversial (Humphreys, 1997; Silberstein, 1998).

The varieties of emergence can be divided into several groups along lines similar to those divisions between the types of reduction (figure 5.2).

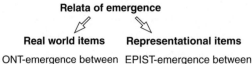

Relata of emergence

Real world items	Representational items
ONT-emergence between	EPIST-emergence between
• parts/wholes	• concepts
• properties	• theories
• events/processes	• models
• causal capacities	• frameworks
• laws	• laws
• entities ⋯	• states of a dynamical system ⋯

Figure 5.2

Ontological relations are objective in the sense that they link ontic items, e.g., properties, independent of any epistemic considerations. Relations of the second sort are epistemic, because they depend on our abilities to comprehend the nature of the links or dependencies among real world items.

At least four major forms of emergence have been championed; each is an elaboration of the failure of its corresponding reduction relation:

- Non-elimination
- Non-identity
- Mereological emergence (holism)
- Nomological emergence

Non-elimination

If a property, entity, causal capacity, kind or type cannot be eliminated from our ontology, then one must be a realist about said item. Obviously, this leaves open the question of what the criteria ought to be for non-elimination in any given case; but they will almost certainly be epistemological/explanatory in nature.

Non-identity

If a property, type or a kind cannot be ultimately identified with a physical (or lower-level) property, type or kind then one must accept that said item is a distinct non-physical (or higher-level) property, type or kind. Again, this leaves open the criteria for non-identifiability and again, such criteria are generally epistemological/explanatory in nature.

Mereological emergence (holism)

These are cases in which objects have properties that are not determined by the intrinsic (non-relational) physical properties of their most basic physical parts. Or, cases in which objects are not even wholly composed of basic (physical) parts at all. British (classical) emergentism held that mereological emergence is true of chemical, biological and mental phenomena (McLaughlin, 1992).

Nomological emergence

These are cases in which higher-level entities, properties, etc., are governed by higher-level laws that are not determined by or necessitated by the fundamental laws of physics governing the structure and behavior of their most basic physical parts. For example, according to Kim (1993), British emergentism held that while there were bridge laws linking the biological/mental with the physical, such bridge laws were inexplicable brute facts. That is, on Kim's view British emergentism did not deny global supervenience. But British emergentism did deny that the laws governing the mental for example were determined (or explained) by the fundamental laws of physics (McLaughlin, 1992; Kim, 1993). A more extreme example of nomological emergence would be where there were no bridge laws whatsoever linking fundamental physical phenomena with higher-level phenomena. In such cases, fundamental physical facts and laws would only provide a necessary condition for higher-level facts and laws. This would imply possible violations of global supervenience. Both Cartwright (1999) and Dupré (1993) *seem* to defend something like this kind of nomological emergence. An even more extreme example is found in cases in which either fundamental physical phenomena or higher-level phenomena are not law-governed at all. This would amount to eliminativism or antirealism regarding nomological or physical necessity; see Van Fraassen (1989) for a defense of this view. It is important to note that in all cases of *nomological emergence*, it is *in principle* impossible to derive or predict the higher-level phenomena on the basis of the lower-level phenomena.

The *epistemological link* must describe how things are related such that one epistemologically emerges from another. At least two major views have been championed:

- Predictive/explanatory emergence
- Representational/cognitive emergence

Predictive/explanatory emergence

Wholes (systems) have features that cannot *in practice* be explained or predicted from the features of their parts, their mode of combination, and the laws governing their behavior. In short, X bears predictive/explanatory emergence with respect to Y if Y cannot (reductively) predict/explain X. More specifically, in terms of types of intertheoretic reduction, X bears predictive/explanatory emergence with respect to Y: if Y cannot *replace* X, if X cannot be *derived* from Y, or if Y cannot be shown to be *isomorphic* to X. A lower-level theory Y (description, regularity, model, schema, etc.), for purely epistemological reasons (conceptual, cognitive or computational limits), can fail to predict or explain a higher-level theory X. If X is predictive/explanatory emergent with respect to Y for *all possible cognizers in practice*, then we might say that X is *incommensurable* with respect to Y. A paradigmatic and notorious example of predictive/explanatory emergence is chaotic, non-linear dynamical systems (Silberstein and McGeever, 1999). The emergence in chaotic systems (or models of non-linear systems exhibiting chaos) follows from their sensitivity to initial conditions, plus the fact that physical properties can only be specified to finite precision; infinite precision would be necessary to perform the required "reduction", given said sensitivity. It does not follow, however, that chaotic systems provide evidence of violations of mereological supervenience or nomological supervenience (Kellert, 1993, pp. 62, 90), e.g., dynamical systems have attractors as high-level emergent features only in the sense that you cannot deduce them from equations for the system. McGinn (1999) and other mysterians hold that folk psychology is predictive/explanatory emergent with respect to the theories of neuroscience.

Representational/cognitive emergence

Wholes (systems) exhibit features, patterns or regularities that cannot be fully represented (understood) using the theoretical and representational resources adequate for describing and understanding the features and regularities of their more basic parts and the relations between those more basic parts. X bears representational/cognitive emergence with respect to Y, if X does *not* bear predictive/explanatory emergence with respect to Y, but nonetheless X represents higher-level patterns or non-analytically guaranteed regularities that cannot be fully, properly or easily represented or understood from the perspective of the lower-level Y. As long as X retains a significant *pragmatic* advantage over Y with respect to understanding the phenomena in question, then X is representational/cognitive emergent with respect to Y. Nonreductive physicalism holds that folk psychology is representational/cognitive emergent with respect to the theories of neuroscience (Antony, 1999).

The Reduction and Emergence Debate Today: Specific Cases Seeming to Warrant the Label of Ontological or Epistemological Emergence

Not since the first half of the twentieth century have emergence and reduction enjoyed so much critical attention. Claims involving emergence are now rife in discussions of philosophy of mind, philosophy of physics, various branches of physics itself including quantum mechanics, condensed matter theory, non-linear dynamical systems theory (especially so-called chaos theory), cognitive-neuroscience (including connectionist/neural network modeling and con-sciousness studies) and so-called complexity studies (Silberstein and McGeever, 1999). To quote Kim:

> we are now seeing an increasing and unapologetic use of expressions like "emergent," "emergent property," and "emergent phenomenon" . . . not only in serious philo-sophical literature but in the writings in psychology, cognitive science, systems theory, and the like (1998, pp. 8–9).

Kim also says that

> the return of emergentism is seldom noticed, and much less openly celebrated; it is clear, however, that the fortunes of reductionism correlate inversely with those of emergentism . . . It is no undue exaggeration to say that we have been under the reign of emergentism since the early 1970s (1999, p. 5).

There are two primary reasons for the return of emergentism. First, regarding nomological emergence, a growing body of literature focusing on actual scientific practice suggests that there really are not many cases of successful intertheoretic reduction in the empiricist tradition of *demonstrating* nomological supervenience.

> Our scientific understanding of the world is a patchwork of vast scope; it covers the intricate chemistry of life, the sociology of animal communities, the gigantic wheel-ing galaxies, and the dances of elusive elementary particles. But it is a patchwork nevertheless, and the different areas do not fit well together (Berry, 2000, p. 3).

Focus on actual scientific practice suggests that either there really are not many cases of successful epistemological (intertheoretic) reduction or that most philosophical accounts of reduction bear little relevance to the way reduction in science actually works. Most working scientists would probably opt for the latter claim.

Often discussed cases of failed or incomplete intertheoretic reduction in the literature include:

1   the reduction of thermodynamics to statistical mechanics (Primas, 1991, 1998; Sklar, 1999)
2   the reduction of thermodynamics/statistical mechanics to quantum mechanics (Hellman, 1999)
3   the reduction of chemistry to quantum mechanics (Cartwright, 1997; Primas, 1983)
4   the reduction of classical mechanics to quantum mechanics (such as the worry that quantum mechanics cannot recover classical chaos) (Belot and Earman, 1997).

Take the case of chemistry and its alleged reduction to quantum mechanics. Currently chemists do not use fundamental quantum mechanics (Hamiltonians and Schrödinger's equation) to do their science. Quantum chemistry cannot be deduced directly from Schrödinger's equation due to multiple factors that include the many-body problem (Hendry, 1998). Quantum mechanical wave functions are not well-suited to represent chemical systems or support key inferences essential to chemistry (Woody, 2000). It is still an open question as to whether quantum mechanics can describe or represent a molecule (Berry, 2000). Indeed, little of current chemistry can be represented by pure quantum mechanical calculations (Primas, 1983; Scerri, 1994; Ramsey, 1997). Chemistry uses idealized models whose relationship to fundamental quantum mechanics is questionable (Primas, 1983; Hendry, 1999). As Cartwright (1997, p. 163) puts it:

> Notoriously, we have nothing like a real reduction of the relevant bits of physical chemistry to physics – whether quantum or classical. Quantum mechanics is important for explaining aspects of chemical phenomena but always quantum concepts are used alongside of sui generis – that is, unreduced-concepts from other fields. They do not explain the phenomena on their own.

Another well-known example is the case of thermodynamics and statistical mechanics. First, there is a variety of distinct concepts of both temperature and entropy that figure in both statistical mechanics and classical thermodynamics. Second, thermodynamics can be applied to a number of very differently constituted microphysical systems. Thermodynamics can be applied to gases, electromagnetic radiation, magnets, chemical reactions, star clusters and black holes. As Sklar (1993, p. 334) puts it:

The alleged reduction of thermodynamics to statistical mechanics is another one of those cases where the more you explore the details of what actually goes on, the more convinced you become that no simple, general account of reduction can do justice to all the special cases in mind.

Third, the status of the probability assumptions that are required to recover thermodynamic's principles within statistical mechanics are themselves problematic or *ad hoc*. For example, the assumption that the micro-canonical ensemble is to be assigned the standard, invariant, probability distribution. Fourth, perhaps the thorniest problem of all, statistical mechanics is time symmetric and thermodynamics possesses time asymmetry.

These are especially important examples because they involve difficulties between different levels of explanation *within physical science*. Some of the four (e.g., the reduction of thermodynamics to statistical mechanics) were once thought of as successes for philosophical accounts of intertheoretic reduction (Sklar, 1993).

Perhaps the most highly advertised case of failed intertheoretic reduction is the attempt to reduce folk psychology to theories of neuroscience. Presently a popular *ontological* version of the mind/body problem goes by the name of "the hard problem of phenomenal consciousness": how and why are brain states conscious? (Chalmers, 1996). As Kim (1998, pp. 102–3) puts it:

> We are not capable of designing, through theoretical reasoning, a wholly new kind of structure that we can predict will be conscious; I don't think we even know how to begin, or indeed how to measure our success . . . In any case it seems to me that if emergentism is correct about anything, it is more likely to be correct about qualia than about anything else.

For more on the problems of phenomenal consciousness and emergence see Silberstein (2001).

In this spirit, philosophers of science and mind have made a cottage industry of collecting many of the cases of incomplete intertheoretic reduction, calling them all "emergence"; see, for example, *Special Issue: Reduction and Emergence, Philosophical Studies*, 95 (1–2), August 1999 and Beckermann et al. (1992). The essays in both volumes span psychology, biology and physics. Each of the essays is an examination of an attempted intertheoretic reduction that is currently having grave difficulties. Taken in toto, these cases seem a barometer of the prospects for unifying the sciences, and therefore *indicative* of the prospects of epistemological and ontological reductionism. There is a movement afoot devoted to arguing this point. The movement is known as the "disunity of science movement" or the "anti-fundamentalism movement" (Dupré, 1993; Cartwright, 1999). However, an *indication* is not an argument, so each case deserves to be examined in its own right.

There is no doubt danger in lumping all these cases together. It is clear, for example, that thermodynamics is predictive/explanatory emergent with respect to statistical mechanics. As of yet, few are ready to conclude that thermodynamical phenomena are, for example, nomologically or mereologically emergent with respect to statistical mechanical phenomena. By way of contrast, when Kim talks about phenomenal consciousness being emergent, he seems to be making a claim about emergent phenomenal consciousness which goes beyond a function of igno-rance interpretation (Kim, 1998, 1999). It is not uncommon for such equivoca-tions on the term "emergence" to appear in the same volume.

This brings me to the second major reason for the return of emergence. There are some people who allege that quantum mechanics *itself* provides examples of mereological emergence:

> In quantum theory, then, the physical state of a complex whole cannot always be reduced to those of its parts, or to those of its parts together with their spatiotem-poral relations, even when the parts inhabit distinct regions of space. Modern science, and modern physics in particular, can hardly be accused of holding reductionism as a central premise, given that the result of the most intensive scientific investigations in history is a theory that contains an ineliminable holism (Maudlin, 1998, p. 55).

By and large, a system in classical physics can be analyzed into parts, whose states and properties determine those of the whole they compose. But the state of a system in quantum mechanics resists such analysis. The quantum state of a system gives a specification of its probabilistic dispositions to display various properties on its mea-surement. Quantum mechanics' most complete such specification is given by what is called a pure state. Even when a compound system has a pure state, its subsystems generally do not have their own pure states. Schrödinger, emphasizing this charac-teristic of quantum mechanics, described such component subsystems as "entan-gled." Such entanglement of systems demonstrates nonseparability – the state of the whole is not constituted by the states of its parts. State assignments in quantum mechanics have been taken to violate state separability in two ways: the subsystems may simply not be assigned any pure states of their own, or else the states they are assigned may fail to completely determine the state of the system they compose.

The quantum state of a system may be either pure or mixed. A pure state is rep-resented by a vector in the system's Hilbert space. It is commonly understood that any entangled quantum systems violate state separability in so far as the vector rep-resenting the state of the system they compose does not factorize into a vector in the Hilbert space of each individual subsystem that could be taken to represent its pure state. A set of entangled quantum systems compose a system whose quantum state is represented quantum mechanically by a tensor-product state-vector which does not factorize into a vector in the Hilbert space of each individual system:

$$\Psi_{1,2,\ldots,R} \neq \Psi_1 \otimes \Psi_2 \otimes \cdots \otimes \Psi_R$$

Now in such a case each subsystem $1, 2, \ldots, n$ may be uniquely assigned what is called a mixed state (represented in its Hilbert space not by a vector but by a so-called von Neumann density operator). But then state separability fails for a different reason: the subsystem mixed states do not uniquely determine the compound system's state.

On the basis of nonseparability, many people have argued that quantum mechanics provides us with examples of systems that have properties that do not always reduce to the intrinsic properties of the most basic parts, i.e., quantum mechanical systems exhibit mereological emergence (Healey, 1991; Hawthorne and Silberstein, 1995; Humphreys, 1997). Such entangled systems appear to have novel properties of their own. Quantum systems that are in superpositions of possible states are behaviorally distinct from systems that are in mixtures of these states and individual systems can be become entangled and thus form a new unified system which is not the sum of its intrinsic parts. From this, some further infer that: "the state of the compound [quantum] system determines the state of the constituents, but not vice versa. This last fact is exactly the reverse of what [mereological] supervenience requires" (Humphreys, 1997, p. 16). The opinion of a growing number of philosophers of physics is expressed by Maudlin (1998, pp. 58–60):

> Quantum holism ought to give some metaphysicians pause. As has already been noted, one popular "Humean" thesis holds that all global matters of fact supervene on local matters of fact, thus allowing a certain ontological parsimony. Once the local facts have been determined, all one needs to do is distribute them throughout all of space-time to generate a complete physical universe. Quantum holism suggests that our world just doesn't work like that. The whole has physical states that are not determined by, or derivable from, the states of the parts. Indeed, in many cases, the parts fail to have physical states at all. The world is not just a set of separately existing localized objects, externally related only by space and time. Something deeper, and more mysterious, knits together the fabric of the world. We have only just come to the moment in the development of physics that we can begin to contemplate what that might be.

At any rate, quantum nonseparability is not restricted to settings such as twin-slit experiments and EPR (non-locality) experiments. Superpositions and entangled states are required to explain certain chemical and physical phenomena such as phase transitions that give rise to superconductivity, superfluidity, paramagnetism, ferromagnetism; see Anderson (1994), Auyang (1998) and Cornell and Wieman (1998).

Some interpretations of quantum mechanics such as Bohr (1934) and Bohm and Hiley (1993) imply mereological emergence (holism) with respect to *entities*: there are physical objects that are not wholly composed of basic (physical) parts. On Bohr's interpretation one can meaningfully ascribe properties such as position or momentum to a quantum system only in the context of some well-defined experimental arrangement suitable for measuring the corresponding property. Although a quantum system is purely physical on this view, it is not composed of distinct happenings involving independently characterizable physical objects such as the quantum system on the one hand, and the classical apparatus on the other. On Bohm's interpretation, it is not just quantum object and apparatus that are holistically connected, but any collection of quantum objects by themselves constitute an indivisible whole. A complete specification of the state of the "undivided universe" requires not only a listing of all its constituent particles and their positions, but also of a field associated with the wave-function that guides their trajectories. If one assumes that the basic physical parts of the universe are just the particles it contains, then this establishes ontological holism in the context of Bohm's interpretation.

For the purposes of this discussion, what is most important is not whether or not quantum mechanics actually does provide cases of mereological emergence, but that the belief that it does, in part, fuels emergentism. Though it must be said, there are some philosophers who are still skeptical about the reality, coherence or importance of quantum holism (Lewis, 1986; Dickson, 1998). Not everyone acknowledges that nonseparability implies mereological emergence. For example, Healey argues that whether or not nonseparability implies mereological emergence is a matter of interpretation (1989, pp. 142–5). Healey's own modal interpretation (1989) does imply mereological emergence, however he stipulates that the formalism of quantum mechanics is open to interpretations that do not. He argues (Healey, 1991) that nonseparability in general and so-called non-locality are best explained by positing mereological emergence.

Questions for Future Research

Recall that the best reason for believing in reductionism is an acceptance of mereological and/or nomological supervenience based in large part on successful intertheoretic reduction (or epistemological reduction). Do the preceding examples of epistemological and ontological emergence indicate emergentism is true? At this juncture, may we even say whether emergentism or reductionism is more probable? What does the current state of disunity within any given science and across the various sciences imply about emergence? Regarding the ultimate fate of mereological and nomological emergence respectively, there are two general possibilities. Either these respective forms of emergence are merely a function of our ignorance or they are real facts about the world. If they are real facts about the world then they may be either universally true or restricted to a particular domain such as microphysics. Of course, the ultimate fate of mereological emergence might be different from that of nomological emergence and vice-versa. For example, the possibilities for nomological emergence are as follows:

There are four *reductive outcomes*:

1 Any claimed emergence is due to philosophical ignorance. A better, more appropriate *philosophical* theory of intertheoretic reduction needs to be con-

structed that will show that the lower-level theory does reductively explain the higher-level theory in question. It is possible (if not probable) that different cases will require different accounts of intertheoretic reduction for their resolution.

2   Any case of emergence is due to *empirical* or *experimental* ignorance. Future discoveries will allow us to see how the lower-level theory does in fact reductively explain the higher-level theory in question.

3   Any claim to emergence relies on lower-level theories that are false or incomplete, and such theories will be replaced or supplemented by correct lower-level theories in order to reductively explain the higher-level theory.

Outcomes 1–3 would all be unqualified wins for epistemological reductionism if not ontological reductionism.

4   The higher-level theory will cease to be predictive/explanatory emergent with respect to the lower-level theory, but for some (indeterminate) length of time the higher-level theory will be representational/cognitive emergent with respect to the lower-level theory.

This is more or less a win for epistemological (if not ontological) reductionism. There are then two *emergent outcomes*:

5   The higher-level theory is predictive/explanatory emergent with respect to the lower-level theory and for whatever reason, due to whatever *epistemological limits*, the lower-level theory and its successors will never be able to reductively explain the higher-level theory. This is a win for epistemological emergence only.

6   The higher-level theory is predictive/explanatory emergent with respect to the lower-level theory (and its successors) *because* the phenomena/laws represented by the higher-level theory are *nomologically emergent* with respect to the phenomena/laws represented by the lower-level theory. The lower-level phenomena only provide a *necessary (but not sufficient) condition* for the *emergence* of the higher-level phenomena. This would be an unqualified loss for *both* epistemological and ontological reductionism.

One important question for the future is to determine, in each specific instance of incomplete intertheoretic reduction (such as the cases discussed earlier), which of these six possibilities actually obtains. However it should be clear that emergentism and reductionism might form a *continuum* and not a dichotomy. This is true in several respects. First, even if mereological emergence is real it does not necessarily imply nomological emergence. Even if the quantum is mereologically emergent, it could still be the case that all higher-level phenomena nomologically supervenes upon it. Second, both mereological and nomological emergence might be restricted to certain domains. For example, mereological emergence might be limited to the quantum and nomological emergence limited to the mental. Third, for any given case we can always divide the question for ontological and episte-mological emergence. Or more generally, it could turn out, for example, that epistemological emergence is inescapable while ontological emergence is rare or nonexistent. Of course, given the former, it is an open question how we would ever discover the latter.

Recent accounts of *intertheoretic reduction*, the more radical versions of the semantic and pragmatic models mentioned earlier, such as *GRR* (Schaffner, 1992, 1998) and the more explicitly pragmatic and ontic *causal mechanical* model (Machamer et al., 2000), explicitly reject microreduction, in part because of the problematic cases mentioned earlier. Such alternative accounts of intertheoretic reduction, in their rejection of microreduction, explicitly acknowledge the con-tinuum between reduction and emergence. For example, the causal mechanical model of intertheoretic reduction focuses on explanations as characterizing complex (nested and inter-connected) causal mechanisms and pathways, such as we find in molecular biology and neuroscience. The emphasis in this model is on causal/mechanical processes as opposed to nomological patterns of explanation. More importantly for our purposes, this model admits of *multilevel* descriptions of causal mechanisms that mix different levels of aggregation from cell to organ back to molecule.

Take the following example from behavioral genetics:

> there is no *simple* [reductive] explanatory model for behavior even in simple organisms. What *C. elegans* [a simple worm] presents us with is a tangled network of influences [causal mechanisms] at genetic, biochemical, intracellular, neuronal, muscle cell, and environmental levels (Schaffner, 1998, p. 237).

This kind of reductive explanation focuses on interlevel causal processes and emphasizes the limits and rarity of logical empiricist accounts of intertheoretic reduction. This approach to reduction is diachronic, emphasizing the gradual, partial and fragmentary nature of many real world cases. This model clearly views intertheoretic reduction as a continuum and not a dichotomy.

One can also find similar web-like and bushy cases of intertheoretic reduction within physics. For example, cases in which two domains (such as quantum mechanics and chemistry) are related by an asymptotic series often require appeal to an intermediate theory (Berry, 1994; Primas, 1998; Batterman, 2000, 2001). In the asymptotic borderlands between such theories, phenomena emerge that are not fully explainable in terms of either the lower-level or the higher-level theory, but require both theories or an intermediary (Batterman, 2000, 2001). Examples of this phenomena can be found in the borders between: quantum mechanics and chemistry, as well as thermodynamics and statistical mechanics (Berry and Howls, 1993; Berry, 1994; 2000; Batterman, 2000). Batterman speaks of the "asymptotic emergence of the upper level properties" in such cases, and he goes on to suggest that "it may be best, in this context, to give up on the various philosophical models of reduction which require the connection of kind predicates in the reduced theory with kind predicates in the reducing theory. Perhaps a more fruitful approach is to investigate asymptotic relations between the different theory pairs. Such asymptotic methods often allow for the understanding of emergent structures which dominate observably repeatable behavior in the limiting domain between the theory pairs" (2000, pp. 136–7).

Intertheoretic reduction *à la* singular asymptotic expansions is not easy to characterize, though it is fair to say that it falls within the semantic approach to intertheoretic reduction. Examples of intertheoretic relations involving singular asymptotic expansions include: Maxwell's electrodynamics and geometrical optics; molecular chemistry and quantum mechanics and; classical mechanics and quantum mechanics (Primas, 1998; Berry, 2000).

There are several things worth noticing about both the preceding models of intertheoretic reduction. Such reductions are not universally valid, they can only be considered on a case-by-case basis. Such reductions require specification of context, the new description or higher-level theory cannot be derived from the lower-level theory. Indeed, such reductions generally start with the higher-level theory/context and work back to the more fundamental theory (Berry, 1994). The lower-level theory (the reducing theory) is not, as a rule, more powerful or universal in its predictive/explanatory value than the higher-level theory (the reduced theory). Indeed, the new ontology and topology generated by the higher-level description cannot be replaced or eliminated precisely because of its more universal explanatory power; and the intertheoretic reductions on such accounts show why this must be the case. Contrary to the standard view, failure of reduction need not imply failure of explanation. A more fundamental theory can explain a higher-level theory ("from below" as it were) without providing a reduction of that theory in the standard senses of the term. Emergent phenomena need not be inexplicable brute facts contrary to classical emergentism. Given such accounts of intertheoretic reduction, there is good reason to think that contra the dreams of the unity of science movement, that unification of scientific theories will be local at best.

Such alternative accounts of intertheoretic reduction suggest that the relationship between "higher-level" and "lower-level" scientific theories is a nested hierarchy as opposed to a pyramid structure. And if we think such accounts of reduction reflect the actual ontology of the world, they suggest that the relationship between the various "levels" (subatomic, atomic, molecular, etc.) is also a nested hierarchy. An even more radical speculation along these lines is that the relationship between higher-level and lower-level scientific theories as well as between the various ontic "levels" themselves looks more like non-Boolean lattices (Primas, 1991). The various domains will have overlapping areas or unions, but they will not be co-extensional. So properties in one domain may be necessary for properties in another domain to emerge, but not sufficient. Such alternative accounts of intertheoretic reduction do not obviously imply or demand either mereological or nomological supervenience.

Humphreys suggests (1997) if there were widespread mereological emergence or nonseparability then lower-level property instances would often "merge" in the formation of higher-level properties such that they no longer exist as separate subvenient entities. Widespread mereological emergence calls into question the very picture of reality as divided into a "discrete hierarchy of levels"; rather it is

> more likely that even if the ordering on the complexity of structures ranging from those of elementary physics to those of astrophysics and neurophysiology is discrete, the interactions between such structures will be so entangled that any separation into levels will be quite arbitrary (Humphreys, 1997, p. 15).

Given widespread mereological emergence, the standard divisions and hierarchies between phenomena that are considered fundamental and emergent, simple and aggregate, kinematic and dynamic, and perhaps even what is considered physical, biological and mental are redrawn and redefined. Such divisions will be dependent on what question is being put to nature and what scale of phenomena is being probed.

But on the face of it, one can embrace these alternative models of intertheoretic reduction while maintaining that all apparent emergence is just a function of ignorance. For example, Schaffner strongly suggests that nothing about such tangled causal processes warrants any claims for either mereological emergence or nomological emergence (such as vital or configurational forces). Rather, at worst, such systems provide us with cases of predictive/explanatory emergence or representational/cognitive emergence (Schaffner, 1998, pp. 242–5).

At present, both the emergentist and reductionist feel that, so far, things are going their way. The emergentist points to failures of ontological and methodological reductionism, and the reductionist points to successes. Regarding the problematic cases of intertheoretic reduction, the perennial reductionist reply is to claim that the future will bring success, just as in the past; emergentists likewise feel that they will be redeemed by the future just as they are by the present. This much is true I think, given the examples of both epistemological and ontological emergence canvassed, there is no reason why the burden of proof should continue to lie exclusively with emergentism. At this juncture, neither view is irrational in light of the evidence and neither view is conclusive. Ultimately, emergentists and reductionists are divided by a deeply held philosophical or aesthetic preference that neither will relinquish easily. For example, many philosophers persist in assuming that nomological and mereological reductionism are true in spite of the actual state of unification within science and in spite of the fact that fundamental physics itself might prove a counter-example to mereological supervenience. Do the past successes of reductionism warrant those assumptions on their part or is the assumption based largely on faith?

We know what questions need to be answered to resolve the debate between emergentism and reductionism, but is it possible to ever answer them? How will we know when we have answered them? It is no doubt prudent to remain agnostic while patiently awaiting the outcome of each "crucial question" for the debate. But unfortunately, not all the problems and questions are empirical. Given that progress on the *ontological* questions of reductionism/emergentism is inextricably bound with progress on the *epistemological* questions of reductionism/emergentism, and vice-versa, there still remains a deeper *conceptual* or *philosophical* problem about how to ultimately adjudicate the evidence at any given point in time. For example, the problem with reducing chemistry to quantum mechanics is not just a computational or calculational one. The explanatory success of chemistry requires both a new *ontology* and a new *topology* (e.g., molecules) beyond that of quantum mechanics (Primas, 1998; Hendry, 1999). Can we therefore conclude that chemical phenomena are ontologically emergent in some important respects? But trying to answer this seemingly straight ontological question will immediately raise the specter of trying to cut the Gordian knot of ontology (e.g., cross-theoretic identities) and epistemology (e.g., intertheoretic reduction). Any answer to the question will require falling back on *philosophical* criteria that are not easily justified. Perhaps the point here is that, in any given case, deciding on the means of intertheoretic reduction (formal or otherwise) and deciding whether or not the attempted reduction is successful (the criteria for successful reduction), is inescapably normative.

For example, is it smooth intertheoretic reduction that motivate and sustain claims of cross-theoretic property identity or the other way around? Likewise, is it the failure of smooth intertheoretic reduction that motivates and sustains claims for failures of cross-theoretic identities, or the other way around? Is there any fact of the matter regarding such questions or are such questions largely normative?

Whichever way we choose, it seems to either lead in circles or raise new and equally hairy problems. If we hold that ontological concerns such as the question of identifying the mental and the physical for example should be completely subordinated to the epistemological question of whether or not the theory of folk psychology can be intertheoretically reduced to some theory of neuroscience, then we need an acceptable and agreed upon account of intertheoretic reduction. As Patricia Churchland puts it, "By making theories the fundamental relata [of the reduction relation], much of the metaphysical bewilderment and dottiness concerning how entities or properties could be reduced simply vanishes" (as quoted in Bickle, 1998, p. 44). But this, of course, brings us back full circle to our problematic cases of intertheoretic reduction. Exactly what we lack at the moment is an acceptable and agreed upon account, method or criteria of intertheoretic reduction in many problematic cases.

Take the case of nonreductive physicalism versus reductive physicalism for example. Both accounts of the mental accept mereological and nomological supervenience, yet the former denies that the mental can be cross-identified with the physical. This is because nonreductive physicalism denies that successful intertheoretic reduction is, in principle or in practice, sufficient for ontological identification of properties (Antony, 1999, pp. 37–43). On this view, mental properties are ontologically distinct while being explicable and predictable in principle from their physical basis. Nonreductive physicalism holds that the identification of one property with another is not a function of successful intertheoretic reduction, but whether or not the higher-level property figures in patterns or causal relations in non-analytically-guaranteed regularities. An entity/property is ontologically non-identifiable if it participates essentially in regularities that are novel from the point of view of the alleged reducing base – a situation not precluded by successful intertheoretic reduction. Truths discovered that are not true by definition about higher-level properties are irreducible to lower-level truths. As Antony (1999) puts it, nonreductive physicalism is "a non-ontologically-reductive materialism, coupled with an insistence on explanatory reduction" (p. 43). Thus, the only thing that really separates reductive from nonreductive physicalism then, is their respective *philosophical* criterion for identifying one natural kind/property with another; there is no disagreement here about the basic ontological and scientific facts. According to nonreductive physicalism the fact that folk psychology is representational/cognitive emergent with respect to neuroscientfic theories of mind, is sufficient to block the cross-theoretic identity of mental properties with physical properties. According to reductive physicalism on the other hand, if folk psychology can in principle be intertheoretically reduced to some theory of neuroscience then that is sufficient for cross-identification of mental properties with physical properties. The question is this: Is there any objective fact of the matter about who is right in such a dispute? In the long run, it is important to try to separate out the normative from the more empirical aspects of the debate between emergentism and reductionism.

References

Anderson, P. W. (1994): *A Career in Theoretical Physics*. Singapore: World Scientific Publishing.

Antony, L. (1999): "Making Room for the Mental. Comments on Kim's 'Making Sense of Emergence,'" *Philosophical Studies*, 95(2), 37–43.

Auyang, S. (1998): *Foundations of Complex-System Theories*. Cambridge: Cambridge University Press.

Batterman, R. W. (2000): "Multiple Realizability and Universality," *British Journal of the Philosophy of Science*, 51, 115–45.

Batterman, R. W. (2001): *The Devil in the Details: Asymptotic Reasoning in Explanation, Reduction, and Emergence*. Oxford: Oxford University Press.

Beckermann, A., Flohr, H. and Kim, J. (eds.) (1992): *Emergence or Reduction? Essays on the Prospects for Nonreductive Physicalism*. Berlin: DeGruyter.

Bedau, M. (1997): "Weak Emergence," in J. E. Tomberlin (ed.), *Philosophical Perspectives (11): Mind, Causation, and World*, Boston: Blackwell, 374–99.

Belot, G. and Earman, J. (1997): "Chaos out of Order: Quantum Mechanics, the Correspondence Principle and Chaos," *Studies in History and Philosophy of Modern Physics*, 2, 147–82.

Berry, M. V. (1991): "Asymptotics, Singularities, and the Reduction of Theories," in D. Prawiz, B. Skyrms and D. Westerståhl (eds.), *Logic, Methodology, and Philosophy of Science IX: Proceedings of the Ninth International Congress of Logic, Methodology and Philosophy of Science, Uppsala, Sweden, August 7–14, 1991,* volume 134 of *Studies in Logic and Foundations of Mathematics,* Amsterdam: Elsevier Science B. V, 597–607.

Berry, M. V. (1994): "Singularities in Waves and Rays," in R. Balian, M. Kléman and J. P. Poirier (eds.), *Physics of Defects (Les Houches, Session XXXV, 1980),* Amsterdam: North Holland, 453–543.

Berry, M. V. (2000): "Chaos and the Semiclassical Limit of Quantum Mechanics (is the Moon there when somebody looks?)," *Proceedings of the Vatican Conference on Quantum Mechanics and Quantum Field Theory,* 2–26.

Berry, M. V. and Howls, C. (1993): "Infinity Interpreted," *Physics World,* 12, 35–9.

Bickle, J. (1998): *Psychoneuronal Reduction: The New Wave.* Cambridge, MA: MIT Press.

Blazer, W., Pearce, D. A. and Schmidt, H.-J. (1984): *Reduction in Science: Structure Examples and Philosophical Problems.* Dordrecht: D. Reidel Publishing Company.

Bohm, D. and Hiley, B. J. (1993): *The Undivided Universe.* London: Routledge.

Bohr, N. (1934): *Atomic Theory and the Descriptive of Nature.* Cambridge: Cambridge University Press.

Cartwright, N. (1997): "Why Physics?" in R. Penrose, A. Shimony, N. Cartwright and S. Hawking (eds.), *The Large, the Small and the Human Mind,* Cambridge: Cambridge University Press, 161–8.

Cartwright, N. (1999): *The Dappled World: A Study of the Boundaries of Science.* New York: Cambridge University Press.

Causey, R. L. (1977): *Unity of Science.* Dordrecht: Reidel.

Chalmers, D. (1996): *The Conscious Mind.* Oxford: Oxford University Press.

Churchland, P. M. (1981): "Eliminative Materialism and the Propositional Attitudes," *Journal of Philosophy,* 78, 67–90.

Churchland, P. M. (1985): "Reduction, Qualia, and the Direct Introspection of Brain States," *Journal of Philosophy,* 82, 8–28.

Churchland, P. M. (1989): *A Neurocomputational Perspective: The Nature of Mind and the Structure of Science.* Cambridge, MA: MIT Press.

Churchland, P. S. (1986): *Neurophilosophy: Toward a Unified Science of the Mind-Brain.* Cambridge, MA: MIT Press.

Cornell, E. A. and Wieman, C. E. (1998): "The Bose-Einstein Condensate," *Scientific American,* 278(3), 40–5.

Dennett, D. (1988): "Quining qualia," in A. J. Marcel and E. Bisiach (eds.), *Consciousness in Contemporary Science,* Oxford: Clarendon Press, 240–78.

Dickson, W. M. (1998): *Quantum Chance and Non-Locality.* Cambridge: Cambridge University Press.

Dupré, J. (1993): *The Disorder of Things: Metaphysical Foundations of the Disunity of Science.* Boston: Harvard University Press.

Feyerabend, P. K. (1962): "Explanation, Reduction and Empiricism," in H. Feigl and G. Maxwell (eds.), *Minnesota Studies in the Philosophy of Science III.* Minnesota: University of Minnesota Press, 231–72.

Feyerabend, P. K. (1977): "Changing Patterns of Reconstruction," *British Journal for the Philosophy of Science,* 28, 351–82.

Giere, R. (1988): *Explaining Science: A Cognitive Approach.* Chicago: University of Chicago Press.

Hawthorne, J. and Silberstein, M. (1995): "For Whom the Bell Arguments Toll," *Synthese,* 102, 99–138.

Healey, R. A. (1989): *The Philosophy of Quantum Mechanics: An Interactive Interpretation.* Cambridge: Cambridge University Press.

Healey, R. A. (1991): "Holism and Nonseparability," *Journal of Philosophy*, 88, 393–421.

Hellman, G. (1999): "Reduction (?) to What (?) Comments on L. Sklar's 'The Reduction (?) of Thermodynamics to Statistical Mechanics'," *Philosophical Studies*, 95(1–2), 200–13.

Hendry, R. (1998): "Models and Approximations in Quantum Chemistry," in N. Shanks (ed.), *Idealization in Contemporary Physics*, Amsterdam: Rodopi, 123–42.

Hendry, R. (1999): "Molecular Models and the Question of Physicalism," *Hyle*, 5(2), 143–60.

Hooker, C. A. (1981): "Towards a General Theory of Reduction. Part I: Historical and Scientific Setting. Part II: Identity in Reduction. Part III: Cross-Categorical Reduction," *Dialogue*, 20, 38–59, 201–36, 496–529.

Humphreys, P. (1997): "How Properties Emerge," *Philosophy of Science*, 64, 1–17.

Kellert, S. (1993): *In the Wake of Chaos.* Chicago: University of Chicago Press.

Kemeny, J. and Oppenheim, P. (1956): "On Reduction," *Philosophical Studies*, 7, 6–17.

Kim, J. (1993): "Multiple Realization and the Metaphysics of Reduction," in J. Kim (ed.), *Supervenience and Mind*, Cambridge: Cambridge University Press, 309–35.

Kim, J. (1998): *Mind in a Physical World: An Essay on the Mind-Body Problem and Mental Causation.* Cambridge, MA: MIT Press.

Kim, J. (1999): "Making Sense of Emergence," *Philosophical Studies*, 95(1–2), 3–36.

Kitcher, P. (1989): "Explanatory Unification and the Causal Structure of the World," in P. Kitcher and W. C. Salmon (eds.), *Minnesota Studies in the Philosophy of Science, vol. 13, Scientific Explanation*, Minneapolis: University of Minneapolis Press, 410–505.

Lewis, D. (1986): "Causation," in D. Lewis (ed.), *Philosophical Papers, Volume II.* Oxford: Oxford University Press, 159–213.

McGinn, C. (1999): *The Mysterious Flame: Conscious Minds in a Material World.* New York: Basic Books.

McLaughlin, B. (1992): "Rise and Fall of British Emergentism," in A. Beckermann, H. Flohr and J. Kim (eds.), *Emergence or Reduction? Essays on the Prospects for Nonreductive Physicalism*, Berlin: DeGruyter, 30–67.

Machamer, P., Darden, L. and Craver, C. F. (2000): "Thinking About Mechanisms," *Philosophy of Science*, 67, 1–25.

Maudlin, T. (1998): "Part and Whole in Quantum Mechanics," in E. Castellani (ed.), *Interpreting Bodies: Classical and Quantum Objects in Modern Physics*, Princeton: Princeton University Press, 46–60.

Nagel, E. (1961): *The Structure of Science.* New York: Harcourt, Brace.

Oppenheim, P. and Putnam, H. (1958): "The Unity of Science as a Working Hypothesis," in H. Feigl and G. Maxwell (eds.), *Minnesota Studies in the Philosophy of Science. Philosophical Studies, vol. 95*, Minneapolis, Minnesota: University of Minnesota, 45–90.

Primas, H. (1983): *Chemistry, Quantum Mechanics, and Reductionism.* Berlin: Springer-Verlag.

Primas, H. (1991): "Reductionism: Palaver Without Precedent," in E. Agazzi (ed.), *The Problem of Reductionsm in Science*, Dordrecht: Kluwer, 161–72.

Primas, H. (1998): "Emergence in the Exact Sciences," *Acta Polytechnica Scandinavica*, 91, 83–98.

Ramsey, J. L. (1997): "Molecular Shape, Reduction, Explanation and Approximate Concepts," *Synthese*, 111, 233–51.

Rorty, R. (1970): "In Defense of Eliminative Materialism," *The Review of Metaphysics*, 24, 112–21.

Rueger, A. (2000a): "Physical Emergence, Diachronic and Synchronic," *Synthese*, 124, 297–322.

Rueger, A. (2000b): "Robust Supervenience and Emergence," *Philosophy of Science*, 67, 466–89.

Sarkar, S. (1998): *Genetics and Reductionism*. Cambridge: Cambridge University Press.

Scerri, E. (1994): "Has Chemistry Been at Least Approximately Reduced to Quantum Mechanics?" in D. Hull, M. Forbes and R. Burian (eds.), *PSA 1994, vol. 1*. East Lansing, MI: Philosophy of Science Association, 160–70.

Schaffner, K. F. (1992): "Philosophy of Medicine," in M. H. Salmon, J. Earman, C. Glymour, J. G. Lennox, P. Machamer, J. E. McGuire, J. D. Norton, W. C. Salmon and K. F. Schaffner (eds.), *Introduction to the Philosophy of Science*, Englewood Cliffs: Prentice Hall, 310–45.

Schaffner, K. F. (1998): "Genes, Behavior, and Developmental Emergentism: One Process, Indivisible?" *Philosophy of Science*, 65, 209–52.

Schaffner, K. F. (2000): "Behavior at the Organismal and Molecular Levels: The Case of *C. elegans*," *Philosophy of Science*, supplement to 67(3), D. A. Howard (ed.), *Part II: Symposia Papers*, S273–S288.

Silberstein, M. (1998): "Emergence and the Mind/Body Problem," *Journal of Consciousness Studies*, 5, 464–82.

Silberstein, M. (2001): "Converging on Emergence: Consciousness, Causation and Explanation", *Journal of Consciousness Studies*, 8(9)(10), 61–98.

Silberstein, M. and McGeever, J. (1999): "The Search for Ontological Emergence," *The Philosophical Quarterly*, 49, 182–200.

Sklar, L. (1967): "Types of Inter-theoretic Reduction," *The British Journal of the Philosophy of Science*, 18, 109–24.

Sklar, L. (1993): *Physics and Chance: Philosophical Issues in the Foundations of Statistical Mechanics*. Cambridge: Cambridge University Press.

Sklar, L. (1999): "The Reduction (?) of Thermodynamics to Statistical Mechanics," *Philosophical Studies*, 95(1–2), 187–99.

Stephan, A. (1992): "Emergence – A Systematic View on its Historical Facets," in A. Beckermann, H. Flor and J. Kim (eds.), *Emergence or Reduction?: Essays on the Prospects of Nonreductive Physicalism*, New York: DeGruyter, 68–90.

Takesaki, M. (1970): "Disjointness of the kms States of Different Temperatures," *Communications in Mathematical Physics*, 17, 33–41.

Van Fraassen, B. C. (1989): *Laws and Symmetry*. Oxford: Oxford University Press.

Wilkes, K. (1988): "Yishi, Duh, Um and Consciousness," in A. J. Marcel and E. Bisiach (eds.), *Consciousness in Contemporary Science*, Oxford: Clarendon Press, 148–77.

Wilkes, K. (1995): "Losing Consciousness," in T. Metzinger (ed.), *Conscious Experience*, Thorverton: Imprint Academic, 35–89.

Wimsatt, W. C. (1976): "Reductive Explanation: A Functional Account," in R. S. Cohen, C. A. Hooker, A. C. Michalos and G. van Evra (eds.), *Philosophy of Science Association*, Dordrecht: Reidel, 671–710.

Woody, A. I. (2000): "Putting Quantum Mechanics to Work in Chemistry: The Power of Diagrammatic Representation," *Philosophy of Science*, supplement 67(3), D. A. Howard (ed.), *Part II: Symposia Papers*, S612–S627.

Chapter 6

Models, Metaphors and Analogies

Daniela M. Bailer-Jones

Metaphor and analogy, often not distinguished very sharply, are manifestations of ways in which information can be expressed and, as some would argue, is processed in our mind. This is so not only in everyday situations, but also in the sciences where information is presented in highly systematic and methodologically specialized ways. Appreciating that metaphor and analogy are pervasive in everyday situations and in the language we use to describe them, it comes as no surprise to encounter them in science also. In tune with this assumption, the analysis of metaphor and analogy has had considerable influence on the analysis of scientific models. Taking models as manifestations of how scientists develop interpretations of empirical information in their subject area, the metaphor approach to scientific models can elucidate how scientists create and formulate interpretations of aspects of the empirical world.

I shall begin by outlining what scientific models are, and then provide a review of the treatment of analogy in the philosophy of science because the use of analogies is central for scientific modeling as well as for metaphorical language use (pp. 108–13). Issues guiding the analysis of metaphor and the application of this analysis in science are then presented (p. 114), while the next section (p. 118) deals with the specific claim that scientific models are metaphors. The final section (p. 121) highlights current issues that have emerged from the discussion of scientific models as metaphors and still need to be resolved. It also summarizes the relationships between model, metaphor and analogy as examined here.

Models

I consider the following as the core idea of what a scientific model is: A model is an interpretative description of a phenomenon that facilitates access to that phenomenon. (I take "phenomenon" to cover objects as well as processes.) This access can be perceptual as well as intellectual. Interpretations may rely, for instance, on idealisations or simplifications or on analogies to interpretative descriptions of other phenomena. Facilitating access usually involves focusing on specific aspects of a phenomenon, sometimes deliberately disregarding others. As a result, models tend to be partial descriptions only. Models can range from being objects, such as a toy aeroplane, to being theoretical, abstract entities, such as the Standard Model of the structure of matter and its fundamental particles. As regards the former, scale models facilitate looking at (perceiving) something by enlarging it (e.g. a plastic model of a snow flake) or shrinking it (e.g. a globe as a model of the earth). This can involve making explicit features which are not directly observable (e.g. the structure of DNA or chemical elements contained in a star). The majority of scientific models are, however, a far cry from consisting of anything material like the rods and balls of molecular models used for teaching; they are highly theoretical. They often rely on abstract ideas and concepts, frequently employing a mathematical formalism (as in the big bang model, for example), but always with the intention to provide access to aspects of a phenomenon that are considered to be essential. Bohr's model of the atom informs us about the configurations of the electrons and the nucleus in an atom, and the forces acting between them; or modeling the heart as a pump gives us a clue about how the heart functions. The means by which scientific models are expressed range from the concrete to the abstract: sketches, diagrams, ordinary text, graphs, mathematical equations, to name just some. All these forms of expression serve the purpose of providing intellectual access to the relevant ideas that the model describes. Providing access means giving information and interpreting it, and expressing it efficiently to those who share in the specific intellectual pursuits. Scientific models are singularly about empirical phenomena (objects and processes), whether these are how metals bend and break or how man has evolved.

One might object that, according to my explication, more or less anything that is used in science to describe empirical phenomena is a model, and indeed this seems to be the case. The tools employed to grant us and others intellectual access are of great diversity, yet this, in itself, is no reason to deny them the status of being constituents of models. Modeling in science is pervasive, and it has become increasingly varied and more and more abstract. The sheer diversity of models makes it unlikely that *all* models are metaphors or rely on analogies, but some do, and these are the focus of this article.

When scientific models are associated with metaphor and analogy, the topic is how scientists develop and convey scientific accounts for empirical phenomena encountered in their research. The idea of viewing scientific models as metaphors appeared in the 1950s (Hesse, 1953; Hutten, 1954) and was taken up in Mary Hesse's (1966) work in the 1960s with the aim to show that metaphorical models and analogy are more than heuristic devices that can be jettisoned once a 'proper' theory is in place. (For a review of the work at that time and before, see Leatherdale (1974).) Work on models and metaphor continues to be discussed to this day (Paton, 1992; Bhushan and Rosenfeld, 1995; Miller, 1996; Bradie, 1998, 1999;

Bailer-Jones, 2000b), and it usually concurs with the rejection of the received opinion that theories prevail over models – a rejection which need not be linked to a metaphor view of models, however (Cartwright, 1999; Giere, 1999; Morgan and Morrison, 1999). Metaphor, in this context, is seen as very closely tied to analogy, just as models are closely tied to analogy (Achinstein, 1968; Harré, 1988). In addition, model, metaphor and analogy have been popular notions in their own right over the past decades. There is a great deal of work on analogy coming from artificial intelligence research (Falkenhainer et al., 1989; Holyoak and Thagard, 1989; Hofstadter, 1995) and from cognitive psychology (Gentner and Markman, 1997; Van Lehn, 1998). Metaphor has been addressed in philosophy of language (Davidson, [1978] 1984; Searle, 1979) as well as in cognitive linguistics (Kittay, 1987; Langacker, 1987; Lakoff and Johnson, 1980, 1999; Lakoff, 1993). Today, the study of scientific models as metaphors or analogies in philosophy cannot be separated from the study of these phenomena in neighboring subjects, nor should they. On the other hand, drawing from a large number of fields with different aims and research questions does not make progress in research on scientific models any swifter. To draw inferences about the use of metaphor and analogy in scientific modeling, one needs to tread carefully when assessing whether findings from neighboring disciplines can be integrated – while integrating them is likely to be an important stepping stone in the analysis of scientific models. In any case, the idea that analogy centrally occurs in human information processing and knowledge generation is common to work done in all these areas.

Analogy

The Greek word analogy ($\alpha\nu\alpha\lambda\text{o}\gamma\iota\alpha$) means 'proportion', e.g. 2 is to 4 as 4 is to 8. To use analogy for illustration is a common occurrence in Greek thought, as when the pre-Socratic Thales of Miletus claims that the earth floats on water like a piece of wood would (Aristotle, *De Caelo*, B 13, 294a28 f.) Analogy is often understood as pointing to a resemblance between relations in two different domains, i.e. *A* is related to *B* like *C* is related to *D*. To give an example, the electrons in an atom are related to the atomic nucleus like the planets are related to the sun. The term "formal" analogy points to relations between certain individuals of two different domains that are identical, or at least comparable. Such an identity of structure does not require a "material" analogy, that is, the individuals of the domains are not required to share attributes (Hesse, 1967). Both the motion of electrons and planets is determined by an attractive force which is why they orbit around the atomic nucleus and the sun respectively, even though the causes of attraction are not the same (gravitational versus electrostatic) which is why the relationships are perhaps more correctly called "comparable" than "identical." Although electrons and planets share the relationship of attraction, they differ hugely in attributes, such as size and physical make-up.

Analogies can exist as formal relationships between phenomena or, rather, between the theoretical treatment of phenomena. Pointing to examples, such as light and sound waves or magnetism and dielectric polarisation, Pierre Duhem stresses that "it may happen that the equations in which one of the theories is formulated is algebraically identical to the equations expressing the other. . . . [A]lgebra establishes an exact correspondence between [the theories]" (Duhem, [1914] 1954, p. 96). To find such a correspondence serves "intellectual economy," and it can also "[constitute] a method of discovery" by "bringing together two abstract systems; either one of them already known serves to help us guess the form of the other not yet known, or both being formulated, they clarify each other" (Duhem, 1954, p. 97). There is "nothing here that can astonish the most rigorous logician"; it is a strategy in perfect agreement with "the logically con-ducted understanding of abstract notions and general judgements" (1954, p. 97), and yet, Duhem judges analogies in science as heuristic, thus as no longer essen-tial once a theory is formulated.

Norman Campbell ([1920] 1957) treats analogy in science as somewhat more crucial for his notion of theory. In his account a theory contains a hypothesis, a set of propositions of which it is not known whether they are true or false about a certain subject (for my purposes, "first subject"). If the ideas expressed in these propositions were not connected to some *other* ideas (associated with a "second subject"), they would be, according to Campbell, no better than arbitrary assump-tions. As the propositions constituting the hypothesis are not themselves testable, they require some kind of confirmation via a translation into *other* ideas, i.e. ideas about a second subject, whereby the latter are known to be true through obser-vational laws. Campbell's example is the kinetic theory of gases (first subject) being the hypothesis formulated in terms of a set of propositions. The second subject would then be "the motion of a large number of infinitely small and highly elastic bodies contained in a cubical box" (Campbell, 1957, p. 128). For the latter, laws are available so that it can be known which propositions concerning the second subject are true. The first subject, Campbell proposes, benefits from this knowledge: Via a "dictionary," the transition from first to second subject can be made, and the knowledge about the small elastic bodies illuminates the case of interest, i.e the first subject, and is employed to test the hypothesis indirectly. Campbell asserts: "[i]n order that a theory may be valuable . . . it must display an analogy. The propositions of the hypothesis must be analogous to some known laws" (Campbell, 1957, p. 129). In modern terms, the small elastic bodies in the box would be considered as a model, though Campbell does not use this term. While Campbell greatly advertises the importance of analogy, he seems little concerned with its advantages for the practice of science, that is for discovery or teaching.[1]

Besides analogies *between* the theoretical treatments of phenomena which Campbell considers, Hesse (1966) also proposed that scientific models are to be viewed as analogues[2] to the aspects of the real world that are their subject (Hesse, 1953, p. 201); see also Hesse (1967) and Harré (1970, p. 35). Rom Harré calls this "[a] behavioural analogy between the behaviour of the analogue of the real productive process and the behaviour of the real productive process itself" (1988, p. 127). In contrast to this, most subsequent discussions about the formal relationship of analogy concern analogies between theoretical treatments of different empirical phenomena and the examination of the potential of the use of analogy for purposes of "intellectual economy" and for "scientific discovery." For this, the study of nineteenth century science, specifically the work of Kelvin, Faraday, and Maxwell, provides numerous case studies (Duhem, [1914] 1954; Campbell, [1920] 1957; Hesse, 1953). Examples include the analogy between heat and electrostatics, where the same equations can be employed in both areas, with temperature corresponding to electrical potential and source of heat to positive charge (North, 1980, p. 123), or Maxwell's approach to electromagnetism by analyzing an electromagnetic ether in terms of vortices along lines of magnetic force (Harman, 1982, 1998). From such examples, analogy can be recognized as a constituent of scientific argument (North, 1980) and can be appreciated as cognitively relevant. It seems plausible that the development of models of new phenomena benefits, in many cases, from considering analogies to other, already existing and more familiar models, even if these appear to belong to quite different phenomena. The key point is precisely that the two phenomena are not the same. Proclaiming one thing to be analogous to another is not simply a statement about what the two subjects have in common. Rather, in the interesting cases of analogy, there are differences between the relations and attributes present in both domains; these are called "disanalogies" or "negative analogies." Electrons and planets are attracted by the atomic nucleus and the sun respectively, but not through the same kind of force. Any positive analogy comes with negative, and also sometimes with neutral, not yet explored analogies (Hesse, 1966, 1967; Harré, 1970, 1988). The effective use of analogy presupposes that its users know, or can explore, what the positive and the negative analogies between two domains are.

Analogy is often thought to occur in science because it supports a central function of models: explanation (Harré, 1960; Nagel, [1960] 1979, p. 107; Hesse, 1966; Achinstein, 1968). According to some authors, models being explanatory mostly coincides with them being developed on the basis of an analogy to some other object or system (Achinstein, 1968, p. 216). Explanation is thus linked to making the transition from something unfamiliar to something more familiar: "The analogies help to assimilate the new to the old, and prevent novel explanatory premises from being radically unfamiliar" (Nagel, [1960] 1979, p. 46). Analogy counts as a plausible candidate for providing explanations because the use of more familiar and already accepted models (models that have led to understanding in different, but comparable situations) appears as a promising strategy in a new context. Correspondingly, Peter Achinstein states:

> Analogies are employed in science to promote understanding of concepts. They do
> so by indicating similarities between these concepts and others that may be familiar
> or more readily grasped. They may also suggest how principles can be formulated

and a theory extended: if we have noted similarities between two phenomena (for example, between electrostatic and gravitational phenomena), and if principles governing the one are known, then, depending on the extent of the similarity, it may be reasonable to propose that principles similar in certain ways govern the other as well (Achinstein, 1968, pp. 208–9).

Achinstein quotes as examples, among others, the analogies between the atom and the solar system, between waves of light, sound and water, between nuclear fission and the division of a liquid drop, between the atomic nucleus and extranuclear electron shells, and between electrostatic attraction and the conduction of heat (Achinstein, 1968, pp. 203–5).

Exploring how people understand is the subject of cognitive psychology, where research into analogy has generated considerable interest over the last twenty years; for overviews, see Gentner and Markman (1997) and Holyoak and Thagard (1997). Results involve that analogy can be analyzed in terms of similarity, similarities of relationships (e.g. encountering interference in water waves *and* in light) and similarities of object attributes (e.g. oxygen and helium being gaseous at room temperature). Correspondingly, analogy can consist of attribute mappings as well as of relationship mappings, but Gentner (1983) produces empirical evidence according to which mappings of relations tend to be favored and considered the "deeper" analogies by those who are confronted with the analogies. Analogies become relevant, in science in particular, when they highlight a "system of connected knowledge, not a mere assortment of facts" (Gentner, 1983, p. 162). Furthermore, there is a preference for comparing items that are similar because their differences are "alignable." Items that are dissimilar have little in common, their differences are not alignable, and they thus have a smaller impact on people's perception of similarity. Gentner and Markman view "the ability to carry out fluent, apparently effortless, structural alignment and mapping [as] a hallmark of human cognitive processing" (1997, p. 53).

Acknowledging the importance of analogy in scientific reasoning (Hesse, 1966; Gentner, 1982; Harré, 1988) makes it tempting to identify scientific modeling with drawing analogies. However, while many models have their roots in an analogy, such as Thomson's plum pudding model of the atom or Bohr's model of the atom based on the solar system, few existing models in science have not developed beyond the boundaries of the analogy from which they originated, and others may simply not have their origin in an analogy at all (Bailer-Jones, 2000b). Moreover, an analogy is a relationship between things or processes while a model is a type of description about some thing or process. If anything, a model could be an analogue, but this is not the issue because the way to evaluate a model is not to judge whether it is analogous to something, but whether it, as it stands (analogous or not), *provides access* to a phenomenon in that it interprets the available empirical data about the phenomenon. An analogy used for modeling can act as a catalyst to aid modeling, even though the aim of modeling has nothing intrinsically to do with analogy. It is, of course, more than reasonable to stress the importance of analogy in the modeling process given that analogy is one of the cognitive strategies available for creative discovery from which scientific models result.

Metaphor

A metaphor is a linguistic expression in which at least one part of the expression is transferred (μεταφερειν) from one domain of application (source domain), where it is common, to another (target domain) in which it is unusual, or was probably unusual at an earlier time when it might have been new. This transfer serves the purpose of creating a specifically suitable description of aspects of the target domain, where there was no description before (e.g. "black hole") or none was judged suitable. Martin and Harré (1982, p. 96) call these "crises of vocabulary." Metaphoric expressions are used for descriptions, and the occasion for the use of metaphor arises when the two domains between which the transfer occurs can be viewed as being related: by similarity of object attributes, or by similarity of relationships (Gentner, 1983). Thus the relationship of analogy is usually an important factor in being able to understand a metaphor. However, establishing the importance of analogy for understanding a metaphor is not to claim that the analogy necessarily precedes the metaphor. One could equally argue that it is the metaphor that prompts the recognition of an analogy – it is feasible that *both* types of cases occur; the latter possibility would still warrant that the metaphor is connected to the analogy (or analogies) suggested by it: "every metaphor may be said to mediate an analogy or structure correspondence" Black ([1977] 1993, p. 30). In astronomical observations, one talks about signal-to-noise ratio. *Signal* is the light emitted from the object one wants to observe; *noise* represents the uncertainty in the signal (and the *background*) due to quantum fluctuations of photon emission and thus represent a limit to the precision with which the signal can be determined. The analogy connected with the noise metaphor is to a sound signal, e.g. emitted from an interlocutor whilst noise from other people talking and perhaps a nearby road needs to be separated from the signal so as to make out the information of interest. As listeners dealing with sound waves, we are quite proficient in filtering out all those unpredictable random frequencies that could prevent us from making out the signal in which we are interested, and a comparable skill would be required for optical waves in astronomy. Without this analogy, the metaphor of noise, as used in astronomy, is incomprehensible.

The claim that scientific models are metaphors is tied to the fact that often an analogy is exploited to construct a model about a phenomenon. Thus, if scientific models are metaphors, then analogy is an important factor in this. "The brain is the hardware for which a child gradually develops suitable software" implies an analogy between data processing in a computer and the cognitive development of a child, just like the liquid drop model of the atomic nucleus suggests an analogy between the atomic nucleus and a liquid drop in that the overall binding energy of the nucleus is, in approximation, proportional to the mass of the nucleus – like in a liquid drop. The view that scientific models are metaphors depends, of course, on what metaphor is taken to be, other than metaphors being connected to analogies. Only in view of that can one assess how the analysis of metaphor translates into an understanding of scientific models. I start by focusing on the first of these issues.

The analysis of metaphor is traditionally conducted by contrasting literal with figurative (or metaphorical) language (a distinction on which I shall shed some doubt below). This requires the reliance on an intuitive or common-sense understanding of "literal," despite the difficulty of pinpointing what makes literal language literal. Of course, we have a sense in which talk about "little green men" appears metaphorical in comparison to "extraterrestrial intelligent life." "Literal" implies, by default, that an expression is not transferred from another domain, i.e. is "more directly" about something and perhaps more "typical," "common," "usual" or "expected." Inevitably, such a classification remains unsatisfactory, partly because we do not tend to find metaphorical statements more difficult to comprehend than so-called literal statements that could stand in their place (Rumelhart, [1979] 1993). Metaphors, moreover, can be perfectly usual and familiar. Nobody stumbles over *processing information* or *developing software* said of the mind, or a *phylogenetic tree* merely because it is no oak, beech, lime or fir. Just as we understand the brain-as-computer metaphor, we understand that a phylogenetic tree displays dependency relations of a group of organisms derived from a common ancestral form, with the ancestor being the trunk and organisms that descend from it being the branches. Most metaphors are understood with ease which indicates that there are no grounds to treat them as deviations of language use. On the contrary, they are pervasive and central (Richards, 1936).

While there may be no clear-cut distinction between literal and metaphorical, one can still observe different "degrees" of metaphoricity, and the conditions under which we are capable of comprehending metaphors can be outlined correspondingly:

A Even though a metaphor is entirely novel to us, we are endowed with the cognitive skill to interpret it just as easily as if we were familiar with that particular use of terminology.

B While we recognize a phrase as metaphorical in principle, we are so familiar with the particular type of metaphor that the metaphor is neither unusual nor unexpected; the brain-as-a-computer metaphor is an example of this. Another is to think of the energy distribution of a system as a landscape with mountains and valleys, and a gravitational force that is responsible for differences in potential energy depending on height, exemplified in phrases such as *potential well* or *tunnelling through a potential barrier*.

C We are so familiar with what once was a metaphor that a special effort would be required to recognize it as such; examples are electric *current*, electric *field*, *excited* state or a chemical *bond forming, breaking, bending, twisting* or even *vibrating*. Such metaphors are "dead"; they are pervasive in our language and they appear to us just like literal expressions (Machamer, 2000), especially as sometimes they are our *only* expression for what they describe. Historical priority would probably be the only grounds on which a *current* of a river or a *field* ploughed by a farmer would be judged more literal than *electric current* or *electric field*.

These degrees of metaphoricity are only partially related to the novelty of the metaphors, because some metaphors, (unlike those in C, or even in B), will always remain recognizable as metaphorical, no matter how familiar and well-known they have become. An example would be "God does not play dice" expressing resistance to indeterminacy in physics.

In the following, I shall focus on metaphors of the first and second kind, namely metaphors that have not become and perhaps never will become entirely ordinary and are consequently not quite so easily taken for literal language. In these cases, it is often presumed that the metaphorical phrase has a special quality in the way it communicates information, sometimes referred to as "cognitive content" (Black, 1954). A "strong cognitive function" is assigned to metaphor when "a metaphorical statement can generate new knowledge and insight by *changing* relationships between the things designated (the principal and subsidiary subjects)" (Black, [1977] 1993, p. 35). This is thought to happen because metaphor inspires some kind of creative response in its users that cannot be rivalled by literal language use. Think of "little green men" as a metaphor for extraterrestrial intelligent life, as it is used in science, not restricted to fantasy. Of course, the original domain of that expression is fantasy, and there "little green men" may mean exactly that: small green people. If the phrase is used in science contexts, however, the implicit reference to fantasy highlights the fact that we have no idea of what extraterrestrial intelligent life might be like. Something naively and randomly specific – little green men – is chosen to indicate that there is no scientific way of being specific about the nature of extraterrestrial intelligent life. Precisely that we do not know what extraterrestrials are like is what we can grasp from the phrase "little green men." No amount of interpreting the literal phrase "little green men" without a system of "associated commonplaces" (Black, 1954); "implicative complex" in Black ([1977] 1993) would enable us to achieve this, thus knowledge of the domain of application is crucial. "Little green men" is transferred from the domain of fantasy to the radically different domain of science where one would not usually expect this expression. Nonetheless, there is no reason to think that the metaphor of little green men can be less reliably interpreted and understood by its recipients than any phrase from the "right" domain of language use, such as "not further specifiable forms of extraterrestrial intelligent life." On the contrary, according to Max Black's interaction view which goes back to Ivor Richards (1936), we even gain insight through the metaphor that no literal paraphrase could ever capture; a metaphor cannot be *substituted* by a literal expression. Neither is it simply a *com-*

parison between the two relevant domains, as in an elliptical simile ("Extraterrestrial intelligent life is (like) little green men"), because, as Black suspects, metaphor can *create similarity*.[3] If this is true, metaphorical meaning can no longer be viewed as a sheer function of the literal meaning of linguistic expressions belonging to a different domain. Instead, the proposal of the interaction view is that the meanings of the linguistic expressions associated with either domain shift. The meanings of the expressions are extended due to new ideas that are generated when the meanings associated with primary and secondary subject interact. The interaction takes place on account of the metaphor which forces the audience to consider the old and the new meaning together.

While the idea that scientific models are metaphors appears in Black (1962),[4] it was further explored in Hesse (1966, pp. 158–9) who draws from the interaction view:

> In a scientific theory the primary system is the domain of the explanandum, describable in observation language; the secondary is the system, described either in observation language or the language of a familiar theory, from which the model is taken: for example, "Sound (primary system) is propagated by wave motion (taken from a secondary system)", "Gases are collections of randomly moving particles."

Hesse postulates a meaning shift for metaphors. Their shift she takes to be in pragmatic meaning that includes reference, use and a relevant set of associated ideas (1966, p. 160). Correspondingly, a shift in meaning can involve change in associated ideas, change in reference and/or change in use. On these grounds, Hesse gets close to dissolving the literal/metaphorical distinction: "the two systems are seen as more like each other; they seem to interact and adapt to one another, even to the point of invalidating their original literal descriptions if these are understood in the new, postmetaphoric sense" (1966, p. 162); see also Hesse (1983). The crucial point is that metaphors can (in spite or because of this) be used to communicate reliably and are not purely subjective and psychological. Not "*any* scientific model can be imposed a priori on *any* explanandum and function fruitfully in its explanation" (1966, p. 161). Scientific models, in contrast to poetic metaphors, have to subject to certain objective criteria, or as Hesse puts it, "their truth criteria, although not rigorously formalizable, are at least much clearer than in the case of poetic metaphor" (1966, p. 169). Correspondingly, one may "[speak] in the case of scientific models of the (perhaps unattainable) aim to find a 'perfect metaphor,' whose referent is the domain of the explanandum" (1966, p. 170). In my formulation, a model is evaluated with regard to whether it provides access to a phenomenon and matches the available empirical data about the phenomenon reasonably well.

Rom Harré and his co-authors also discuss models and metaphor together. They claim that both could be interpreted successfully with the same tool, namely their type-hierarchy approach (Aronson et al. 1995, p. 97). Yet, the role of metaphor in science is different (Harré, 1960, 1970, p. 47; Martin and Harré, 1982). According to Martin and Harré (1982), metaphorical language is used in the sciences to fill gaps in the scientific ordinary language vocabulary. Examples are metaphorical expressions that have acquired very specific interpretations, like *electric field*, *electric current* or *black hole*, which is why they should be understood "without the intention of a point-by-point comparison" (Martin and Harré, 1982, p. 100). Such metaphorical terms can, however, be viewed as a "spin off" of scientific models (1982, p. 100). Martin and Harré (1982, p. 100) explain:

> The relationship of model and metaphor is this: if we use the image of a fluid to explicate the supposed action of the electrical energy, we say that the fluid is functioning as a model for our conception of the nature of electricity. If, however, we then go on to speak of the "rate of flow" of an "electrical current", we are using metaphorical language based on the fluid model.

It seems that many examples lend force to the view that metaphorical scientific terminology, even if hardly recognizable as such any longer, can be a "spin off" of models (without the claim that models themselves are metaphorical) and I shall discuss one example below. That Martin and Harré, different from myself, consider models simply as analogues has no bearing on this specific point.

Metaphorical Models

I now single out the features of models discussed in association with the claim that scientific models are metaphors. The listed points presuppose that the metaphors in question are connected to analogies and that something like the cognitive claim attached to the interaction view holds.

Familiarity and understanding

Models and metaphors exploit the strategy of understanding something in terms of something else that is better understood and more familiar; they exploit the analogy relationship suggested by a metaphor or explored in a model. Of course, being familiar does not equate with being understood, but familiarity can be a factor in understanding. This is also not to suggest that understanding can be reduced to the use of analogy, but having organized information in one domain (source) of exploration satisfactorily can help to make connections to and do the same in another domain (target). The aim is to apply the same pattern in the target domain as in the source domain, with the same assumptions of structural relationships in both the source and the target domains. For instance, to think of the energy generation process in quasars in terms of energy generation in binary stars is helpful because it was by studying binary star systems that the importance of accretion of mass as a power source was first recognized. Moreover, turning the gravitational energy into the "internal" energy of a system is perhaps the only way to account for the enormous energies that must be present in quasars. The proposed conversion process of gravitational energy is, in turn, inspired by disks in planet or star formation. Piecing together these ideas based on analogies to already better-analyzed empirical phenomena paved the way to the formulation of the accretion disk model that is constitutive in explaining energy present in quasars and radio galaxies. For more examples, see Cornelis (2000).

Material for exploration

Models and metaphors can be hypothetical and exploratory. Besides a positive analogy which may have given rise to the formulation of a model or metaphor, there are negative and neutral analogies that can be explored. This exploration furthers creative insight, as the interaction view proposes, because sometimes negative and neutral analogies offer a pool of ideas of what can be tested about the target domain. Metaphorical models nevertheless have to stand up to empirical reality, which is why Hesse talks about "clearer truth criteria than for poetic metaphors" and "the (perhaps unattainable) aim to find a 'perfect metaphor'" (1966, p. 170), i.e. a perfect description, one that provides an empirically adequate description of a phenomenon. An example for metaphorical exploration is artificial neural networks as used in computing for pattern recognition. Digital computers are serial processors and good at serial tasks such as counting or adding up. They are less good at tasks that require the processing of a multitude of diverse items of information, tasks such as vision (a multitude of colors and shapes etc.) or speech recognition (a multitude of sounds) at which the human brain excels. The example of the brain demonstrates how to cope with such tasks through many simple processing elements that work in parallel and "share the job." This makes the system tolerant to errors; in such a parallel distributed processing system, a single neuron going wrong has no great effect. The idea of artificial neural networks was therefore to transfer the idea of parallel processing to the computer so as to take advantage of the processing features of the brain. Moreover, the assumption that learning occurs in the brain when modifications are made to the effective coupling between one cell and another at a synaptic junction is simulated in artificial systems through positive or negative reinforcement of connections. Artificial neural networks produce impressive results in pattern recognition, even though there remain considerable negative analogies between them and the human brain. Not only do the number of connections differ hugely from the brain, but the nodes in artificial neural networks are highly simplified in comparison to neurons in the brain. Explaining the neural network metaphor involves becoming aware of its appropriate applications as well as its limits.

Coping with negative analogies

Metaphors, analyzed as being connected to analogies, usually involve the statement of negative analogies; these do not tend to hinder the use of the metaphor, however. Scientific models, in contrast, require attention to so-called negative analogies. Even though models claim no more than to be partial descriptions, to use them efficiently, their users need to be aware of the negative analogies, those "things" not described by the model and that do not stand up to empirical tests. Knowing what the model is not a model of is part of the model. As shown, an artificial neural network does not simulate the structure of the human brain in every respect, but we need to know in what respects it does. Not spelling out disanalogies explicitly in a model can have detrimental effects. Some metaphors, especially if even the positive analogy is questionable, can be positively misleading, e.g. the common interpretation of entropy as a measure of disorder. Consider the example of a partitioned box of which one half contains a gas and the other is empty. When the partition is removed, the gas spreads over both halves of the box. This constitutes an increase of entropy because it is extremely unlikely that all gas molecules will ever return to one half of the box spontaneously. It is not intuitive why the second situation should be viewed as a state of higher disorder than the first; a more precise way of modeling entropy is to talk about the number of available microstates per macrostate. For some examples from chemistry, see Bhushan and Rosenfeld (1995).

New terminology

Metaphorical models are "new vocabulary" in terms of which empirical data can be described. This "vocabulary" makes possible a description intended to provide interpretations of data. In a narrower sense of a vocabulary, metaphorical terminology is employed to meet the problem of catachresis, i.e. to provide scientific terminology where none existed previously (Boyd, 1993). Sometimes, such new terminology has its root in the analogues that inspired the formulation of the model to which the terminology belongs (a "spin off" of the model); Martin and Harré (1982). An example of this is simulated annealing, a method used in optimization for determining the best fit parameters of a model based on some data. The physical process of annealing is one in which a material is heated to a high temperature and then slowly cooled. This process increases the chance that the material relaxes to a low energy state rather than getting stuck in a higher energy metastable state. Annealing is about avoiding "local minima" of energy states so as to reach the "global minimum." Having found a local minimum, it may be necessary to expend some energy first, i.e. to "climb over" an "energy mountain," to find a more global minimum, the low energy state. This can be interpreted to correspond to finding an "optimal fit" in the search for the best fitting parameters for a data model; one wants to avoid terminating the search early at what seems a good fit in a local search area. In the computational method of simulated annealing, not only the equations from statistical physics, such as the Boltzmann equation, are adopted almost exactly, but the descriptive terminology is also taken over. Terms such as *temperature, specific heat capacity* and *entropy* are applied to optimization in a meaningful way.

Literal versus metaphorical

Finally, the converse of the metaphor claim needs to be considered briefly: What would "being literal" mean in the context of scientific modeling? What would be the consequences of a literal–metaphorical distinction for scientific models if they are, as the claim goes, metaphors? In the context of science, this would presumably mean that there are "proper," "precise" or "literal" ways of describing an empirical phenomenon and other, metaphorical and less direct ways, the latter being exemplified by scientific models. One would then have to ask what "literal" ways of describing empirical phenomena are. Of course, there may just not be any alternative description for certain phenomena, e.g. for "electron spin," although this is what talk about literal and metaphorical language implies. One answer that may be put forward by some is that *theories* are the literal descriptions. However, theories cannot range as an alternative to models, if, as my claim goes, they are not descriptions of phenomena (Cartwright, 1999). Instead, theories may be employed in models and applied to empirical phenomena only *through* scientific models. In this case they cannot be viewed as an independent mode of description of phenomena, i.e. as literal in contrast to models that are metaphorical.

Current Issues

There is a range of general issues concerning scientific models which I have barely covered. These include how models work, why they are needed and to what extent they are used. Other issues are, for instance, how models relate to theories or whether simulations are models. Yet, in this last section, I want to focus only on points that are specifically raised in the context of comparing scientific models with metaphors.

Creativity

Employing metaphors can sometimes be a creative use of language, both in formulation and interpretation. Consequently, explaining human creativity, in science and elsewhere, is a common agenda of those seeking to know how metaphor works. Metaphor is a popular answer to any exploration of creativity, but it is also a fairly impenetrable keeper of its secret because it is not so easily analysed itself. Especially in the light of metaphorical models, I wonder whether one ought not dispel the opinion that metaphors are formed by sudden strikes of genius. (Note that this would, in any case, only apply to new metaphors on first use, not on the vast number of worn and trite specimens.) Certainly the majority of scientific models are developed by laborious and continued efforts that stretch over years and require enormous endurance on the side of those who seek these results (Bailer-Jones, 2000a). Ingenuity does not equate with effortlessness, and progress often occurs in very small steps. Correspondingly, one should examine the thesis that metaphors also only gradually become widely accepted and uniformly understood, just like scientific insight rarely strikes scientists out of the blue. In any case, it is worth challenging the myth of sudden, inexplicable insights in science that is often associated with creativity-generating metaphor (e.g. Kekulé's notorious dream of the snake biting its tale leading him to the conception of the six-carbon benzene ring). Creativity in science deserves to be investigated in its own right, and not only in the framework of metaphor; see issues 4-3 and 4-4, 1999, of *Foundations of Science* on Scientific *Discovery and Creativity*.

Acquisition of new meaning

Formulating a metaphor is about gaining a "newish" expression connected with a "newish" description of interpretation of a subject matter. "Newish" is supposed to take account

(a) of a gradual development of metaphorical expressions, and
(b) of those aspects of the expression that are familiar by way of the analogue which the metaphor exploits.

The interaction view proposes something like a sudden discovery of a new metaphor ensued possibly by a gradual shift in meaning of the primary and the secondary subject in response to that discovery. The implication is that the metaphorical expression is likely to give up one meaning and slowly to acquire another. However, even if we now talk about electric currents or artificial neural networks, we have not lost the capacity of using "current" to describe a river or "neural network" to describe neurons connected in the brain. Expansion of the domain of application can be observed because the expression now occurs metaphorically, but the use of the expression in its source domain need not disappear. Whether talk is about electric fields or ploughing a field, both are understood equally well, and there is no obvious reason why metaphor should function differently from literal language use in describing (Rothbart, 1984; Machamer, 2000). At some point, "field" has obviously acquired an additional meaning that permits it to be applied in the context of electromagnetism. Understanding it is a matter of working out which meaning a term happens to have in which context. Similarly, ideas for models can be employed in different domains: models of statistical mechanics do not disappear because they have a new application in the technique of simulated annealing. Yet, even if one denies a fundamental difference between literal and metaphorical language, questions remain about how we succeed in the complex task of interpreting linguistic expressions, picking among the range of possible interpretations in view of context and associations. For scientific models, this means that even when we surmount viewing them as either literal or metaphorical, we still need to examine how precisely they provide information about the empirical world.

Metaphorical models and metaphorical terminology

In future work, it is essential to distinguish carefully whether it is a model as a whole that is portrayed as metaphorical or whether it is merely a case of metaphorical ordinary language being used in describing a model (metaphorical or not), or both. Several different combinations seem possible, and it remains to be examined what effect these different levels of metaphorical penetration have on scientific thinking:

- Certain models are considered metaphorical in the sense that a transfer from one domain to another has taken place, but where no specific metaphorical terminology is used in this model, e.g. Bohr's model of the atom.
- In other cases, a structural relationship is made out between two domains that warrants a transfer leading to the formulation of a model in the target domain. In addition, this transfer gives rise to metaphorical language use accompanying the use of the model, e.g. *temperature* in simulated annealing or *noise* in observational astronomy.
- Then there are models where the descriptive terminology employed is metaphorical, but the two domains involved in this metaphorical terminology are not related in structure. An example is gravitational lensing. All a gravitational lens has in common with an optical lens is that it bends a light ray. The *bending* of a light ray due to gravitation is, unlike the case of an optical lens, not interpreted in terms of the optical phenomenon of refraction, so the metaphor is not connected to any deeper structural analogy between gravitational and optical lenses.
- Finally, yet other scientific metaphors that can be found in popular culture are without impact on scientific modeling which is why they can be disregarded for the current purposes. Examples are *litmus tests* in politics, a *critical mass* of participants needed before ideas can be *generated*, a military *nerve centre*, learning by *osmosis*, being *tuned in* or *turned off*, somebody being an Elvis *clone*, etc. (Hutchinson and Willerton, 1988).

The relationships between models, metaphors and analogies

I briefly recapitulate the connections between scientific models, metaphors and analogies to highlight the central research question resulting from their confrontation.

A model is an interpretation of an empirical phenomenon. As such, it is a description, although a partial description not intended to cover all aspects of the phenomenon in question, just like metaphors are, although the latter need not be interpretations. The task of scientific models is to facilitate (perceptual as well as intellectual) access to phenomena. While metaphors may also facilitate access to phenomena, their main characteristic is not this, but a transfer of at least one part of an expression from a source domain of application to a target domain. The implication is that the use of the expression in the source domain may be more familiar and/or better understood than its use in the target domain. Some scientific models can be analyzed as metaphors because their formulations involve a transfer of conceptions from a different domain (artificial neural networks, simulated annealing, Bohr's model of the atom). However, such a transfer is only of interest in the context of models insofar as it assists the purpose of the model, namely to interpret an empirical phenomenon.

Insightful metaphors are those that point to an analogy between phenomena of two different domains. The development of scientific models also often relies on analogies. Both the interpretation of models and that of metaphors frequently benefits from the analogies associated with them. Analogy deals with resemblances of attributes, relations or processes in different domains, exploited in models and highlighted by metaphors. Note that neither metaphors nor models *are* analogies – they are descriptions. This raises the question whether, at the cognitive level, there is anything involved in the metaphor claim concerning scientific models that can *not* be reduced to analogy. Is there something, e.g. the importance of context or associations, that lifts the cognitive force of metaphor above that of analogy? Much of this question seems to rest on not only the study of analogy, but also on whether there exist alternative strategies for knowledge formation.

To summarize, neither metaphors nor models are mysteriously creative or otherwise mysterious in how they contribute to our thinking about phenomena, although this is not to suggest that we understand everything about metaphor or about scientific modeling. Both can be, and need to be, however, subject to research. Many cognitive and creative claims about metaphors and metaphorical models appear reliant on the relationship of analogy, but whether analogy really deserves to be considered as the base category in developing interpretative descriptions equally requires further investigation. Finally, beyond the commonalties of scientific models and metaphor already highlighted, there is one other: scientific models appear to be, contrary to past research traditions, as central in scientific practice for describing and communicating aspects of the empirical world as metaphors are in ordinary language.

Acknowledgments

For their comments on aspects of this article I would like to thank Peter Machamer, Stephan Hartmann, Michael Bradie, Andreas Bartels and Coryn Bailer-Jones.

Notes

1 Mellor (1968) argues that Campbell requires analogy largely to overcome the abyss between theory and observation, and that, were it not for Campbell's strict theory–observation distinction, his account would not differ significantly from Duhem's.
2 "Analogy" refers to the relationship between two objects; an "analogue" is the object itself that is seen to be in the relationship of analogy to something.
3 This claim put forward very carefully is later reaffirmed: "I still wish to contend that some metaphors enable us to see aspects of reality that the metaphor's production helps to constitute" (Black, [1977] 1993, p. 38).
4 Black ([1977] 1993, p. 30) later contends: "I am now impressed, as I was insufficiently when composing *Metaphor*, by the tight connections between the notions of models and metaphors. Every implication-complex supported by a metaphor's secondary subject, I now think, is a *model* of the ascriptions imputed to the primary subject: Every metaphor is the tip of a submerged model."

References

Achinstein, P. (1968): *Concepts of Science*. Baltimore, Maryland: John Hopkins Press.
Aronson, J. L., Harré, R. and Way, E. C. (1995): *Realism Rescued: How Scientific Progress is Possible*. Chicago, Illinois: Open Court.
Bailer-Jones, D. M. (2000a): "Modeling Extended Extragalactic Radio Sources," *Studies in History and Philosophy of Modern Physics*, 31B, 49–74.
Bailer-Jones, D. M. (2000b): "Scientific Models as Metaphors," in F. Hallyn (ed.), *Metaphor and Analogy in the Sciences*, Dordrecht: Kluwer Academic Publishers, 181–98.
Bhushan, N. and Rosenfeld, S. (1995): "Metaphorical Models in Chemistry," *Journal of Chemical Education*, 72, 578–82.
Black, M. (1954): "Metaphor," *Proceedings of the Aristotelian Society*, 55, 273–94.
Black, M. (1962): *Models and Metaphors*. Ithaca, New York: Cornell University Press.
Black, M. ([1977] 1993): "More about Metaphor," in A. Ortony (ed.), *Metaphor and Thought*, Cambridge: Cambridge University Press, 19–41.
Boyd, R. (1993): "Metaphor and Theory Change: What is 'Metaphor' a Metaphor for?" in A. Ortony (ed.), *Metaphor and Thought*, Cambridge: Cambridge University Press, 481–532.
Bradie, M. (1998): "Models and Metaphors in Science: The Metaphorical Turn," *Protosociology*, 12, 305–18.

Bradie, M. (1999): "Science and Metaphor," *Biology and Philosophy*, 14, 159–66.

Campbell, N. R. ([1920] 1957): *Foundations of Science* (formerly titled: *Physics, The Elements*). New York: Dover Publications.

Cartwright, N. (1999): *The Dappled World*. Cambridge: Cambridge University Press.

Cornelis, G. C. (2000): "Analogical Reasoning in Modern Cosmological Thinking," in F. Hallyn (ed.), *Metaphor and Analogy in the Sciences*, Dordrecht: Kluwer Academic Publishers, 165–80.

Davidson, D. ([1978] 1984): "What Metaphors Mean," in *Inquiries into Truth and Interpretation*, Oxford: Clarendon Press, 245–64.

Duhem, P. ([1914] 1954): *The Aim and Structure of Physical Theory*. Translated from the French 2nd edn, Princeton, New Jersey: Princeton University Press.

Falkenhainer, B., Forbus, K. D. and Gentner, D. (1989): "The Structure-Mapping Engine: Algorithm and Examples," *Artificial Intelligence*, 41, 1–63.

Gentner, D. (1982): "Are Scientific Analogies Metaphors?" in D. S. Miall (ed.), *Metaphor: Problems and Perspectives*, Sussex: The Harvester Press, 106–32.

Gentner, D. (1983): "Structure Mapping: A Theoretical Framework for Analogy," *Cognitive Science*, 7, 155–70.

Gentner, D. and Markman, A. B. (1997): "Structure Mapping in Analogy and Similarity," *American Psychologist*, 52, 45–56.

Giere, R. (1999): *Science Without Laws*. Chicago: University of Chicago Press.

Harman, P. M. (1982): *Energy, Force and Matter*. Cambridge: Cambridge University Press.

Harman, P. M. (1998): *The Natural Philosophy of James Clerk Maxwell*. Cambridge: Cambridge University Press.

Harré, R. (1960): "Metaphor, Model and Mechanism," *Proceedings of the Aristotelian Society*, 60, 101–22.

Harré, R. (1970): *The Principles of Scientific Thinking*. London: Macmillian.

Harré, R. (1988): "Where Models and Analogies Really Count," *International Studies in the Philosophy of Science*, 2, 118–33.

Hesse, M. (1953): "Models in Physics," *British Journal for the Philosophy of Science*, 4, 198–214.

Hesse, M. (1966): *Models and Analogies in Science*. Notre Dame: University of Notre Dame Press.

Hesse, M. (1967): "Models and Analogy in Science," in P. Edwards (ed.), *The Encyclopedia of Philosophy*, New York: Macmillian, 354–9.

Hesse, M. (1983): "The Cognitive Claims of Metaphor," in J. P. van Noppen (ed.), *Metaphor and Religion*, Brussels: Study Series of the Vrije Universiteit Brussel, 27–45.

Hofstadter, D. (1995): *Fluid Concepts and Creative Analogies*. London: Penguin.

Holyoak, K. and Thagard, P. (1989): Analogical Mapping by Constraint Satisfaction," *Cognitive Science*, 13, 295–355.

Holyoak, K. and Thagard, P. (1997): "The Analogical Mind," *American Psychologist*, 52, 35–44.

Hutchinson, B. and Willerton, C. (1988): "Slanging with Science," *Journal of Chemical Education*, 65, 1048–9.

Hutten, E. (1954): "The Role of Models in Physics, "*British Journal for the Philosophy of Science*, 4, 284–301.

Kittay, E. F. (1987): *Metaphor. Its Cognitive Force and Linguistic Structure*. Oxford: Clarendon.

Lakoff, G. (1993): "The Contemporary Theory of Metaphor," in A. Ortony (ed.), *Metaphor and Thought*, Cambridge: Cambridge University Press, 202–51.

Lakoff, G. and Johnson, M. (1980): *Metaphors We Live By*. Chicago: University of Chicago Press.

Lakoff, G. and Johnson, M. (1999): *Philosophy in the Flesh: The Embodied Mind and Its Challenge to Western Thought*. New York: HarperCollins Publishers.

Langacker, R. W. (1987): *Foundation of Cognitive Grammar. Vol. 1: Theoretical Prerequisites*. Stanford, California: Stanford University Press.

Leatherdale, W. H. (1974): *The Role of Analogy, Model and Metaphor in Science*. Amsterdam: North Holland.

Machamer, P. (2000): "The Nature of Metaphor and Scientific Descriptions," in F. Hallyn (ed.), *Metaphor and Analogy in the Sciences*, Dordrecht: Kluwer Academic Publishers, 35–52.

Martin, J. and Harré, R. (1982): "Metaphor in Science, in D. S. Miall (ed.), *Metaphor*, Sussex: The Harvester Press, 89–105.

Mellor, D. H. (1968): "Models and Analogies in Science: Duhem *versus* Campbell?" *Isis*, 59, 282–90.

Miller, A. (1996): *Insights of Genius*. New York: Springer-Verlag.

Morgan, M. and Morrison, M. (eds.) (1999): *Models as Mediators*. Cambridge: Cambridge University Press.

Nagel, E. ([1960] 1979): *The Structure of Science*. Indianapolis, Indiana: Hackett Publishing Company.

North, J. D. (1980): "Science and Analogy," in M. D. Grmek, R. S. Cohen and G. Cimino (eds.), *On Scientific Discovery*, Boston Studies in the Philosophy of Science, Dordrecht: D. Reidel Publishing Company, 115–40.

Paton, R. C. (1992): "Towards a Metaphorical Biology," *Biology and Philosophy*, 7, 279–94.

Richards, I. A. (1936): *The Philosophy of Rhetoric*. New York: Oxford University Press.

Rothbart, D. (1984): "The Semantics of Metaphor and the Structure of Science," *Philosophy of Science*, 51, 595–615.

Rumelhart, D. E. ([1979] 1993): "Some Problems with the Notion of Literal Meanings," in A. Ortony (ed.), *Metaphor and Thought*, Cambridge: Cambridge University Press, 71–82.

Searle, J. (1979): *Expression and Meaning*. Cambridge: Cambridge University Press.

VanLehn, K. (1998): "Analogy Events: How Examples Are Used During Problem Solving," *Cognitive Science*, 22, 347–88.

Chapter 7

Experiment
and Observation[1]

James Bogen

Introduction

People once believed a fabulous engine called the Scientific Method harvests empirical evidence through observation and experimentation, discards subjective, error ridden chaff, and delivers objective, veridical residues from which to spin threads of knowledge. Unfortunately, that engine is literally fabulous. Lacking a single method whose proper application always yields epistemically decisive results, real-world scientists make do with messy, quirky techniques and devices for producing and interpreting empirical data which proliferate as investigators improvise fixes for practical and theoretical problems which bedevil their research.[2] Their evolution is punctuated rather than linear – marked as much by abandonment and modification of previously accepted tools and techniques as by conservation and accumulation.

Failing as they did to take into account the diversity and malleability of observational and experimental practice, twentieth century philosophers of science who tried to derive highly general a priori epistemic directives from theories of logic, rationality, judgment, and the like, have been unable to answer important questions about the design and conduct of scientific research. This chapter's moral is that because of this failure, philosophers of science should pay more attention to nuts and bolts details of observation and experimentation.

Although experiment and observation are undertaken to further a great many different purposes (including discovering new effects for scientists to explain, filling in, and correcting details of theories, developing, calibrating, and figuring out fruitful applications of equipment) I will be concentrating on just one – the production and interpretation of data for use in testing theoretical claims and practical ideas about their applications.[3]

I'll use a single term "empirical" in connection with both observation and experiment (along with their equipment, techniques, epistemics, etc.). William

Herschel (discoverer of Uranus) draws a useful distinction between them. Observation, he says, is a matter of

> noticing facts as they occur without any attempt to influence the frequency of their occurrence . . . (Herschel, 1966, p. 76)

It can be likened to passively

> listen[ing] to a tale, told us, perhaps obscurely, piecemeal, and at long intervals of time, with our attention, more or less awake (p. 77) [where in many cases the] . . . tale is told slowly and in broken sentences (p. 78).

Experiment, by contrast, is a matter of

> putting in action causes and agents over which we have control, and purposely varying their combinations, and noticing what effects take place (Herschel, p. 76)

Herschel likens this to cross-examining a

> witness and by comparing one part of his evidence with the other, while he is yet before us . . . reasoning upon it in his presence [so that we can] . . . put pointed and searching questions, the answer to which may at once enable us to make up our minds (Herschel, 1966, pp. 66, 77).

In experiments, natural or artificial systems are studied in artificial settings designed to enable the investigators to manipulate, monitor, and record their workings, shielded, as much as possible from extraneous influences which would interfere with the production of epistemically useful data. Investigators who cannot produce the data they need in this way can sometimes rely on "experiments of nature." That's the term Bernard used in connection with diseases which provided evidence for the study of the physiology of the affected organs and systems (Bernard, 1949, p. 10). Like observable instances of astronomical regularities, experiments of nature are interactions in which natural mechanisms conspire, without contrivance by the investigators, to produce effects of interest to the investigator. Obtaining and interpreting data from such occurrences sometimes involves enough equipment and elaborate arrangements to blur Herschel's distinction. (Eddington's use of eclipse photographs to calculate the deflection of starlight is an example (Dicke, 1964, p. 2).)

But even so, there are epistemically significant differences between observing without interfering, and setting up an experiment, altering its components, and intervening in their workings as needed to investigate the significance of the data the experiment produces. Medical research can dramatize this point. Suppose a hideous disease is observed more frequently in people who eat certain foods than in people who do not. It may not be at all obvious from the statistical distribution of these and other observed factors whether the diet and the disease are con-

nected causally or only accidentally. Someone might be able to find out by manipulating people's diets, life styles, and environments as required to eliminate or
control for confounding factors which are not apparent in the statistics, but this
would be morally impermissible. To deal with cases like this (and cases where the
experiments needed to settle a question would be too costly or too difficult to
perform), philosophers and statisticians have been trying to develop formal
methods of causal analysis with which to infer causal relations from statistical distributions. The technical difficulties they confront provide a detailed mathematical picture of certain fascinating epistemic differences between experiment and
observation (Glymour, 1997, pp. 233–42).

Robert Boyle's dialogue, *The Skeptical Chymist,* contrasts using empirical evidence to evaluate, and using it to illustrate a theory. One of the characters is
Themistius, a peripatetic who believes theories can be evaluated only by argument
from undisputed *a priori* principles, and accordingly that the only legitimate use
of experiment is "to illustrate, rather than to demonstrate" after the manner of
astronomers who use cardboard spheres to explain their theories to laymen who
don't have enough mathematics to follow demonstrations of their truth (Boyle,
1661, pp. 20, 21). For example, he burns some green wood and uses the resulting ashes, moisture, smoke, and fire used to illustrate the peripatetic doctrine that
all non-elemental stuffs are composed of earth, water, air and fire. To make the
illustration work, he uses peripatetic theory to explain to the observers what they
have seen. Since (according to the theory) elemental earth is heavy and dry the
dryness and weight of the ashes prove that they are composed of earth. Since all
elements tend to move toward their natural places, the smoke proves itself to be
air by "ascending to the top of the chimney and . . . vanishing into air, like a river
losing itself in the sea" (Boyle, 1661, p. 21), and so on. Carneades, the dialogue's
skeptical chemist, maintains to the contrary that scientific claims should be tested
by experiments in which factors which cannot be fruitfully studied (and may not
even occur) in natural settings, are produced, isolated, and tortured[4] until they
confess truths about nature (Boyle, 1661, p. 10).

Although Carneades is right about this, it would be hard to teach science if
empirical methods couldn't also be used to provide illustrations; and shabby as it
is, Themistius' illustration exemplifies a point whose importance cannot be over-
stated: Natural and artificial empirical results are typically very different from the
things scientists use them to investigate. Thus, what Jean Perrin wanted to learn
about were unobservable atoms, not the motions of resin beads he observed in
hopes of learning about them (Perrin, 1990, chs III, IV). What neuro-cognitive
psychologists want to find out about are cognitive processes and neuronal mechanisms which support them, not the scores on psychological tests, the functional
images of the brain, and the other empirical evidence they use to study them. To
think their main goal is to understand the scores, and other empirical evidence,
rather than the brain functions is like thinking that chefs make sauces for the
purpose of using whisks and mixing bowls. A moral to draw from Themistius and
Carneades is that what empirical results can teach depends upon what they can

legitimately be interpreted as indicating. *The Skeptical Chymist* is filled with examples of what can be involved in deciding whether a proposed interpretation is legitimate. Themistius' illustration assumes that burning decomposes wood into its component elements. Carneades objects that for all they know, heating produces new stuffs, or that the ash and other residues came from the air, the container in which the sample was heated, or impurities in the sample (Boyle, 1661, p. 27ff). As we'll see, a number of twentieth century controversies in the philosophy of science amount to versions of the question whether reasoning from experimental or observational outcomes can reveal more about what goes on in nature than Themistius' question begging illustrations.

Neglecting Experiment; Distorting Observation

The logical positivists[5] are the giants whose shoulders we stand on. They made invaluable contributions to philosophy of science, and are largely responsible for its establishment as an academic discipline. Their work was important to the linguistic turn which founded analytic philosophy, but they are largely responsible for a neglect of experiment and observation which blinded twentieth century philosophers of science to facts about the production and interpretation of empirical data which bear importantly on their epistemological and metaphysical concerns. The logical positivists taught their followers to treat scientific theories as if they were deductively closed collections of propositions, including observation reports expressed in a vocabulary which includes terms which signify observables, theoretical propositions expressed in a vocabulary whose terms do not signify observables, and correspondence rules of mixed vocabulary which can be used to derive predictions and explanations of observables from theoretical propositions, and to test theoretical propositions against observation reports (Nagel, 1961, pp. 90–117). This conception models scientific prediction, explanation, and theory testing in terms of inferential relations among sentence-like structures. But so conceived, what does science have to do with the natural world of non-sentential, extra-linguistic things, features, events, processes, etc., scientists investigate? Hempel (1935, pp. 50–1) responded to related questions by claiming that no theorist who

> supports a cleavage between statements and reality is able to give a precise account of how a comparison between statements and facts may possibly be accomplished.

In keeping with this, and with their appreciation of the power of newly developed logical tools for the study of the systems of sentences they used to model scientific theories, analytic philosophers of science downplayed the cleavage and devoted themselves to investigating the syntax and semantics of observation and theoretical languages. In sharp contrast to nineteenth century figures – like

Whewell (1991) and Duhem (1991) – and earlier empirically minded thinkers – like Bacon (1994), Boyle (1661), and Hooke (1968) – who engaged in what amounts to philosophy of science before the subject became institutionalized in its present form, the logical positivists and their sympathizers treated observation and experiment as black boxes which outputted observation sentences in relatively mysterious ways of next to no philosophical interest.[6] Decades would pass before philosophers of science began to appreciate how much the epistemic value of empirical data as evidence for or against a scientific claim depends upon the way it was produced, and the degree to which some features of scientific practice can be illuminated by considering facts about data production instead of logical relations between theoretical claims and descriptions of empirical results. But you didn't have to sympathize with the logical positivists to ignore empirical methods. In the late 1950s, a group of their critics including Kuhn, Hanson, and Feyerabend developed new distractions associated with Hanson's slogan, "seeing is a theory laden enterprise" (Hanson, 1958, p. 19) to direct philosophers' attention away from empirical practice. As a result of all of this, Alan Franklin could complain 30 years later that someone who told philosophers of science of the death of Lummer and Prigsheim would "get the same reaction – total unconcern – that the ambassador from England gets at the end of *Hamlet* when he announces the death of Rosenkrantz and Guildenstern" (Franklin, 1989, p. 1). (Lummer's and Prigsheim's investigations of black body radiation featured "some of the most important experiments in the history of physics" (*ibid.*)).[7] Feyerabend (1985), Kuhn (1970), Hanson, and their followers understood theory loading in different ways. The most common understandings resembled one or more of the following substantially different versions Kuhn developed of his own idea that paradigms[8] influence observation to such an extent that observers who work in different paradigms cannot "see" the same things. (Kuhn, 1970, pp. 111–23.)

K1: Perceptual

Bruner and Postman found that on short exposures, subjects looking at normal and anomalous playing cards described them all as if they were normal, failing, e.g. to report that a black four of hearts was black. It took repeated, longer exposure for them to learn to describe the anomalous cards correctly (Kuhn, 1970, p. 63). Kuhn interprets this as indicating that someone who lacks the concept of a deck containing a red club or a black diamond cannot have (or notice having) the same visual experience as an observer who has it. He goes on to suggest that scientific paradigms determine observers' concepts to such an extent that when investigators with conflicting paradigms look at the same thing, their observations will differ. In particular, an investigator's paradigm may prevent her from observing what would otherwise support a competing paradigm's theoretical claims (Kuhn, 1970, pp. 111, 113–14, 115, 120–1).

K2: Semantical

Whether or not the parties to a scientific disagreement can have the same perceptual experiences, the theoretical commitments of their paradigms influence the meanings of crucial descriptive terms to such an extent that they will be unable to accept each others observation reports unless they understand them to mean different things (Kuhn, 1970, p. 127ff).

K3: Salience

Paradigms determine what experiments and observations investigators will carry out, and what features of their results they will attend to or take seriously. The paradigms investigators work in may thus prevent them from obtaining significant empirical evidence or appreciating its bearing on their positions (Kuhn, 1970, pp. 64, 121–38).

Such ideas encouraged philosophers to ignore the study of real-world empirical methods and direct their energies to disputes about theory loading and its implications for theory evaluation and scientific progress. Some philosophers were thus led to worry, in effect, whether empirical research can deliver anything more epistemically respectable than Themistian illustrations of the investigator's theoretical commitments. But let's take a quick look at the merits of K1 and K2 before we turn to this.

Now it is notorious that observers' mental sets can lead them to sincerely report having seen what was not there to be seen.[9] But as a generalization, K1 is both implausible and false. Priestley and Lavoisier performed similar experiments using similar equipment. They watched such things as burning candles, levels of water in graduated tubes, and small animals asphyxiating in bell jars. Despite their conflicting paradigms, there is no evidence that the water levels, the candles burning out, the animals keeling over, or the chronometer readings they used to time them looked significantly different to them. Opposed as their paradigms were, they frequently reported the same observations. What separated them was not their perceptions, but the conclusions they drew from their evidence (Conant, 1957).[10] Cases like this are as troublesome for K2 as for K1. Even if (as K2 supposes) the investigators understood their theoretical terms differently, that is no reason to think they couldn't understand the numbers and other symbols they used to record temperature, pressure, or weight readings, etc., in the same way. Furthermore K2 doesn't even apply to the many data which consist of drawings, photographs, tracings, sound recordings, and other non-verbal records.

A more sophisticated version of K2 has it that different background beliefs enable investigators to use different theoretical terms to describe what they observe in significantly different ways. For example, observers who identify musical pitch with air pressure oscillations can use oscillation talk to report what they hear. Just as observers who lacked the relevant theory of pitch wouldn't report their observations this way, people who use phrases like "440 Hertz" but not letters, A–G, to report pitches wouldn't report pitches the way we do (Churchland, 1992, p. 53). Alternatively, some say that an experimentalist who believes the visual display she looks at contains "information transmitted without interference" from neutrino emitting interactions in the interior of the sun "to the appropriate receptor" in her laboratory can say that in looking at the display, she sees the interior of the sun, while investigators who lack the relevant background beliefs would have to say instead, e.g. that they see solar neutrino fluxes, or just Geiger counter splodges (Shapere, 1982, p. 492). This might seem to help explain how observation reports can bear on theoretical claims about things we always thought were unobservable. If you don't understand how the visual display the observer watches can have anything to do with unobservable solar neutrino fluxes, just describe her as seeing the fluxes! K2 might seem to explain how scientists with different theoretical commitments can have honest disagreements over the significance of an empirical outcome. Of course, they disagree if their theories prevent them from accepting each others' observation reports.

Such stories talk past the issues of empirical epistemology.[11] Most empirical work is aimed at detecting and answering questions about things, facts, events, processes, and their features, all of which I'll refer to as effects. Some effects are instances of phenomena which occur in nature (e.g. astronomical regularities) or in the laboratory (e.g. Compton scattering, laser effects) with sufficient regularity, and result from uniform enough operations of sufficiently simple systems of causal influences to make them susceptible to the derivation of quantitative predictions and detailed, systematic explanations based from highly general theoretical principles.[12] Other effects, produced less tidily and occurring less regularly, may also be explained or predicted, but only by appeal to causal interactions which depend too heavily upon, and vary too much with, local conditions to be accounted for by models as simple, and generally applicable as those used in connection with phenomena. Examples include evolutionary phenomena like the color change in moths in Manchester during the industrial revolution (Mitchell, 1987, p. 354). Investigators use what they can find out about effects of both kinds to test their theories and devise practical applications of them; to design, calibrate and assess the reliability of their equipment; to design experiments and devise tactics for making observations and producing data. Effects are studied by reasoning from data. Data are sentential or non-sentential records of things which investigators perceive, or which register on their equipment. Numerical records of measurements and tests scores, drawings, photographs, EKG, and seismic tracings are examples of the latter. The crucial epistemic questions of empirical epistemology have to do with how conclusions about effects are supported by reasoning from data. Among the most important of these are questions I'll call the three Rs:

1 *Relevance*: What bearing does the effect the investigators believe their data reveals have on the theoretical or practical issues they use them to pursue?

2 *Reality*: Is the effect whose occurrence, or whose features, the data seem to indicate real or spurious? and

3 *Reliability*: Are the data imprecise, inaccurate, or otherwise epistemically defective *with respect to the features upon which the investigator's reasoning to conclusions about the effect of interest depends?* Equipment and methods of data production are *not* required to be reliable, and data are *not* required to be true, approximately true, accurate, precise, etc., to any degree, or with respect to any features which are not essential to the evaluation of the reasoning which uses it. That is why, for example, the visually obvious disparities between the shapes, relative sizes, and relative positions of lunar mountains and craters as they are, and as Galileo's drawings depict them do not discredit Galileo's use of the drawings to argue that the surface of the moon is opaque and irregular, rather than smooth and crystalline (Galileo, 1989, pp. 41–7).

The investigator's observations of items whose relevant features are easily perceived and perceptually discriminated need not be at all problematic. But even when they are problematic, the fact that one can describe an investigator as having observed what she can learn about only by reasoning from data sheds no light on the epistemic legitimacy of inferences from data, or the epistemic significance of the data they depend upon. In short, I don't think empirical epistemologies need concern themselves with K1 or K2. K3, however, is another matter as we'll see.

The Socio-Theoretical Turn

With regard to the second R, most philosophers, historians, and other students of science would now agree with Peter Galison, that unlike sound deductive arguments by which mathematicians and logicians can hope to settle their disputes decisively enough to permanently close debate, experimentalists typically cannot demonstrate once and for all "the reality – or artificiality – of an effect"[13] (Galison, 1987, p. 2). The same holds for disputes about Relevance and Reality. At the same time, it is undeniable that scientists can often end disputes and achieve long lasting consensus over the three Rs. A number of philosophers began attending to experiment and observation in response to the attempts of a diverse, multi-disciplinary group of theorists who made it their project to explain consensus as the product of social interactions shaped by the influence of a variety of social and behavioral factors. Although they tend to disagree too much among themselves to want to be lumped together, and although no one uses the label I've chosen for them, I'll call them Social Theorists. All of them argue for the relatively innocuous view that regardless of how well a scientific claim is confirmed by the available evidence, the consensus generating processes by which the claim comes to be accepted or rejected by any particular group of scientists is always significantly influenced in a variety of ways by a variety of social, political, and cultural factors. Much more controversially, many Social Theorists think that social, political, and cultural values determine, not just whether a more or less well confirmed claim will be accepted, but also, that whether or to what degree it is supported[14] by the evidence (Shapin, 1994, pp. 193–309). Many think that such epistemic virtues as truth, precision, accuracy, and rationality, can be completely accounted for by appeal to the very same sorts of factors. For example, Bloor scoffs at Durkheim for saying that

> [I]n the early stages of cultural evolution . . . a [scientific] belief might be deemed true because it is socially accepted . . . for us it is only socially acceptable on condition that it is true (Bloor, 1983, p. 3).

and maintains, to the contrary, that truth itself is the product of social acceptance.

Different groups of investigators do in fact disagree over whether a body of empirical evidence is sufficiently accurate, or whether an instrument or a procedure for producing data is sufficiently reliable. For example, some cognitive neuroscientists are happy to base their conclusions on functional brain images whose accuracy with regard to levels and locations of neuronal activity are considered inadequate by others (Steinmetz and Seitz, 1991, Mazziota et al., 1995). And Newton and some of his contemporaries were happy to argue from measurements and estimates of speeds, distances and times which seem significantly inaccurate to us (Newton, 1999, pp. 797–801, 803). Social Theorists take such variability to mean that the meaning and the authority of an epistemic norm depends upon, and is relative to the practices of the groups who embrace it. Because different groups rely upon different procedures and standards for measurement, description, and mathematical analysis, what it is for a given result to meet a given standard of accuracy, precision, etc., may also vary from group to group. Social Theorists conclude from this that truth, degree of accuracy, error, and the like are not features which empirical results possess independently of their evaluation. Instead, they are constituted by the processes by which investigators agree to assign them. Like the epistemic norms they figure in, they are relative to the practices of investigators, none of which is intrinsically any better epistemically than any other. Some Social Theorists treat the production and maintenance of consensus as the products of social interactions in which individuals function (whether or not they realize it) to promote their or their groups' interests (Barnes, 1997; Collins, 1999).

Social Theorists apply a variety of strategies to the study of the relevant interactions and the consensus generating mechanisms they belong to. Latour employs the model of political contests in which competitors employ rhetorical and other devices to gain and consolidate support, and neutralize opposition, treating laboratory equipment, chemical reagents, experimental animals, books, papers, and other naturally occurring and artificially produced objects on analogy to potential friends and foes (Latour, 1987, pp. 30–59, 63–94). With Woolgar, he used ethnographical techniques to study the folkways of investigators at work in a laboratory (Latour and Woolgar, 1986, chs 1, 2). Shapin explores the roles trust and authority play in resolving empirical controversies, and the influence on them of the investigators' class, social position and public persona (Shapin, 1994, pp. 3–125; Lloyd, 1993). A number of Social Theorists think of investigators as technicians who employ their material and conceptual resources to produce what their intended audiences will accept as credible and useful empirical results. So conceived, an important part of the scientist's work is to manipulate theoretical considerations and laboratory effects to obtain a satisfactory and sustainable fit. Understanding the tricks of this trade requires close, case by case studies of problems posed, and opportunities afforded by the skills and limitations of the investigators, the peculiarities of their methods, and the behaviors of their equipment and the items they apply it to. See for example, Pickering (1986), Clarke and Fujimara (1992, chs 3, 5, 6), Gooding (1992), Hacking (1991, pp. 186–209) and Galison (1987; 1997, chs 2–8). The Social Theorists' pursuits of such ideas helped redirect philosophers' attention to real world empirical practices. Both supporters and opponents of their positions[15] can learn a great deal from the wealth of information about empirical practice their work provides.

Some Issues for Empirical Epistemologists

The following are brief illustrations of issues which are now engaging, and should continue to engage philosophers interested in experiment and observation.

Evaluating reality, relevance, and reliability

Philosophers of science have proposed a number of highly abstract, general, purportedly exceptionless epistemic standards for use in connection with the three Rs. Some are supposed to determine the acceptability of data. Others are supposed to determine the acceptability of reasoning from data. Although many examples of good, real world scientific work accord with the most plausible of standards, their claims to universality have fared badly. Here are some of the most influential of these received, but discredited proposals.

Contrary to the assumption (a) that unreplicated evidence is always epistemically defective, neuroscientists, evolutionary and other biologists, particle physicists, cosmologists, engineers, and many other scientists rely on unreplicated, poorly replicated, and in come cases, unreplicable data and effects (Bogen, 2001; Galison,

1987, pp. 180–97). Galileo's moon drawings, Eddington's calculations of starlight deflection (Earman and Glymour, 1980), and Millikan's oil drop experiments (see below) are a few of many counterexamples to the common assumption that (b) both data and claims about effects derived from data are unacceptable unless there is good reason to think they are true, or that they reach some high threshold of accuracy, approximation to the truth, or probability. Although it has seemed obvious that (c) in interpreting data one must not assume the correctness of central components of the theory it is being used to test, just such assumptions figure in the use of Michelson–Morley data to argue against Fresnel's aether theory (Laymon, 1988, p. 250). Contrary to the assumption that (d) descriptions of effects calculated from data must be logically consistent with the claims they are used to argue for, Newton appealed to Kepler's laws in his demonstration of Universal Gravitation even though they are incompatible with Universal Gravitation when applied to the solar system (Duhem, 1991, pp. 190–5; Laymon, 1983). Thus purportedly exceptionless received epistemic standards to which many examples of good scientific research accord are violated by others.

Bayesian confirmation theory (BC), is the most widely accepted recent alternative to received accounts of theory testing. But it provides no better epistemic standards for the evaluation of reasoning from data than the received accounts. According to BC, empirical evidence, e, confirms a claim, h, only if the probability of h conditional on e and background knowledge, k, is higher than the probability of h conditional on k alone (and, according to some Bayesians, higher than some threshold probability above 0.5) (Earman and Salmon, 1992, pp. 89–100). But many effects calculated from data which investigators accept as good evidence fail to meet this test, as do most of the data investigators rely upon. Often they fail by default because there are no non-arbitrary, objective ways to determine the prior probabilities required for the calculation of the relevant conditional probabilities. For example, consider the oil drop experiment Millikan used to argue for h_m. The magnitudes of "all static electrical charges both on conductors and insulators' are multiples of the fixed and unvarying magnitude of the charge on the electron" (Millikan, 1935, pp. 72–3, 76) is a case in point.

The effect he calculated from his data in support of h_m was a mathematical relationship among the magnitudes of charges on oil drops moving or held in suspension between charged plates in a closed chamber, and the charges on ions produced by irradiating the air in the chamber: (e_m). Every such charge is a multiple of the smallest charge on an ion (Millikan, 1935, pp. 75, 76). Millikan's data included:

D1   stop watch readings used to time the motions of oil drops carrying static electrical charges as they moved under the influence of gravity, the electric field between the charged plates, and additional charges they picked up (Millikan supposed) from ions they captured on collision

D2   measurements of the charges on the plates under various experimental conditions, and

D3 measurements of air pressure, temperature, and other non-electrical influences on the drops motion.

To produce (D1), an investigator would align a drop with the top cross hair of a
low power telescope, stop the first hand of a stop watch when the drop reached
the middle hair, and the second when the drop reached the bottom hair. Before
attempting to calculate charges from his data, Millikan eliminated an impressive
number of stopwatch data points. Some were outriders. Some, he assumed,
reflected the influences on the oil drop motions of convection currents, encoun-
ters with dust particles and other extraneous causal influences he could not
otherwise correct for. According to Franklin, some could not be retained without
running afoul of independently accepted principles (Franklin, 1989, p. 150). A
few other data points were thrown out without explanation. Next, Millikan esti-
mated and corrected for error due to idiosyncrasies of the observers' reaction times
and visual acuities, and peculiarities of the equipment, and calculated the range of
random error. For details, see Millikan (1935, pp. 57–124) and Franklin (1989,
138–64).

Whether or not it can tell us anything about the evidential bearing of e_m on h_m,
this example bodes ill for BC as a general account of reasoning from data. According to Bayes' theorem, the probability of h_m, conditional on raw data (of kinds
D1, D2, D3) and background knowledge, k_m, must be calculated from the probability of h_m conditional on k_m, the probability of the raw data conditional on h_m
and k_m, and the probability of the raw data conditional on (k_m and h_m + k_m and
h_m)! It should be obvious just from the details of Millikan's data reduction that
any attempt to write down all of the numbers needed for the calculation of those
probabilities would be whimsical at best. Thus BC cannot explain why Millikan's
data was relevant to the evaluation of h_m. The same holds for Millikan's use of his
data in support of e_m.

If e_m is a real effect, and not an artifact of data production and interpretation
it is an instance of h_m, and as such, it clearly counts in favor of it. The crucial question for an empirical epistemologist to ask is not "how probable is h_m or e_m conditional on the data?", but rather, "is e_m a real effect?". The answer to that question
depends upon the results of detailed evaluations of local factors which are idiosyncratic to the workings of the equipment, the design and conduct of the experiment, Millikan's treatment of his data (including his decisions about what to
throw out) and so on. The point of the evaluations would not be to decide whether
Millikan's cleaned up data and the quantities he calculated from it to argue for e_m
were highly accurate (there's little chance they were) or whether Millikan's procedures for cleaning up the data and calculating velocities and magnitudes of
charges were generally reliable (some of them certainly were not). The point is to
decide (local reliability) whether the quantities Millikan recorded and calculated
are mistaken *in ways which discredit the argument* for e_m. This turns on considerations as different from one another as the error generating characteristics of the
stopwatches, the influences of convection currents and dust particles, the statisti-

cal significance of Millikan's data reduction procedures, and such physical issues as whether capture by an oil drop changes the magnitude of the charge on a trapped ion.

As this example illustrates, the epistemic worth of an empirical result used to argue for or against a scientific claim typically involves the application of a variety of different ideas and techniques from different areas of mathematics, natural science, engineering, etc. to a variety of different and independent details of experimental design and execution. These mixes are far too heterogeneous to be informatively modeled along the lines of received confirmation theories as instances of the uniform application of general principles as simple and general as Bayes' theorem, or the rules of a predicate calculus. This is by no means to deny that Bayesian techniques, along with the deductive and inductive techniques traditional philosophers of science have favored are used in individual steps that may be taken toward the determination, e.g., of whether a given effect is likely to have been an experimental artifact, or the estimation of the error characteristics of a particular process for producing data. The moral to draw from experiments like Millikan's is that rather than providing perfectly general accounts of the acceptability of data and its relevance to theory evaluation, they are among the tools which may be applied (in different ways for different purposes) to particular investigations. Detailed studies of real world evaluations of reality and local reliability and the motley of formal and empirically based standards and principles they employ have directed the interests of epistemologists of science to accounts of causality which can be used to understand how investigations of the causal influences of components of data generating mechanisms figure in the detection and investigation of effects (Cartwright, 1989; Spirtes et al. 1993; Woodward, 1997, 2000; Pearl, 2000) and to the statistical and probability techniques employed in real world science (Earman, 1992; Mayo, 1996). Whether further investigations along these lines can lead to anything more general than piecemeal, case by case epistemological accounts is an open question.

Laboratory effects

Like the motions of Millikan's oil drops, laser effects, Compton scattering, the effects of knockout genes on neuronal interactions, and the evolution and peculiarities of new strains of fruitflies studied by geneticists, experimentalists often study effects which "seldom or never" occur in a pure state before people have brought them under surveillance. Hacking thinks it isn't much of an exaggeration to say the effects studied by "sciences . . . whose claims to truth answer primarily to work done in the laboratory . . . are created in the laboratory" (Hacking, 1992, p. 33). Exaggeration or not, this confronts philosophers with the task of finding out how, and to what extent, data obtained from effects produced or isolated in the laboratory can be indicative of facts about natural phenomena.

Salience

K3 correctly suggests that the salience and availability of empirical evidence can be heavily influenced by the investigator's theoretical and ideological commitments, and by factors which are idiosyncratic to the education and training, and research practices which vary with, and within different disciplines.

In an experiment designed to produce data bearing on the evolutionary significance of female orgasms, female macaques in the company of males were rigged with battery operated equipment to record muscle contractions, temperatures, and other physiological items relevant to sexual activity. The equipment was miniaturized to allow the macaques to behave normally but limitations of battery storage capacity precluded continuous monitoring. To save power, the investigators arranged things so that the equipment would be turned on when the males were sexually aroused – an odd choice because female macaque orgasms typically occur without male arousal during masturbation and same sex play. When asked about this, the principal investigators said they just wanted to record the important orgasms! As Elisabeth Lloyd argues, this is best understood as a reflection of gender-based cultural preconceptions about female sexuality which were accepted without question by researchers in primatology and evolutionary biology (Lloyd, 1993, pp. 139–42, 149–50).

In the 1960s, experimentalists were confronted with a "zoo" of competing relativistic theories of gravitation (Earman, 1992, p. 174). The "extreme weakness of the gravitational interaction, only 10^{-40} of the strength of the strong interaction" (Dicke, 1964, p. 2) and the limitations of available empirical equipment and techniques to frustrate attempts to produce locally reliable laboratory data relevant to the evaluation of GTR and its competitors. To make the most of what little empirical evidence was available, physicists decided to reduce the number of alternatives it would be required to help evaluate. To this end, they drew up a list of values they wanted a gravitational theory to possess, and eliminated theories lacking one or more of them. To make the cut, a theory had to be logically consistent, to deliver approximations of Newtonian predictions for slow speeds, to accommodate the use of co-variant equations to describe gravity, and to allow space-time to be treated as a four-dimensional manifold. Although the available evidence was inconclusive with regard to the original set of alternatives, it favored GTR over a number of alternatives which met those conditions (Earman, 1992, p. 176–81). Thus reducing the field of alternatives to be judged gave old evidence new epistemic significance. This illustrates one way in which an investigators' theoretical commitments (e.g., to Newtonian predictions for slow speeds, to the use of co-variant equations) influence the significant of empirical evidence, and the conclusions they will find it reasonable to draw from them.

The macaque experiment is remarkable for showing how strong an influence cultural factors can have, not just on biology theorizing (that's no surprise) but on the treatment of a low-level problem in experimental design arising from a factor as ideologically and culturally neutral as battery life.

The epistemic significance of a bit of evidence can depend on the entrenchment of[16] a technique or piece of equipment, i.e., the extent to which the investigators depend on it to produce or interpret data, and the extent to which they treat it as epistemically unproblematic.

William Labov's critique of the arguments by which educational psychologists tried to demonstrate severe language deficits in inner city black children is an illustration. In one experiment, the investigator (a white adult male) showed a black child a toy airplane and asked "what would you say this looks like?", "what color is it?" "what would you use it for?" and "where do you think we could get another one of these?". The data (audio tapes of the interviews) exhibited hesitant, monosyllabic replies broken by pauses of up to 20 seconds. These were the features of the evidence on which the educational psychologists based their arguments. Labov was impressed instead by the children's intonation – an equally audible feature which was not salient to the psychologists. In accordance with standard socio-linguistic practices, he transcribed the interviews in a notation designed to indicate intonation patterns. Thus an utterance of "I don't know" is written:

$$^{3}\text{`o'}$$
$$^{2}\text{a} \qquad ^{2}\text{know}$$

Labov argued that because such intonation patterns are typical of black children's responses in comparable situations they find threatening the data may be indicative, not of verbal incapacity, but of defensive behavior elicited by fear of the investigator and the apparent pointlessness of his questions (Labov, 1972, pp. 206–7). Labov's consideration of intonation led also to the design of new experiments which had not occurred to the psychologists (Labov, 1972, pp. 208, 214). Because different techniques were entrenched in their research practices a feature of the data which was salient to the socio-linguists was ignored by the psychologists. As this example illustrates, philosophers should concern themselves with the epistemic import of entrenchment.

Technology and ideology

Recall that in Lloyd's primatology example, increased battery storage would have enabled the investigators to record long enough to produce data on typical, as well as atypical female orgasms. This suggests how important a role the nature of the equipment and technology available to the researcher can play in the interactions which bring cultural biases to bear on data production. Galison's account of symbiosis in the development of spark and bubble chamber experimental technologies and the conceptual resources of particle physics is a well-documented example of the epistemically significant interplay between theoretical commitments and data production (Galison, 1997, pp. xvii–63).[17]

Hidden possibilities

If an empirical result agrees with a theory, does that entitle us to conclude anything more than that the result agreed with the theory? Bacon is supposed to have thought so for cases of crucial experiments in which the result also rules out the theory's known competitors.[18] But Duhem argues that, to prove the correctness of a theory, a crucial experiment would have to test every possible alternative, and scientists are never in a position to know what alternatives remain to be discovered (Duhem, 1991, pp. 180–90). Thus, Duhem would deny that empirical evidence which favored GTR over all of its known competitors establishes its correctness. The most it establishes is that GTR did better than the alternatives with which it was compared. That wouldn't be so bad if the evidence told the investigators what probability to assign to GTR. But as long as the possibility space remains uncharted, it would seem that probability could be assigned only relative to the known possibilities, or by assigning a prior probability to the disjunction of unknown alternative possibilities. The latter alternative is unattractive because it is as hard to see how the priors could be assigned non-arbitrarily, as to see what should be made of probabilities which depend on arbitrarily stipulated priors. The former conjures up memories of defective theories which were accepted on evidence which would have been unimpressive if alternative theories had been developed (Earman, 1992, ch. 7). Thus a pressing question for empirical epistemology is the issue of how empirical results can contribute to the exploration and limitation of possibility spaces.

Replacing empiricism

Old-fashioned empiricists thought the test of a scientific claim is how well it stacks up against evidence delivered by the senses – unaided, or aided by magnifying or amplifying devices. But what about evidence produced by equipment like Geiger counters and galvanometers which are attuned to signals which the senses cannot pick up? It doesn't take too much of a stretch for empiricists to extend their notion of empirical evidence to apply to what registers only on observational equipment. But some of the evidence scientists rely on presents a more severe challenge to empiricism. For example, PET and fMRI imaging are so far the best, non-invasive techniques for comparing levels of cognitively significant neuronal activity in anatomically different regions of the brain during the performance of a mental task. The quantities exhibited by PET images are usually calculated from signals emitted by an oxygen isotope which has been introduced into the circulatory system. fMRI calculates its quantities from weak radio signals which vary with levels of oxygenated hemoglobin in the blood emitted when the brain is subjected to a strong magnetic field. Both quantities are indicative of blood volumes which vary with the electro-chemical neuronal activities of interest to the investigator. The empirical evidence produced by PET and fMRI consists of images resulting from elaborate computations embodying assumptions and techniques from physics, biology, statistics, and elsewhere (Corbetta 1998; Haxby et al., 1998). The images are not so much data as graphic representations of interpretations of radiation or radio signals. Unlike the distinction between what can be perceived by humans and what can register only on experimental equipment, the distinction between producing and interpreting data is too central to empiricism for its adherents to give up. It will take an alternative to empiricism to help us understand evidence like this which blurs the distinction between data production and interpretation.

Notes

1 Thanks to Peter Machamer for helpful suggestions, some of which I followed.
2 For some early history, see Eamon (1994) and Dear (1995).
3 This is not to endorse Popper's idea that experimental and observational results are scientifically significant only if they were produced for the purpose of testing a theory, let alone the hypothetico-deductive account according to which theory testing is always a matter of trying to falsify it by producing an experimental or observational result which is incompatible with a prediction the theory entails (Popper, 1959, pp. 27–48). For some influential alternatives, see Glymour (1980, pp. 33–9, 110–75) and Howson and Urbach (1993, pp. 6–11, 117–70).
4 The torture image comes from Bacon who has been accused in recent feminist science criticism of supporting the violation of nature by white, male science and technology; see Merchant in Zimmerman et al. (1993, pp. 272–5).
5 To save space, I use this term for logical empiricists as well as the Vienna Circle, and various of their sympathizers.
6 For samples of issues that occupied philosophers of science as they neglected experiment, see Schlick (1959, pp. 209–27), Neurath (1983, ch. 7, 8), Russell (1940, pp. 124–57) and Carnap (1953).
7 In addition to Franklin (1989), philosophers Ian Hacking (1991), and Deborah Mayo (1996), historians Peter Galison (1987, 1997), and Mary Jo Nye (1972) and sociologists and anthropologists of science including Andrew Pickering (1986), Harry Collins (1999), Bruno Latour and Steve Woolgar (1986) are among the writers who helped correct the neglect of experiment.
8 By "paradigm" Kuhn says he sometimes means a "constellation" of shared "beliefs, values, techniques, and so on," and sometimes "one sort of element in that constellation, the concrete puzzle-solutions which, employed as models or examples" on which scientists base their investigations (Kuhn, 1970, p. 175).
9 For example, see Stuewer's story of painfully conscientious observers reporting nonexistent flashes on a scintillation screen (Stuewer, 1985, pp. 284–9).

10 Surprisingly, Priestley v Lavoisier is one of the examples Kuhn argues from! (Kuhn, 1970, pp. 56, 118).

11 For the case of neutrinos, see Franklin (2001, pp. 249–318).

12 This and the distinction between phenomena and data are discussed in Bogen and Woodward (1988, 1992) and Woodward (1989).

13 Galison came to this point by paying close attention to experimental practice. Surprisingly, Popper was able to appreciate a linguistic version of the same problem (Popper, 1959, Part I, chp. v) even though he thought it "may be of little concern to the practical research worker" (Popper, 1959, p. 93, Part I, ch. v).

14 On the distinction between support and acceptance, see Hempel (1965, pp. 90–3).

15 For an excellent overview and samples of some Social Theorist positions, see Pickering (1992, ch. 1–6).

16 See Griesemer on entrenchment in Clarke and Fujimara (1992, pp. 52–60).

17 For further examples, see Mitman and Fausto-Sterling, and Holmes in Clarke and Fujimara (1992).

18 But see Urbach (1993, pp. 18, 168).

References

Bacon, F. (1994): *Novum Organum*, trans., P. Urbach and J. Gibson (eds.), Chicago: Open Court.

Barnes, B. (1997): *Interests and the Growth of Knowledge*. London: RKP.

Bernard, C. (1949): *An Introduction to the Study of Experimental Medicine*, tr. H. C. Greene, USA: Henry Schuman, Inc.

Bloor, D. (1983): *Knowledge and Social Imagery*. Chicago: University of Chicago Press.

Bogen, J. (2001): "Two as Good as One Hundred – Poorly Replicated Evidence in 19th Century Neuroscience," *Studies in the History and Philosophy of Biology and Biomedical Science*, forthcoming.

Bogen, J. and Woodward, J. (1988): "Saving the Phenomena," *Philosophical Review*, July, 303–52.

Bogen, J. and Woodward, J. (1992): "Observations, Theories and the Evolution of the Human Spirit," *Philosophy of Science*, 59, 590–611.

Boyle, R. (1661): *The Skeptical Chymist,* reprint Kila, Mt: Kessinger Publishing Co.

Carnap, R. (1953): "Testability and Meaning," in H. Feigl and M. Brodbeck (eds.), *Readings in the Philosophy of Science*, New York:Appleton-Century-Crofts, 47–92.

Cartwright, N. (1989): *Nature's Capacities and Their Measurement*. Oxford: Oxford University Press.

Churchland, P. (1992): *A Neurocomputational Perspective*. Cambridge: MIT Press.

Clarke, A. E. and Fujimara, J. H. (eds.), (1992): *The Right Tools for the Job*. Princeton: Princeton University Press.

Collins, H. M. (1999): *Changing Order*. Chicago: University of Chicago.

Conant, J. B. (1957): "The Overflow of the Phlogiston Theory: The Chemical Revolution of 1775–1789," in J. B Conant and L. K. Nash (eds.), *Harvard Case Histories in Experimental Science, vol. 1*, Cambridge: Harvard University Press, 65–116.

Corbetta, M. (1998): "Functional Anatomy of Visual Attention in the Human Brain: Studies with Positron Emission Tomography," in Parasuraman (1998, 95–122).

Dear, P. (1995): *Discipline and Experience, the Mathematical Way in the Scientific Revolution.* Chicago: University of Chicago Press.

Dicke, R. H. (1964): *The Theoretical Significance of Experimental Relativity.* New York: Gordon and Gordon.

Duhem, P. (1991): *The Aim and Structure of Physical Theory*, transl., P Wiener, Princeton: Princeton University Press.

Eamon, W. (1994): *Science and the Secrets of Nature, Books of Secrets in Medieval and Early Modern Culture.* Princeton: University of Princeton Press.

Earman, J. (1992): *Bayes or Bust?* Cambridge: MIT Press.

Earman, J. and Glymour, C. (1980): "Relativity and the Eclipses," *Historical Studies in the Physical Sciences*, 11, 49–85.

Earman, J. and Salmon, W. (1992): "The Confirmation of Scientific Hypotheses," in M. H. Salmon (ed.), *Introduction to the Philosophy of Science*, Inglewood Cliffs: Prentice Hall, 42–103.

Feyerabend, P. K. (1985): "Attempt at a Realistic Interpretation of Experience," in P. K. Feyerabend *Philosophical Papers, vol. 1*, Cambridge: Cambridge University Press, 17–36.

Franklin, A. (1989): *The Neglect of Experiment.* Cambridge: Cambridge University Press.

Franklin, A. (2001): *Are There Really Neutrinos?* Cambridge: Perseus.

Galileo, G. (1989): *Siderius Nuncius,* trans., A. Van Helden (ed.), Chicago: University of Chicago Press.

Galison, P. (1987): *How Experiments End.* Chicago: University of Chicago Press.

Galison, P. (1997) *Image and Logic.* Chicago: University of Chicago Press.

Glymour, C. (1980): *Theory and Evidence.* Princeton: Princeton University Press.

Glymour, C. (1997): "A Review of Recent Work on the Foundations of Causal Inference," in V. R. Kim and S. P. Turner (eds.), *Causality in Crisis*, South Bend: University of Notre Dame Press, 201–48.

Gooding, D. (1992): "Putting Agency Back Into Experiment," in Pickering (1992, pp. 65–112).

Hacking, I. (1991): *Representing and Intervening.* Cambridge, Cambridge University Press.

Hacking, I. (1992):"The Self Vindication of the Laboratory Sciences," in Pickering (1992, pp. 29–65).

Hanson, N. R. (1958): *Patterns of Discovery.* Cambridge: Cambridge University Press.

Haxby, J. V., Courtney, S. M. and Clark, V. P. (1998): "Functional Magnetic Resonance Imaging and the Study of Attention," in Parasuraman (1998, pp. 123–42).

Hempel, C. G. (1935): "On the Logical Positivist's Theory of Truth," *Analysis*, 2(4), 49–59.

Hempel, C. G. (1965): *Aspects of Scientific Explanation and Other Essays in the Philosophy of Science.* New York: Free Press.

Herschel, J. F. W. (1966): *Preliminary Discourse on the Study of Natural Philosophy.* New York: Johnson Reprint Corporation.

Hooke, R. (1968): "Micrographia," in R. T. Gunther (ed.), *Early Science in Oxford, vol. XIII*, London: Dawsons of Pall Mall.

Howson, C. and Urbach, P. (1993): *Scientific Reasoning, The Bayesein Approach*, 2nd edn, Chicago: Open Court.

Kuhn, T. S. (1970): *The Structure of Scientific Revolutions.* Chicago: University of Chicago Press.

Labov, W. (1972): *Language in the Inner City*. Philadelphia: University of Pennsylvania Press.

Latour, B. (1987): *Science in Action*. Cambridge: Harvard University Press.

Latour, B. and Woolgar, S. (1986): *Laboratory Life*. Princeton: Princeton University Press.

Laymon, R. (1983): "Newton's Demonstration of Universal Gravitation and Philosophical Theories of Confirmation," in J. Earman (ed.), *Testing Scientific Theories, Minnesota Studies in the Philosophy of Science, vol. X*, Minneapolis: University of Minnesota Press, 179–200.

Laymon, R. (1988): "The Michelson–Morley Experiment and the Appraisal of Theory," in A. Donovan, L. Laudan and R. Laudan (eds.), *Scrutinizing Science*, Baltimore: Johns Hopkins, 245–66.

Lloyd, E. (1993): "Pretheoretical Assumptions in Evolutionary Explanations of Female Sexuality," *Philosophical Studies*, 69, 139–53.

Mayo, D. G. (1996): *Error and the Growth of Empirical Knowledge*. Chicago: University of Chicago Press.

Mazziota, J. C., Toga, A. W., Evans, A., Fox, P. and Lancaster, J. (1995): "A Probabilistic Analysis of the Human Brain: Theory and Rationale for Its Development," *Neuroimage*, 2, 89–101.

Millikan, R. A. (1935): *Electrons (+ and –), Protons, Photons, Neutrons, and Cosmic Rays*. Chicago: University of Chicago Press.

Mitchell, S. D. (1987): "Competing Units of Selection? A Case of Symbiosis," *Philosophy of Science*, 54, 351–67.

Nagel, E. (1961): *The Structure of Science*. New York: Harcourt, Brace, World.

Neurath, O. (1983): *Philosophical Papers*. Dordrecht: D. Reidel.

Newton, I. (1999): *The Principia*, transl., I. B. Cohen and A. Whitman (eds.), Berkeley: University of California Press.

Nye, M. J. (1972): *Molecular Reality*. London: MacDonald.

Parasuraman, R. (ed.) (1998): *The Attentive Brain*. Cambridge: MIT Press.

Pearl, J. (2000): *Causality Models, Reasoning, and Inference*. Cambridge: Cambridge University Press.

Perrin, J. (1990): *Atoms*, tr., D. L. l. Hammick, Woodbridge: Oxbow Reprints.

Pickering, A. (1986): *Constructing Quarks*. Chicago: University of Chicago Press.

Pickering, A. (ed.) (1992): *Science as Practice and Culture*. Princeton: Princeton University Press.

Popper, K. R. (1959): *The Logic of Scientific Discovery*. New York: Basic Books, first published in German, 1934.

Russell, B. (1940): *An Inquiry into Meaning and Truth*. London: Allen and Unwin.

Schlick, M. (1959): in A. J. Ayer (ed.), *Logical Positivism*, New York: Macmillan, 209–27.

Shapere, D. (1982): "The Concept of Observation in Science and Philosophy," *Philosophy of Science*, 49, 485–525.

Shapin, S. (1994): *A Social History of Truth, Civility and Science in Seventeenth-Century England*. Chicago: University of Chicago Press.

Spirtes, P., Glymour, C. and Scheines, R. (1993): *Causation, Prediction, and Search*. New York: Springer Verlag.

Steinmetz, H. and Seitz, R. J. (1991): "Functional Anatomy of Language Processing: Neuroimaging and the Problem of Individual Variability," *Neuropsychologia*, 29(12) 1149–61.

Stuewer, R. H. (1985): "Artificial Disintegration and the Cambridge–Vienna Controversy," in P. Achinstein and O. Hannaway (eds.), *Observation, Experiment, and Hypothesis in Modern Physical Science*, Cambridge: MIT Press, 239–308.

Urbach, P. (1993): *Francis Bacon's Philosophy of Science*. La Salle: Open Court.

Whewell, W. (1991): *Selected Writings on the History of Science*, Y. Elkana (ed.), Chicago: University of Chicago Press.

Woodward, J. (1989): "Data and Phenomena," *Synthese*, 79, 393–472.

Woodward, J. (1997): "Explanation, Invariance, and Intervention," *Philosophy of Science*, 64, S26–S46.

Woodward, J. (2000): "Data, Phenomena, and Reliability," in D. Howard (ed.), *Proceedings of the 1998 Meetings of the Philosophy of Science Association*, S163–S169.

Zimmerman, M. E., Callicott, J. B., Sessions, G., Warren, K. J. and Clark, J. (eds.) (1993): *Environmental Philosophy From Animal Rights to Radical Ecology*. Englewood Cliffs: Prentice Hall.

Chapter 8

Induction and Probability

Alan Hájek and Ned Hall

We will discuss induction and probability in that order, aiming to bring out the deep interconnections between the two topics; we will close with a brief overview of cutting-edge research that combines them.

Induction: Some Preliminaries

Arguably, Hume's greatest single contribution to contemporary philosophy of science has been the problem of induction (Hume, 1739). Before attempting its statement, we need to spend a few words identifying the subject matter of this corner of epistemology. At a first pass, induction concerns *ampliative* inferences drawn on the basis of evidence (presumably, evidence acquired more or less directly from experience) – that is, inferences whose conclusions are not (validly) entailed by the premises. Philosophers have historically drawn further distinctions, often appropriating the term "induction" to mark them; since we will not be concerned with the philosophical issues for which these distinctions are relevant, we will use the word "inductive" in a catch-all sense synonymous with "ampliative." But we will follow the usual practice of choosing, as our paradigm example of inductive inferences, inferences about the future based on evidence drawn from the past and present.

A further refinement is more important. Opinion typically comes in *degrees*, and this fact makes a great deal of difference to how we understand inductive inferences. For, while it is often harmless to talk about the conclusions that can be rationally *believed* on the basis of some evidence, the force of any given evidence is more accurately measured by noting its effects on the rational assignment of *degrees* of belief. The usual assumption – one that directly connects the two topics of this chapter – is that rationally warranted degree of belief can be modeled as a species of *probability*, inductive inference itself being modeled by some rule for changing probabilities in light of evidence.

The strength of the support that some evidence gives some hypothesis is *not* measured by the degree of belief in the hypothesis that is warranted in light of that evidence. There is widespread disagreement among philosophers of science as to how exactly it should be measured (Milne, 1996); but all parties agree that the notion is an essentially *comparative* one, in that it depends not only on how likely an hypothesis is, in light of the given evidence, but also on how likely it would be, in light of *different* evidence.

The First Problem of Induction

One serious problem that inductive inferences present us with is that of saying with any precision what distinguishes the *good* ones from the *bad* ones. We will approach that problem shortly, and because we take it up second will call it the "second problem of induction." It should be distinguished from the first problem, which is that of saying why the "good" ones *deserve* this label.

For various reasons (whose elucidation, space does not permit), we think it is best to formulate this problem as a problem about settling a conflict between two rival inductive methods – rival sets of rules for adjusting degrees of belief in light of evidence. Let us personify two such rivals in the form of Billy and Suzy, two friends. Suzy is a paragon of cognitive virtue: she always evaluates the impact of evidence on hypotheses in accordance with the "right" inductive principles. Billy evaluates the force of evidence in accordance with different principles, a fact which shows up in the following bizarre behavior. He regularly sticks his fingers in light sockets, always getting a nasty shock when he does so. To be clear, Billy doesn't *like* these shocks *at all*. It's just that each time, his inductive methods lead him to the conclusion, given the evidence available to him, that it is overwhelmingly likely that the sensation will be exquisitely pleasant. Billy and Suzy disagree about these predictions, and since – let us stipulate – they have exactly the same evidence, their disagreement traces to the different principles they adhere to in evaluating the force of such evidence. Is it possible to provide any compelling argument for the conclusion that the inductive principles to which Suzy adheres are rationally superior to Billy's?

There are two relevant parties here, and we need to consider the possibility that there is an argument compelling to one but not the other. Let us stipulate that such an argument must make use of acceptable premises that do not beg the question against the party to be compelled. We will take it that such premises will at least include all propositions detailing the evidence available to Billy and Suzy. Let us further stipulate that the premises must, in some sense, *support* the given conclusion, and that they can do so in one of only two ways: either they entail the conclusion that Suzy's inductive principles are rationally preferable to Billy's, or they provide some measure of inductive support for this conclusion.

It might seem that no argument of the first kind that would be compelling to *either* Billy or Suzy is possible, especially if we limit our attention to arguments that proceed only from the available evidence, and that attempt to establish the superiority of Suzy's inductive methods over Billy's by way of the intermediate conclusion that her methods will, in the future, yield correct predictions more often than Billy's. Here, the point, familiar since Hume, that the past places no logical constraints on the future renders such an intermediate conclusion inaccessible. But there is always the possibility of finding additional, non-question-begging premises, or of finding some other route to the conclusion – loopholes that, as we'll see below, Reichenbach's "pragmatic" justification of induction attempts to exploit.

Moreover, there is at least one clear way in which such an argument could be constructed: namely, if Billy's inductive rules undermine themselves by predicting, given the evidence, that they will systematically issue *false* predictions in the future. If Suzy's principles do not undermine themselves in this way, then they will clearly be rationally preferable; what's more, this conclusion validly follows from premises perfectly acceptable to both. Still, there is little point in hoping for such an argument, as it turns out to be far too easy – and costless – to construct inductive methods that are immune from self-undermining.[1] So we might as well build into our description of the Billy/Suzy scenario that Billy is adhering to just such a method.

The Inductive Justification of Induction

What then of the possibility of a compelling *inductive* argument? None could succeed in convincing Billy. For either the evidence available to Billy inductively warrants, by his lights, the conclusion that Suzy's inductive methods are preferable to his own, or it doesn't. But if the evidence warrants this conclusion then his principles undermine themselves, and we have stipulated that they are immune from such self-undermining. So consider whether some inductive argument could be produced that would provide *Suzy* with compelling grounds for holding her inductive methods to be rationally preferable to Billy's.

To give her the best possible case, let us suppose that Suzy's own inductive principles strongly endorse, in light of the evidence available to her, the conclusion that those very principles will yield wildly successful predictions in the future, whereas Billy's will yield an unbroken string of falsehoods. It seems, then, that she has a compelling and powerful argument for the target conclusion.

But there has been a bit of sleight of hand. The problem is *not* that her "inductive justification of induction" is circular or question-begging, for given that *she* is its target it manifestly *isn't* (van Cleve, 1984); any lingering sense that it *is* can be explained away by noting that no such inductive argument could convince

Billy.) Rather, we need to remember that strength of inductive support is a comparative notion. In the case at hand, the "track record" of Suzy's inductive methods provides, by the lights of these methods, *extra* reason to have faith in them only to the extent that other track records were possible that would have yielded a more pessimistic prediction – i.e., only if it was at least possible for the evidence to produce a self-undermining verdict. But it is *prima facie* quite desirable to adhere to an inductive method that is immune from the possibility of self-undermining, particularly given that this is both easy and costless to do. Assuming that Suzy *is* following such a method, the track record of its successes, however spectacular, contributes nothing *at all* to the case in its favor.

In any case, even at the beginning of inquiry – before any evidence has been amassed – there remains just as clear and intuitive a difference in the rational acceptability of Suzy's and Billy's inductive methods, one which need wait on no "supporting evidence" to become visible: The difference is that hers are superior. Not, then, because the evidence favors them.

The Pragmatic Justification

It seems transparent that no valid argument could be produced whose premises are obviously correct and non-question-begging and whose conclusion is that our particular inductive practices are rationally warranted. Famously, Reichenbach (1938, 1949) attempts to produce just such an argument.

The more formal version of the argument seeks to justify the use of the "straight rule" in making predictions about the limiting relative frequency of outcomes in an infinitely repeated experiment (the rule predicts this limit to equal the frequency observed so far); it has been so thoroughly discussed in the literature that we will pass over it (Salmon, 1967). The less formal version is more clever, and less talked-about. For the sake of definiteness, let us use the label "the scientific method" (SM) as a name for whatever method it is we use – at least, when we are at our cognitive best – for drawing inductive conclusions. And let us suppose that we have some standard of success for an inductive method – say, that the long-range frequency of correct inferences drawn on its basis must be sufficiently high. For any given inductive method, there is of course no guarantee that it will succeed in this sense: the world must cooperate, and it is a contingent matter whether it will do so. Reichenbach thus grants that we can have no a priori guarantee of the success of SM. But he argues that we are "pragmatically" justified in our continued adherence to SM, since if *any* method will succeed, *it* will. For suppose the world is such that some rival method RM will succeed, but SM will not. Well, a central component of SM consists in projecting observed regularities into the future, and in a world where RM succeeds, a relevant such regularity simply *is* the pattern of RM's successes. If, for example, a successful method for predicting the future in world w is to consult an oracle, then SM will eventually establish that the oracle is reliable – and so SM will itself ultimately yield the advice that one should consult the oracle when making predictions. The key claim – that if the world is nice enough to allow for the success of some inductive method, then it is nice enough to allow for the success of SM – is simply a generalization of this example. So, we have a demonstrative argument that SM will be successful if any method will; hence, it seems, a demonstrative argument that we are rationally warranted in adhering to SM.

But the example of the oracle is entirely misleading. For a long string of successful predictions by such an oracle surely constitutes a very *salient* regularity. Suppose the success of the rival inductive method is not nearly so visible; why should we have any confidence that SM will "latch onto" this string of accurate predictions?

We should have *no* such confidence if the following condition holds: For every proposition *A* about the future, there are rival inductive methods that have been highly successful, and equally successful, in the past, but that disagree widely as to the likely truth of *A* (given the available evidence). If that condition is met, then the argument for the crucial premise fails disastrously – for *which* string of successes should SM latch onto? It is easy to see what went wrong: the argument involved a bit of misdirection in getting us to agree implicitly that SM was, in the imagined world w, up against just *one* rival RM. More plausible is that it would be up against a battery of rivals so extensive that they fail as a group to agree on any prediction of substance. It is wishful thinking to suppose that SM could somehow pick the "winner".

Hume's "Skeptical Solution"

Hume's own "skeptical" solution to his problem of induction foreshadowed an important movement in contemporary epistemology, which seeks to "naturalize" the subject (Kornblith, 1985). For Hume took it that no rational basis for induction is possible, while adding that perfectly legitimate empirical psychological questions remain about how exactly it is that deliberating agents draw inductive conclusions from evidence. Hume's own answer emphasized the central role of the apparently brute psychological disposition he called "custom" or "habit"; contemporary fans of this kind of naturalized approach to inductive epistemology could presumably be expected to draw on much more sophisticated theories of human cognitive psychology.

A serious worry is that it is unclear that the naturalizing move in epistemology leaves room for a legitimate, coherent sub-discipline of *normative* epistemology, a discipline that seeks to articulate the principles according to which we *ought* to form our beliefs. That is unfortunate, since our natural or "untutored" cognitive abilities in the inductive domain are notoriously and systematically unreliable, particularly when we're in a situation that forces us to attend sensitively to the probabilistic bearing of evidence on hypotheses (Kahneman et al., 1982). It would seem to require careful *a priori* reflection to distinguish rational inductive inferences from mistakes. At the very least, the defender of a purely naturalized epistemology of induction owes us an account of how *else* we might systematically identify and guard against inductive error.

Popper's Falsificationism

Popper (1968) argues for a different way of dismissing the problem of induction: while agreeing with Hume that no rational justification of induction can be found, he insists that this result is innocuous, simply because induction forms no part of the practice of science. According to Popper, scientists propose "conjectures," and then subject these conjectures to severe observational tests in an effort to falsify them. He claims that we are never rationally warranted in considering such an hypothesis to be probable, given that it has passed such tests. And since, according to the simplest version of falsificationism, deductive relations are all that we need attend to in order to check that an hypothesis has been refuted by some evidence, the problem of induction poses no threat to the rationality of scientific practice.

As a descriptive claim about what scientists, *qua* scientists, actually *do* – let alone about what they *believe* about what they do – Popper's view strikes us as absurd. But even as a normative claim it fares little better. The simplest and most devastating point was nicely emphasized by Putnam (1974): Popper seems willfully blind to the fact that we use evidence from the past and present as a basis for making practical *decisions*, decisions whose rationality is hostage to the rationality of the inferences drawn about their likely consequences on the basis of the given evidence. What would Popper say, for example, about the disagreement between Billy and Suzy? That Billy's behavior is somehow rationally permissible, even in light of the extensive and painful evidential record? Even were it understood merely as a claim about the rationality of belief, falsificationism would be unpalatable; but belief and action are too inextricably linked to sustain such an understanding. The consequences of the position for the rationality of decision provide it, ironically enough, with a decisive refutation.

The Dogmatic Response

The final approach to the first problem of induction that we will consider we call the "dogmatic" response – not as an insult, but because the word nicely summarizes its main features. For according to the dogmatic response, induction is perfectly rational – certain ways of adjusting degrees of belief in the light of evidence are rationally warranted, and certain other ways are irrational – but absolutely no justification can be given of this claim, not even a justification of the kind that would only be compelling to the likes of Suzy (Strawson, 1952). It is rather that the fact that certain inductive inferences are rational and certain others irrational (and perhaps still others neither rational nor irrational) is a brute epistemological fact, incapable of further philosophical explanation or defense.

The principal merits of the view are clear enough. It allows us to maintain, contra Hume and other skeptics about induction, a vigorous distinction between rational and irrational inductive methods and inferences, and it acquires at least some measure of plausibility from the dismal failure of more ambitious attempts to give a justification of induction. Still, the view should only be seen as a kind of philosophical last resort. For there are too many interesting questions about which the dogmatic response falls silent. Notably, whereas any attempt to provide a substantive justification of induction can be expected, to the extent that it succeeds, also to provide insights into what distinguishes good from bad inductive inferences, the dogmatic response is hopeless in this regard. And, as noted above, the problem of providing a clear explication of the distinction between rational ("good") and irrational ("bad") inductive inferences is a deep and central one in its own right. We turn now to a brief discussion of some of the main philosophical approaches in this area.

The Second Problem of Induction: Syntactic Approaches

Traditionally, logic aims to distinguish valid from invalid arguments by virtue of the syntactic form of the premises and conclusion (e.g., any argument that has the form *p and q, therefore p* is valid in virtue of this form). But the distinction between valid and invalid is not fine enough: after all, many invalid arguments are perfectly good, in the sense that the premises provide strong inductive support for the conclusion. Carnap (1950) described this relation of support as the *logical probability* that an argument's conclusion is true, given that its premises are true – hoping that logic, more broadly conceived, could give it a *syntactic* analysis. We will discuss Carnap's approach in more detail below. Other, less ambitious approaches tried to find syntactic criteria for "qualitative confirmation" – criteria, that is, that would identify at least some instances in which evidence raised the probability of an hypothesis, to at least some degree. (See Hempel (1945a,b) for an excellent overview of work in this area.)

These attempts to design a "logic" of induction on the model of formal deductive logic did not succeed. The decisive problem concerns the language-dependence that any such "logic" would have to exhibit. Consider, for example, a language used to represent the outcomes of random draws from an urn filled with colored balls; let the language contain the color predicates "blue" and "green," and also, in the spirit of Goodman (1983), the predicate "grue," where

x is grue at draw i

is equivalent to

x is green at draw i and i ≤ 1,000,000 or x is blue at draw i and i > 1,000,000

Surely the "logical strength" of the argument, "the first million draws are green, therefore the next draw will be green" is greater than $\frac{1}{2}$; if syntax is all that matters, then so too is the logical strength of the argument, "the first million draws are grue, therefore the next draw will be grue." But the two conclusions contradict each other, and so cannot both receive probability greater than $\frac{1}{2}$.

One might try to specify a canonical language, to sentences of which the syntactic rules, whatever they are, are meant to apply – a language free of such monstrosities as "grue". But not only does traditional logic find no need for such a procedure, it is also extraordinarily difficult to see how one could carry it out, at least if we want to analyze the inductive strength of any argument of real interest. By the middle of the seventeenth century, the available evidence strongly supported Keplerian over Ptolemaic astronomy; but what would be the canonical language in which to translate this evidence and these hypotheses, so as to analyze the differential support *syntactically?*

The Second Problem: Modest Probabilism

One might agree with Carnap that induction should be modeled using the tools of probability theory, while denying that syntactic analysis alone can provide or even constrain the values of the relevant probability function. And indeed, what we will call "modest probabilism" about induction and confirmation has become increasingly popular since the demise of logical empiricism. We call this approach "probabilism" because it sees the inductive support or degree of confirmation that evidence E gives hypothesis H as measured by somehow comparing $P(H)$, the probability of H, with $P(H|E)$, the conditional probability of H, given E. (We will have more to say about these quantities shortly.) Perhaps this inductive support is measured by the difference $P(H|E) - P(H)$; perhaps by the ratio $P(H|E)/P(H)$; perhaps in some other way (Good, 1985). But the approach is modest to the extent that it is agnostic about the nature or source of the "confirmation-probability" in question. Its agnosticism notwithstanding, modest probabilism is able to achieve some remarkable successes. For example, it explains straightaway the success (such as it is) of the hypothetico-deductive account of confirmation. For if H implies E, and if $P(E) < 1$, then it follows at once that $P(H|E) > P(H)$

(for this condition is equivalent to $P(E|H) > P(E)$). More interestingly, modest probabilism neatly explains away the Raven's Paradox, and can be easily adapted to illuminate the confirmation of hypotheses that are themselves probabilistic; see Earman (1992) for a fuller discussion.

Partly to flesh out the resources of and problems for this probabilistic approach, we will now switch gears slightly, and take up the second of our topics: an investigation of probability theory and the most important attempts at explicating its conceptual foundations. We begin with an overview of the widely accepted *mathematical* foundations.

Kolmogorov's Axiomatization

Probability theory was inspired by games of chance in seventeenth century France and inaugurated by the Fermat–Pascal correspondence. However, its axiomatization had to wait until Kolmogorov's classic book (1933). Let Ω be a non-empty set ("the universal set"). A *sigma-field* (or *sigma-algebra*) on Ω is a set F of subsets of Ω that has Ω as a member, and that is closed under complementation (with respect to Ω) and countable union. Let P bc a function from F to the real numbers obeying:

1 $P(A) \geq 0$ for all $A \in F$.
2 $P(\Omega) = 1$.
3 $P(A \cup B) = P(A) + P(B) \; \forall \; A, B \in F$ such that $A \cap B = \varnothing$.

Call P a *probability function*, and (Ω, F, P) a *probability space*.

We could instead attach probabilities to members of a collection of *sentences* of a formal language, closed under truth-functional combinations.

It is controversial whether probability theory should include Kolmogorov's further axiom:

4 (Continuity) $E_n \downarrow \varnothing$ implies $P(E_n) \to 0$ (where $E_n \in F \; \forall n$)

Equivalently, we can replace the conjunction of axioms 3 and 4 with a single axiom:

3′ (Countable additivity) If $\{A_i\}$ is a countable collection of (pairwise) disjoint sets, each $\in F$, then

$$P\left(\bigcup_{n=1}^{\infty} A_n\right) = \sum_{n=1}^{\infty} P(A_n)$$

The conditional probability of X given Y is standardly given by the ratio of unconditional probabilities:

$$P(X|\Upsilon) = \frac{P(X \cap \Upsilon)}{P(\Upsilon)}$$

provided $P(\Upsilon) > 0$. We can now prove versions of *Bayes' theorem*:

$$P(A|B) = \frac{P(B|A).P(A)}{P(B)}$$

$$= \frac{P(B|A).P(A)}{P(B|A).P(A) + P(B|\neg A).P(\neg A)}$$

More generally, suppose we have a partition of hypotheses $\{H_1, H_2, \ldots, H_n\}$, and evidence E. Then we have, for each i:

$$P(H_i|E) = \frac{P(E|H_i)P(H_i)}{\sum_{j=1}^{n} P(E|H_j)P(H_j)}$$

The $P(E|H_i)$ terms are called *likelihoods*, and the $P(H_i)$ terms are called *priors*.

If $P(X|\Upsilon) = P(X)$ – equivalently, if $P(\Upsilon|X) = P(\Upsilon)$; equivalently, if $P(X \cap \Upsilon) = P(X)P(\Upsilon)$ – then X and Υ are said to be *independent*. Two cautions: first, the locution "X is independent of Υ" is somewhat careless, encouraging one to forget that independence is a relation that events or sentences bear *to a probability function*. Second, this technical sense of "independence" should not be identified unreflectively with causal independence, or any other pretheoretical sense of the word, even though such identifications are often made in practice. If $P(X|\Upsilon) > P(X)$ – equivalently, if $P(\Upsilon|X) > P(\Upsilon)$ – then X and Υ are *positively correlated*. A cornerstone of any probabilistic approach to induction is the idea that evidence about the observed is positively correlated with various hypotheses about the unobserved.

We now turn to the so-called "*interpretations*" of probability. The term is misleading twice over. Various quantities that intuitively have nothing to do with "probability" obey Kolmogorov's axioms – for example, length, volume, and mass, each suitably normalized – and are thus "interpretations" of it, but not in the intended sense. Conversely, the majority of the most influential "interpretations" of P violate countable additivity, and thus are not genuine interpretations of Kolmogorov's full probability calculus at all. Be that as it may, we will drop the scare quotes of discomfort from now on.

The Classical Interpretation

The classical interpretation, which owes its name to its early and august pedigree – notably the *Port-Royal Logic*, Arnauld (1662) and Laplace (1814) – purports to determine probability assignments in the face of no evidence at all, or symmetrically balanced evidence. In such circumstances, probability is shared equally among all the possible outcomes, so that the classical probability of an event is simply the fraction of the total number of possibilities in which the event occurs – a version of the so-called *principle of indifference*. Unless more is said, it is also arguably the interpretation furthest removed from considerations of induction, reflecting as it does a certain *a prioristic* innocence: in typical applications, the number of possibilities, and thus the share that each gets of the total probability, remain the same (e.g. ⅗) whatever the outcomes in the actual world happen to be. Unfortunately, the classical interpretation can apparently yield contradictory results when there is no single privileged set of possibilities, as Bertrand (1889) brought out in his paradoxes. Classical probabilities are only finitely additive (de Finetti, 1974).

The Logical Interpretation

Logical theories of probability retain the classical interpretation's guiding idea that probabilities can be determined *a priori* by an examination of the space of possibilities. However, they generalize it in two important ways: the possibilities may be assigned *unequal* weights, and probabilities can be computed whatever the evidence may be, symmetrically balanced or not. Indeed, the logical interpretation, in its various guises, seeks to codify in full generality the degree of support or confirmation that a piece of evidence E confers on a given hypothesis H, which we may write as $c(H, E)$.

Early proponents of logical probability include Keynes (1921), W. E. Johnson (1932), and Jeffreys (1939). However, by far the most systematic study of logical probability was by Carnap. His formulation of logical probability begins with the construction of a formal language. He considers (Carnap, 1950) a class of very simple languages consisting of a finite number of logically independent monadic predicates (naming properties) applied to countably many individual constants (naming individuals) or variables, and the usual logical connectives. The strongest (consistent) statements that can be made in a given language describe all of the individuals in as much detail as the expressive power of the language allows. They are conjunctions of complete descriptions of each individual, each description itself a conjunction containing exactly one occurrence (negated or unnegated) of each predicate of the language. Call these strongest statements *state descriptions*.

Any probability measure $m(-)$ over the state descriptions automatically extends to a measure over all sentences, since each sentence is equivalent to a disjunction of state descriptions; m in turn induces a confirmation function $c(-,-)$:

$$c(H,E) = \frac{m(H \,\&\, E)}{m(E)}$$

There are obviously infinitely many candidates for m, and hence c, even for very simple languages. Carnap argues for his favored measure "m^*" by insisting that the only thing that significantly distinguishes individuals from one another is some qualitative difference, not just a difference in labeling. A *structure description* is a maximal set of state descriptions, each of which can be obtained from another by some permutation of the individual names. m^* assigns each structure description equal measure which, in turn, is divided equally among their constituent state descriptions. It gives greater weight to homogenous state descriptions than to heterogeneous ones, thus "rewarding" uniformity among the individuals in accordance with putatively reasonable inductive practice. It can be shown that the induced c^* allows inductive learning from experience – as, annoyingly, do infinitely many other candidate confirmation functions. Carnap claims that c^* nevertheless stands out for being simple and natural.

He later generalizes his confirmation function to a continuum of functions c_λ. Define a *family* of predicates to be a set of predicates such that, for each individual, exactly one member of the set applies, and consider first-order languages containing a finite number of families. Carnap (1963) focuses on the special case of a language containing only one-place predicates. He lays down a host of axioms concerning the confirmation function c, including those induced by the probability calculus itself, various axioms of symmetry (for example, that $c(H, E)$ remains unchanged under permutations of individuals, and of predicates of any family), and axioms that guarantee undogmatic inductive learning, and long-run convergence to relative frequencies. They imply that, for a family $\{P_n\}$, $n = 1, \ldots, k, k > 2$:

$$c_\lambda \text{ (individual } s+1 \text{ is } P_j, \ s_j \text{ of the first } s \text{ individuals are } P_j) = \frac{s_j + \lambda/k}{s + \lambda}$$

where λ is a positive real number.

The higher the value of λ, the less impact evidence has: induction from what is observed becomes progressively more swamped by a classical-style equal assignment to each of the k possibilities regarding individual $s + 1$.

Significantly, Carnap's various axioms of symmetry are hardly logical truths. More seriously, we cannot impose further symmetry constraints that are seemingly just as plausible as Carnap's, on pain of inconsistency (Fine, 1973, p. 202). Goodman taught us: that the future will resemble the past in some respect is trivial; that it will resemble the past in all respects is contradictory. And we may continue: that a probability assignment can be made to respect some symmetry is trivial; that one can be made to respect all symmetries is contradictory. This threatens the whole program of logical probability.

Frequency Interpretations

Frequency interpretations can be thought of as elevating a methodological rule for induction – the straight rule – to the status of a definition of probability.

Empiricist in inspiration, and originating with Venn (1876), they identify an event's probability with the *relative frequency* of events of that type within a suitably chosen reference class. The probability that a given coin lands "heads," for example, might be identified with the relative frequency of "heads" outcomes in the class of all tosses of that coin. But there is an immediate problem: observed relative frequencies can apparently come apart from true probabilities, as when a fair coin that is tossed ten times happens to land heads every time. Von Mises (1957) offers a more sophisticated formulation based on the notion of a *collective*, rendered precise by Church (1940): a hypothetical infinite sequence of "attributes" (possible outcomes) of a specified experiment, for which the limiting relative frequency of any attribute exists, and is the same in any recursively specified subsequence. The probability of an attribute A, relative to a collective ω, is then defined as the limiting relative frequency of A in ω. Limiting relative frequencies violate countable additivity (de Finetti, 1974).

A notorious problem for any version of frequentism is the so-called *problem of the single case*: we sometimes attribute non-trivial probabilities to results of experiments that occur only once. The move to hypothetical infinite sequences of trials creates its own problems: There is apparently no fact of the matter as to what such a hypothetical sequence would be, nor even what its limiting relative frequency for a given attribute would be, nor indeed whether that limit is even defined; and the limiting relative frequency can be changed to any value one wants by suitably permuting the order of trials. In any case, the empiricist intuition that facts about probabilities are simply facts about patterns in the actual phenomena has been jettisoned.

Propensity Interpretations

Attempts to locate probabilities "in the world" are also made by variants of the *propensity* interpretation, championed by such authors as Popper (1959), Mellor (1971), and Giere (1973). Probability is thought of as a physical propensity, or disposition, or tendency of a given type of physical situation to yield an outcome of a certain kind, or to yield a long run relative frequency of such an outcome. This view is explicitly intended to make sense of single-case probabilities. According to Popper, a probability p of an outcome of a certain type is a propensity of a repeatable experiment to produce outcomes of that type with limiting relative frequency p. Given their intimate connection to limiting relative frequencies, such propensities presumably likewise violate countable additivity. Giere explicitly allows single-case propensities, with no mention of frequencies: probability is just a propensity of a repeatable experimental set-up to produce sequences of outcomes. This, however, creates the problem of deriving the desired connection between probabilities and frequencies. This quickly turns into a problem for inductive inference: it is unclear how frequency information should be brought to bear on hypotheses about propensities that we might entertain.

The Subjectivist Interpretation: Orthodox Bayesianism

Degrees of belief

Subjectivism is the doctrine that probabilities can be regarded as degrees of belief, sometimes called *credences*. It is often called "Bayesianism" thanks to the important role that Bayes' theorem typically plays in the subjectivist's calculations of probabilities, although this is yet another misnomer since *all* interpretations of probability are equally answerable to the theorem, and subjective probabilities can be defined without any appeal to it. Unlike the logical interpretation (at least as Carnap originally conceived it), subjectivism allows that different agents with the very same evidence can rationally give different probabilities to the same hypothesis.

But what is a degree of belief? A standard analysis invokes betting behavior: an agent's degree of belief in X is p iff she is prepared to pay up to p units for a bet that pays 1 unit if X, 0 otherwise (de Finetti, 1937). It is assumed that she is also prepared to sell that bet for p units. Thus, *opinion* is conceptually tied to certain *behavior*. Critics argue that the two can come apart: an agent may have reason to misrepresent her opinion, or she may not be motivated to act according to her opinion in the way assumed.

Bayesians claim that ideally rational degrees of belief are (at least finitely additive) probabilities. "Dutch Book" arguments are one line of defense of this claim. A Dutch Book is a series of bets, each of which the agent regards as fair, but which collectively guarantee her loss. De Finetti (1937) proves that *if* your degrees of belief are not finitely additive probabilities, *then* you are susceptible to a Dutch Book. Equally important, and often neglected, is Kemeny's (1955) converse theorem. A related defense of Bayesianism comes from *utility theory*. Ramsey (1926) and Savage (1954) derive both probabilities and utilities (desirabilities) from preferences constrained by certain putative "consistency" assumptions.

Updating probability

Suppose that your degrees of belief are initially represented by a probability function $P_{initial}(-)$, and that you become certain of E (where E is the strongest such proposition). What should be your new probability function P_{new}? The favored updating rule among Bayesians is *conditionalization*; P_{new} is related to $P_{initial}$ as follows:

(Conditionalization) $\quad P_{new}(X) = P_{initial}(X|E)$ (provided $P_{initial}(E) > 0$)

Conditionalization is supported by a "diachronic" Dutch Book argument (Lewis, 1998): on the assumption that your updating is rule-governed, you are subject to a Dutch book if you do not conditionalize. Equally important is the converse theorem (Skyrms, 1987). *Jeffrey conditionalization* allows for less decisive learning experiences in which your probabilities across a partition $\{E_1, E_2, \ldots\}$ change to $\{P_{new}(E_1), P_{new}(E_2), \ldots, \}$, where none of these values need be 0 or 1:

$$P_{new}(X) = \sum_i P_{initial}(X|E_i)P_{new}(E_i)$$

(Jeffrey, 1965). It is again supported by a Dutch book argument (Armendt, 1980). See Diaconis and Zabell (1986) for further probability revision rules.

Orthodox Bayesianism can now be characterized by the following maxims:

B1   Rationality requires an agent's 'prior' (initial) probabilities to conform to the probability calculus.

B2   Rationality requires an agent's probabilities to update by the rule of (Jeffrey) conditionalization.

B3   Rationality makes no further requirements on an agent's probabilities.

If orthodox Bayesianism is correct, then there is a sense in which Hume's problem of induction is immediately solved. Inductive inferences based on observational evidence are justified by the appropriate prior subjective probability assignments, suitably updated on that evidence. For example, by B3, rationality permits you to assign:

$P_{initial}$(the sun will rise on day 10001|the sun rises on days 1, 2, ..., 10000)
= 0.9999

Suppose your evidence is:

the sun rises on days 1, 2, ..., 10000

Then conditionalizing on that evidence, as rationally requires according to B2, gives:

P_{new}(the sun will rise on day 10001) = 0.9999

Similarly, if your prior is of the right form, rationality requires you to assign extremely high probability to all marbles being green after a suitable course of experience with green marbles. And, in general, the problem of justifying our inductive practices factors, according to Bayesians, into the problem of justifying the choice of prior, and the problem of justifying conditionalization; and they claim to have made good on both. So far, so good. However, non-Bayesians will find this a Pyrrhic probabilistic victory. For orthodox Bayesianism equally allows priors that would license counterinductive and grue-some inferences, based on the same evidence.

But Bayesianism is a theme that admits of many variations.

Unorthodox Bayesianism

Each of B1–B3 has its opponents. It will prove convenient to revisit them in reverse order.

The suspicion just raised is that orthodox Bayesianism is too permissive: it imposes no constraints on the assignment of priors, besides their conformity to the probability calculus. Rationality, the objection goes, is not so ecumenical. A standard defence – see, for example Savage (1954) or Howson and Urbach (1993) – appeals to famous "convergence-to-truth," and "merger-of-opinion" results. Roughly, their content is that, in the long run, the effect of choosing one prior rather than another is attenuated: successive conditionalizations on the evidence will, with probability one, make a given agent eventually converge to the truth, and thus initially discrepant agents eventually come to agreement. In an important sense, at least this much inductive logic is implicit in the probability calculus itself.

Unfortunately, these theorems tell us nothing about how quickly the convergence occurs. In particular, they do not explain the unanimity that we in fact often reach, and often rather rapidly. We will apparently reach the truth "in the long run"; but, as Keynes quipped, "in the long run, we shall all be dead."

Against B3, then, there are more stringent Bayesians who hold this truth to be self-evident: Not all priors are created equal. They thus impose further constraints on priors.

One such constraint is that they be *regular*, or *strictly coherent*: if $P(X) = 1$, then $X = \Omega$ (X is necessary/a logical truth); see Shimony (1955). It is meant to guard against the sort of dogmatism that no course of learning by (Jeffrey) conditionalization could cure.

We might also want to recognize the role that certain objective facts, or that certain "expert" opinions, might have in constraining one's subjective probabilities. Call probability function Q an *expert function for P* if the following condition holds:

$$\text{for all } X, P(X|Q(X) = x) = x \tag{8.1}$$

For example, one might conform one's subjective probabilities to corresponding relative frequencies. With Q being the "relative frequency" function, (8.1) becomes a version of the so-called *principle of direct probability*. Or one might think that whatever objective chances might be, they are characterized by their role in conditionally constraining rational credence. With Q being the "objective chance" function, (8.1) becomes a version of a principle that, suitably finessed, becomes Lewis' (1980) *Principal Principle*. Or one might argue, as van Fraassen (1995) does, that epistemic integrity requires one ideally to regard one's future opinions as being trustworthy – perhaps because of their having arisen from a rational process of learning. With Q being one's probability function at some future time, (8.1) becomes a version of van Fraassen's *Reflection Principle*. Q could also encapsulate the opinions of simply *an expert* – a person whom one trusts, for whatever reason.

There have been various proposals for resuscitating symmetry constraints on priors, in the spirit of the classical and logical interpretations. More sophisticated versions of the principle of indifference have been explored by Jaynes (1968). The guiding idea is to maximize the probability function's *entropy*, which for an assignment of positive probabilities $p_1, \ldots, p_n$ to n possibilities equals $-\Sigma_i\, p_i\log(p_i)$.

A set of events (or sentences) is *exchangeable* with respect to a given probability function if every event has the same probability, every conjunction of two events has the same probability, every conjunction of three events has the same probability, and so on. See Skyrms (1994) for an excellent discussion of generalizations of exchangeability, and their use in formulating various Goodmanian theses about projectability. Indeed, commonsense often (but not invariably) seems to require one's probabilities to be exchangeable over "green"-like hypotheses, but not "grue"-like hypotheses.

So there are many motivations for rejecting B3. But the suspicion at the end of the last section may still remain: Bayesianism, even with various of these bells and whistles added, is still too permissive. What is wanted is a justification of the "good" inductive inferences, *and no parallel justification of the "bad" ones*. These principles do not seem to distinguish the good from the bad. Some of them, on the contrary, only seem to nurture the bad inferences – for example, where we might have hoped to kill off grue-like hypotheses, regularity keeps them all alive. Other principles can play both sides with equal ease: exchangeability, for instance, is characterized purely syntactically, so it can be deployed to vindicate grue-like inferences as well as green-like inferences. Still others, such as the Reflection Principle, seem to be neutral with respect to issues of induction.

B2 also has its opponents. Some authors allow, and even insist upon, other rules for the updating of probabilities besides conditionalization. Jaynes advocates revision to the probability function that maximizes entropy, subject to the relevant constraints.

And some Bayesians drop the requirement that rational probability updating be rule-governed altogether; see van Fraassen (1990a) and Earman (1992). Note, however, that in a sense this only makes the problem of induction *worse*. Given that the suggested constraints on the priors do not solve the problem, one might have hoped that the updating rule could take up the slack (and according to the proponents of the convergence results mentioned above, in the long run it does). But if the very requirement of an updating rule is abandoned, then it begins to look as if anything goes: if you want suddenly to jump to a probability distribution that assigns overwhelming probability to all marbles being grue, then you are apparently beyond reproach.

The rejection of B1 is a large topic, and it motivates and can be motivated by some of the non-Kolmogorovian theories of probability, to which we now turn.

Non-Kolmogorovian Theories of Probability

A number of authors would abandon the search for an adequate interpretation of Kolmogorov's probability calculus, since they abandon some part of his axiomatization.

Some authors question its set-theoretic underpinnings. Note that the usual justifications of the probability axioms – Dutch Book arguments and so on – take for granted the sigma-field substructure, rather than justifying it as well. Fine (1973) argues that the requirement that the domain of the probability function be a sigma-field is overly restrictive. Some dispute the requirement that probabilities have numerical values. Fine sympathetically canvasses various theories of comparative probability, exemplified by statements of the form "*A* is at least as probable as *B*." Then there are advocates of *indeterminate* or of *vague* probabilities, who represent probabilities not as single numbers, but as intervals, or more generally sets of numbers, e.g. Levi (1980), Jeffrey (1983) and van Fraassen (1990b). Such vagueness might be indicated by a set of constraints that go beyond those of the probability calculus, but that fall short of the Carnapian ideal of fixing a unique probability function.

Some dispute the usual constraints on the numerical values. Kolmogorov's probability functions are real-valued. A number of philosophers – e.g. Lewis (1980) and Skyrms (1980) – allow probabilities to take values from the real numbers of a *nonstandard model* of analysis; see Robinson (1966) or Skyrms (1980) for the construction of such a model. In particular, they allow probabilities to be *infinitesimal*: positive, but smaller than every positive (standard) real number. This can be motivated by a desire to respect both regularity and certain symmetries in infinite probability spaces. Meanwhile, physicists such as Dirac, Wigner, and Feynman have countenanced *negative* probabilities, and Feynman and Cox have flirted with *complex-valued* probabilities; see Mückenheim (1986) for references. Renyi (1970) allows probabilities to attain the "value" ∞. We may also want to allow logical/necessary truths to be assigned probability less than one, perhaps to account for the fact that mathematical conjectures may be confirmed to varying degrees; see, for example Polya (1968). Thus, mathematics too might be susceptible to induction (to be distinguished from "mathematical induction," a *deductive* argument form!).

Kolmogorov's most controversial axiom is undoubtedly *continuity* – that is, the "infinite part" of countable additivity. He regarded it as an idealization that finessed the mathematics, but that had no empirical meaning. As we have seen, according to the classical, frequency, and certain propensity interpretations, probabilities violate countable additivity. De Finetti marshals a battery of arguments against it (in the name of subjectivism, but his arguments may be regarded as more general).

Various *non-additive* theories of probability that give up even finite additivity have been proposed – for example, Dempster–Shafer theory, which some regard as codifying the notion of "weight of evidence" (Shafer, 1976). So-called "Baconian probabilities" represent another non-additive departure from the probability calculus. The Baconian probability of a conjunction is equal to the minimum of the probabilities of the conjuncts. L. J. Cohen (1970, 1977) regards them as appropriate for measuring inductive support. See Ghirardato (1993) for a survey of non-additive measures of uncertainty, and Howson (1995) for further references.

Lastly, various authors, rather than axiomatizing unconditional probability and defining conditional probability therefrom, take conditional probability as primitive and axiomatize it directly; see Spohn (1986).

Some Future Avenues of Research

Having discussed various landmarks of past work in induction and probability, we find ourselves now in the curiously reflexive position of predicting what future work in these areas will look like. Suitably cautioned by the very nature of our subject, and with appropriate degrees of uncertainty, here are some of our best bets.

We think that there is still much research to be done within a broadly Bayesian framework. There are already signs of the rehabilitation of logical probability, and, in particular, the principle of indifference, by authors such as Stove (1986), Festa (1993), Maher (2000, 2001), and Bartha and Johns (2001). This will surely resonate with developments in the theory of infinitesimals, for example within the system of "surreal numbers" (Conway, 1976; Ehrlich, 2001). Relevant here will also be advances in information theory, randomness and complexity theory (Fine, 1973; Li and Vitanyi, 1997), and approaches to statistical model selection, and in particular the "curve-fitting" problem that attempt to codify simplicity – e.g. the Akaike Information Criterion (Forster and Sober, 1994), the Bayesian Information Criterion (Kieseppä, 2001), Minimum Description Length theory (Rissanen, 1999) and Minimum Message Length theory (Wallace and Dowe, 1999). These may also shed light on the time-honored but all-too-nebulous intuition that "green"-like hypotheses are somehow "simpler" than "grue"-like hypotheses.

Probability theory traditionally presupposes classical set theory/classical logic. There is more work to be done on "non-classical" probability theory. Bayesians may want to enrich their theory of induction to encompass logical/mathematical learning in response to the so-called "problem of old evidence" (Zynda, 1995), and to allow for the formulation of new concepts and theories. We also see fertile connections between probability and logic that have been explored under the rubric of "probabilistic semantics" or "probability logic" – see Hailperin (1996) and Adams (1998). Roeper and Leblanc (1999) develop such probabilistic semantics for primitive conditional probability functions. More generally, we envisage increased attention to the theory of such functions; see, for instance, Festa (1999) for a treatment of Bayesian confirmation theory which takes such functions as primitive, and Hájek 2001 for general arguments in favor of such functions). Further criteria of adequacy for subjective probabilities will be developed – perhaps refinements of "scoring rules" (Winkler, 1996), and more generally, candidates for playing a role for subjective probability analogous to the role that truth plays for belief. There will be more research on the theory of expert functions – for example, in the aggregation of opinions and preferences of multiple experts. This problem is well known to aficionados of the *risk-assessment* literature, which has yet to be mined by philosophers (Kaplan, 1992).

We expect that non-Bayesian research programs will also flourish. Non-additive probabilities are getting impetus from considerations of "ambiguity aversion" (Ghirardato, 2001) and "plausibility theory" (Hild, 2001). Formal learning theory (Kelly, 1996) is also gaining support, and more broadly, philosophers will find much interesting work on induction and learning in the computer science and artificial intelligence literature. And there is a need for more cross-fertilization between Bayesianism and classical statistics, and its recent incarnation in the theory of error statistics (Mayo, 1996). For example, hypothesis testing at a constant significance level has long been known to be inconsistent with Bayesian inference and decision theory. Recent work by Schervish et al. (2002) shows that such "incoherence" is a matter of degree. Moreover, in light of work in the economics literature on "bounded rationality," the study of degrees of incoherence is likely to bear fruit. We foresee related attempts to "humanize" Bayesianism – for example, the further study of vague probability and vague decision theory. And classical statistics, for its part, with its tacit tradeoffs between errors and benefits of different kinds, needs to be properly integrated into a more general theory of decision.

Decision theory and the theory of induction will profit from insights in the causal modeling literature. For example, the so-called "reference class problem" arises because a given event-token can typically be placed under indefinitely many event-types; this is what gives the various problems of induction their teeth. But progress can be made when the relevant causes are identified, and techniques along the lines of those developed by Pearl (2000) and Spirtes et al. (1993) can be appealed to. These techniques will doubtless be finessed. More generally, in this brave new world of inter-disciplinarity and rapid communication, inferential methods developed within one field are increasingly likely to be embraced by practitioners of another.[2]

Notes

1 Space prevents us from giving the details, but the key idea is to build into one's inductive principles a version of van Fraassen's Reflection Principle, which we will discuss below.

2 We thank especially Branden Fitelson, Matthias Hild, Chris Hitchcock, Jim Joyce, and Tim Maudlin for helpful comments.

References

Adams, E. (1998): *Probability Logic.* Stanford University: CSLI.

Armendt, B. (1980): "Is There a Dutch Book Argument for Probability Kinematics?" *Philosophy of Science*, 47, 583–9.

Arnauld, A. (1662): *Logic, or, The Art of Thinking ("The Port Royal Logic")*, tr. J. Dickoff and P. James, Indianapolis: Bobbs-Merrill, 1964.

Bartha, P. and Johns, R. (2001): "Probability and Symmetry," Philosophy of Science, Supplemental volume, forthcoming. Available at http://hypatia.ss.uci.edu/lps/psa2k/bartha.pdf.

Bertrand, J. (1889): *Calcul des Probabilités, 1st edn.* Paris: Gauthier-Villars. Reprinted as 3rd edition, New York: Chelsea Publishing Company, 1972.

Carnap, R. (1950): *Logical Foundations of Probability.* Chicago: University of Chicago Press.

Carnap, R. (1963): "Replies and Systematic Expositions," in P. A. Schilpp (ed.), *The Philosophy of Rudolf Carnap*, La Salle, Ill: Open Court, 966–98.

Church, A. (1940): "On the Concept of a Random Sequence," *Bulletin of the American Mathematical Society*, 46, 130–5.

Cohen, L. J. (1970): *The Implications of Induction.* London: Methuen.

Cohen, L. J. (1977): *The Probable and the Provable.* Clarendon Press: Oxford.

Conway, J. (1976): *On Numbers and Games.* London: Academic Press.

De Finetti, B. (1937): "Foresight: Its Logical Laws, Its Subjective Sources," translated in Kyburg and Smokler (1964, pp. 53–118).

De Finetti, B. (1974): *Theory of Probability.* New York: John Wiley. Reprinted 1990.

Diaconis, P. and Zabell, S. (1986): "Some Alternatives to Bayes's Rule", in B. Grofman and G. Owen (eds.), *Information Pooling and Group Decision Making, Proceedings of the Second University of California, Irwine, Conference on Political Economy*, Greenwich, CT: Jai Press Inc, 25–38.

Earman, J. (1992): *Bayes or Bust? A Critical Examination of Bayesian Confirmation Theory.* Cambridge, MA: MIT Press.

Ehrlich, P. (2001): "Number Systems with Simplicity Hierarchies: A Generalization of Conway's Theory of Surreal Numbers," *The Journal of Symbolic Logic*, 66(3), September.

Festa, R. (1993): *Optimum Inductive Methods : A Study in Inductive Probability, Bayesian Statistics, and Verisimilitude.* Dordrecht: Kluwer Academic Publishers, Synthese Library Series, vol. 232.

Festa, R. (1999): "Bayesian Confirmation," in M. C. Galavotti and A. Pagnini (eds.), *Experience, Reality, and Scientific Explanation*, Dordrecht: Kluwer, 55–87.

Fine, T. (1973): *Theories of Probability.* New York: Academic Press.

Forster, M. and Sober, E. (1994): "How to Tell when Simpler, More Unified, or Less Ad Hoc Theories will Provide More Accurate Predictions," *British Journal for the Philosophy of Science*, 45, 1–35.

Ghirardato, P. (1993): "Non-additive Measures of Uncertainty: A Survey of Some Recent Developments in Decision Theory," *Rivista Internazionale di Sciencze Economiche e Commerciali*, 40, 253–76.

Ghirardato, P. (2001): "Coping with Ignorance: Unforeseen Contingencies and Non-Additive Uncertainty," *Economic Theory*, 17, 247–76.

Giere, R. N. (1973): "Objective Single-Case Probabilities and the Foundations of

Statistics", in P. Suppes, L. Henkin, G. C. Moisil and A. Joja (eds.), *Logic, Methodology and Philosophy of Science, IV*, Amsterdam: North Holland, 467–83.

Good, I. J. (1985): "Weight of Evidence: A Brief Survey," in J. Bernardo, M. DeGroot, D. Lindley and A. Smith (eds.), *Bayesian Statistics 2*, Amsterdam: North Holland, 249–69.

Goodman, N. (1983): *Fact, Fiction, and Forecast, 4th edn.* Cambridge, MA: Harvard University Press.

Hájek, A. (2001): "What Conditional Probability Could Not Be," *Synthese*, forthcoming.

Hailperin, T. (1996): *Sentential Probability Logic.* Bethlehem, PA: Lehigh University Press.

Hempel, C. G. (1945a): "Studies in the Logic of Confirmation, Part I," *Mind*, 54, 1–26.

Hempel, C. G. (1945b): "Studies in the Logic of Confirmation, Part II," *Mind*, 54, 97–121.

Hild, M. (2001): "Non-Bayesian Rationality," available at www.hild.org.

Howson, C. (1995): "Theories of Probability," *British Journal of Philosophy of Science*, 46, 1–32.

Howson, C. and Urbach, P. (1993): *Scientific Reasoning: The Bayesian Approach, 2nd edn.* Chicago: Open Court.

Hume, D. (1739): *A Treatise of Human Nature*, L. A. Selby-Bigge (ed.), Oxford: Clarendon Press. 2nd edn, 1985.

Jaynes, E. T. (1968): "Prior Probabilities," *Institute of Electrical and Electronic Engineers Transactions on Systems Science and Cybernetics*, SSC-4, 227–41.

Jeffrey, R. (1965): *The Logic of Decision.* Chicago: University of Chicago. 2nd edn, 1983.

Jeffrey, R. (1983): "Bayesianism With a Human Face," in J. Earman (ed.), *Testing Scientific Theories, Minnesota Studies in the Philosophy of Science, vol. X*, Minneapolis: University of Minnesota Press, 133–56. Reprinted in Jeffrey (1992, pp. 77–107).

Jeffrey, R. (1992): *Probability and the Art of Judgment*, Cambridge Studies in Probability, Induction, and Decision Theory, Cambridge: Cambridge University Press.

Jeffreys, H. (1939): *Theory of Probability.* Oxford: Oxford University Press. Reprinted in Oxford Classics Texts in the Physical Sciences Series, 1998.

Johnson, W. E. (1932): "Probability: The Deductive and Inductive Problems," *Mind*, 49, 409–23.

Kahneman, D., Slovic, P. and Tversky, A. (eds.) (1982): *Judgment Under Uncertainty: Heuristics and Biases.* Cambridge University Press, Cambridge.

Kaplan, S. (1992): " 'Expert Information' Versus 'Expert Opinions.' Another Approach to the Problem of Eliciting/Combining/Using Expert Knowledge in PRA," *Reliability Engineering and System Safety*, 35, 61–72.

Kelly, K. (1996): *The Logic of Reliable Inquiry.* Oxford: Oxford University Press, forthcoming.

Kemeny, J. G. (1955): "Fair Bets and Inductive Probabilities," *Journal of Symbolic Logic*, 20, 263–73.

Keynes, J. M. (1921): *Treatise on Probability.* London: Macmillan. Reprinted 1962, Harper and Row, New York.

Kieseppä, I. A. (2001): "Statistical Model Selection Criteria and Bayesianism," *Philosophy of Science*, Supplement, forthcoming.

Kolmogorov, A. N. (1933): *Grundbegriffe der Wahrscheinlichkeitrechnung.* Ergebnisse Der Mathematik. Translated as *Foundations of Probability.* New York: Chelsea Publishing Company, 1950.

Kornblith, H. (1985): *Naturalizing Epistemology.* Cambridge, MA: MIT Press.

Kyburg, H. E., Jr. and Smokler, H. E. (eds.) (1964): *Studies in Subjective Probability*. New York: John Wiley and Sons. 2nd edn, 1980.

Laplace, P. S. de (1814): *Essai Philosophique sur les Probabilités*. Paris. Translated into English as *A Philosophical Essay on Probabilities*. New York: Dover, 1951.

Levi, I. (1980): *The Enterprise of Knowledge*. Cambridge, Massachussetts: MIT Press.

Lewis, D. (1980): "A Subjectivist's Guide to Objective Chance," in R. C. Jeffrey (ed.), *Studies in Inductive Logic and Probability, vol II*, Berkeley CA.: University of California Press, 263–93. Reprinted in Lewis (1986).

Lewis, D. (1986): *Philosophical Papers, vol. II*, Oxford: Oxford University Press.

Lewis, D. (1998): *Papers in Philosophical Logic*. Cambridge: Cambridge University Press.

Ming Li and Vitanyi, P. (1997): *An Introduction to Kolmogorov Complexity and its Applications, 2nd edn*. New York: Springer-Verlag.

Maher, P. (2000): "Probabilities for Two Properties," *Erkenntnis*, 52, 63–91.

Maher, P. (2001): "Probabilities for Multiple Properties: The Models of Hesse and Carnap and Kemeny," *Erkenntnis*, forthcoming.

Mayo, D. G. (1996): *Error and the Growth of Experimental Knowledge*. Chicago: University of Chicago Press.

Mellor, D. H. (1971): *The Matter of Chance*. Cambridge: Cambridge University Press.

Milne, P. (1996): "Log[P(h/eb)/P(h/b)] is the one true measure of confirmation," *Philosophy of Science*, 63, 21–6.

Mückenheim, W. (1986): "A Review of Extended Probabilities," *Physics Reports*, 133(6), 337–401.

Pearl, J. (2000): *Causality*. Cambridge: Cambridge University Press.

Polya, G. (1968): *Patterns of Plausible Inference, 2nd edn*. Princeton: Princeton University Press.

Popper, K. R. (1959): "The Propensity Interpretation of Probability," *British Journal for the Philosophy of Science*, 10, 25–42.

Popper, K. R. (1968): *The Logic of Scientific Discovery* revised edn. London: Hutchinson & Co.

Putnam, H. (1974): "The 'Corroboration' of Theories," reprinted in his *Mathematics, Matter, and Method*, Cambridge: Cambridge University Press, 1979, 250–69.

Ramsey, F. P. (1926): "Truth and Probability," in R. B. Braithwaite (ed.), *Foundations of Mathematics and other Essays*. London: Routledge & Kegan Paul, 1960. Reprinted in Kyburg and Smokler (1964, pp. 61–92), and in D. H. Mellor (ed.), *F. P. Ramsey: Philosophical Papers*, Cambridge: Cambridge University Press, 1990.

Reichenbach, H. (1938): *Experience and Prediction*. Chicago: University of Chicago Press.

Reichenbach, H. (1949): *The Theory of Probability*. Berkeley: University of California Press.

Renyi, A. (1970): *Foundations of Probability*. San Francisco: Holden-Day, Inc.

Rissanen, J. (1999): "Hypothesis Selection and Testing by the MDL Principle," *Computer Journal*, 42(4), 260–9.

Robinson, A. (1966): *Non-Standard Analysis*. Amsterdam: North Holland Publishing Company.

Roeper, P. and Leblanc, H. (1999): *Probability Theory and Probability Logic*, Toronto Studies in Philosophy, Toronto: University of Toronto Press.

Salmon, W. (1967): *The Foundations of Scientific Inference*. Pittsburgh, PA: University of Pittsburgh Press.

Savage, L. J. (1954): *The Foundations of Statistics*. New York: John Wiley.

Schervish, M. J., Seidenfeld, T. and Kadane, J. B. (2002): "How Incoherent is Fixed-Level Testing?" *PSA 2000*, forthcoming. Also available at http://hypatia.ss.uci.edu/lps/psa2k/incoherent-fixed-level.pdf.

Shafer, G. (1976): *A Mathematical Theory of Evidence*. Princeton: Princeton University Press.

Shimony, A. (1955): "Coherence and the Axioms of Confirmation," *Journal of Symbolic Logic*, 20, 1–28.

Skyrms, B. (1980): *Causal Necessity*. New Haven, Conn.: Yale University Press.

Skyrms, B. (1987): "Dynamic Coherence and Probability Kinematics," *Philosophy of Science*, 54, 1–20.

Skyrms, B. (1994): "Bayesian Projectibility," in D. Stalker (ed.), *Grue!*, Chicago: Open Court, 241–62.

Spirtes, P., Glymour, C. and Scheines, R. (1993): *Causation, Prediction, and Search*. New York: Springer-Verlag.

Spohn, W. (1986): "The Representation of Popper Measures," *Topoi*, 5, 69–74.

Stove, D. C. (1986): *The Rationality of Induction*. Oxford: Oxford University Press.

Strawson, P. (1952): *Introduction to Logical Theory*. New York: John Wiley & Sons.

van Cleve, J. (1984): "Reliability, Justification, and the Problem of Induction," *Midwest Studies in Philosophy*, 9, 555–68.

van Fraassen, B. (1990a): "Rationality Does Not Require Conditionalization," in E. Ullman-Margalit (ed.), *Science in Reflection: The Israel Colloquium: Studies in History, Philosophy and Sociology of Science*, Dordrecht: Kluwer.

van Fraassen, B. (1990b): "Figures in a Probability Landscape," in J. M. Dunn and A. Gupta (eds.), *Truth or Consequences*, Dordrecht: Kluwer, 345–56.

van Fraassen, B. (1995): "Belief and the Problem of Ulysses and the Sirens," *Philosophical Studies*, 77, 7–37.

Venn, J. (1876): *The Logic of Chance, 2nd edn*. London: Macmillan and Co.

von Mises, R. (1957): *Probability, Statistics and Truth, 2nd English edn*. New York: Macmillan.

Wallace, C. S. and Dowe, D. L. (1999): "Minimum Message Length and Kolmogorov Complexity," *Computer Journal* (special issue on Kolmogorov complexity), 42(4), 270–83.

Winkler, R. L. (1996): "Scoring Rules and the Evaluation of Probabilities," *Test*, 5(1), 1–60.

Zynda, L. (1995): "Old Evidence and New Theories," *Philosophical Studies*, 77, 67–95.

Chapter 9

Philosophy of Space–Time Physics

Craig Callender and Carl Hoefer

Philosophy of space–time physics, as opposed to the more general philosophy of space and time, is the philosophical investigation of special and general relativity. Relativity theory stimulated immediate and deep philosophical analysis, both because of its novel implications for the nature of space, time and matter, and because of more general questions philosophers have about the nature of its claims. With nearly one hundred years of sustained research to draw on, this chapter cannot hope to survey all the topics that have arisen, even all the major ones. Instead, we concentrate on four topics, two with a historical and philosophical pedigree, namely, relationalism and conventionalism, and two that arise in general relativity and cosmology, namely, singularities and the so-called horizon problem. This selection should give the reader a representative taste of the field as it stands today.

Many fascinating topics, however, will not be covered. Notable examples are the topics of time travel, presentism, supertasks, and the Lorentz interpretation of relativity. For up-to-date references and discussions of these topics, the reader can turn to, respectively, Arntzenius and Maudlin (2000), Savitt (2000), Earman and Norton (1996), and Brown and Pooley (2001).

Relationism, Substantivalism and Space–time

Perhaps the most fundamental question one has about space–time is: what *is* it, really? At one level, the answer is simple; at a deeper level, the answer is complex and the continuing subject of philosophers and physicists' struggle to obtain a plausible and intelligible understanding of space–time. In large measure, this struggle can be seen as a continuation of the classical dispute, sparked by the famous Leibniz–Clarke correspondence, between *relational* and *absolutist* conceptions of space – though the terms of the debate have turned and twisted dramatically in the twentieth century.

History

The general theory of relativity's (GTR) simple answer to our question is that space–time is

(a) a four-dimensional differentiable manifold M
(b) endowed with a semi-Riemannian metric g of signature (1,3)
(c) in which all events and material things (represented by stress-energy T) are located, and
(d) in which g and T satisfy Einstein's field equations (EFE).

Had gravitational physics and scientific cosmology begun with Einstein's theory rather than Newton's, this simple answer might seem perfectly natural. Attempting to obtain a deeper understanding of the theory, philosophers struggle to understand GTR's space–time in terms of ideas found in previous theories, ideas whose roots lay in experience, metaphysics and Modern philosophy and physics. The questions that arise from these grounds *seem* to make good sense, independent of their roots: Is space–time a kind of *thing* which, though different from material things and energy-forms, is in some sense just as substantial and real? Do Einstein's equations describe a sort of causal interaction between space–time and matter; or is the relation one of reductive subsumption (and if so, which way)? Can space–time exist without any matter at all? Is motion purely relational in GTR, that is, always analyzable as a change in the relative configuration of bodies, or is it absolute, that is, always defined relative to some absolute structure? Or is it partly relative and partly absolute?

None of these questions receives a clear-cut answer from GTR, which is why absolutely inclined and relationally inclined thinkers can each find grist for their ontological mills in the theory. The complexity and ambiguity of the situation leads some philosophers to argue that it is pointless to try to impose the categories of seventeenth century metaphysics on a theory that has outgrown them (Rynasiewicz, 1996, 2000). Below, we briefly survey some of the key features of GTR that intrigue and frustrate philosophical interpretation, and return at the end to the question of whether the old categories and questions still have value.

No prior geometry In all earlier theories of mechanics and/or gravitation that contained definite doctrines about the nature of space and time (or space–time), space and time were taken as "absolute" structures, fixed and unchanging. As Earman (1989) shows, even the views of the traditional relationist thinkers involved some significant prior geometric and/or temporal structures. The Euclidean structure of space, for example, was universally assumed, as well as some absoluteness of temporal structure.[1]

Not so General Relativity. The background arena in GTR is just M, which can have any of a huge variety of topologies, and whose only "absolute" features are 4-dimensionality and continuity. The rest of the spatio-temporal properties, geometric and inertial and temporal, are all encoded by g, which is not fixed or prior but rather variable under the EFE.[2] This looks extraordinarily promising from a relationist viewpoint: absolute space has finally been banished!

Or has it? Although absolute space or space–time, in the sense of a pre-defined and invariant background, is absent, it is not clear that this amounts to satisfaction of a relationist's desires. Motion has not become "purely relative" in any clear sense; rather, motion is defined relative to the metric, and the metric is by no means definable on the basis of relations between material things. In fact, the EFE turns matters the other way around: given the metric g, the motions of material things (encoded in T) are determined. If material processes affect the structure of spacetime, perhaps this is so much the better for a substantival view of GTR's space–times.

The differing roles of M and g correspond to two different strands of argument for substantivalism, which it will be useful to distinguish. The first strand notes that it is indispensable to the mathematical apparatus of GTR that it *start* from M and build the spatio-temporal structure g on it.[3] Then, invoking the Quinean doctrine that the real is that over which we ineliminably quantify in our best scientific theories, M is claimed to represent a real, existing manifold of space–time points in the world. The second strand looks at g itself, argues that it represents a real structure in the world not reducible to or derivable from material bodies and their relations, and concludes that we have a descendent of Newton's absolute space in GTR.

Manifold and Metric   Above we have indicated that M is the only "fixed background" in GTR, and that only in the sense of dimensionality and continuity, not global shape. But motions (particularly acceleration, but also velocity and position in some models) are defined by g. Which one, then, represents space–time itself? Or must we say it is a combination of both? These questions open a new can of interpretive worms.

A manifold is a collection of space–*time* points, not space points. In other words, the points do not have duration; each one is an ideal point-event, a representative of a spatial location at a single instant of time. They do not exist *over* time and hence serve as a structure against which motion may be defined, as Newton's space points did. If space–time substantivalism is understood as the claim that these points are substantial entities themselves, then the so-called *hole problem* arises (Earman and Norton, 1987). The general covariance of the EFE, interpreted in an active sense, allows one to take a given model $M_1 = <M, g, T>$ and construct a second *via* an automorphism on the manifold: $M_2 = <M, g^*, T^*>$ which also satisfies the EFE. Intuitively, think of M_2 as obtained from M_1 by sliding *both* the metric and matter fields around on the point-manifold (Figure 9.1).

If M_2 and M_1 agree or match for all events before a certain time t, but differ for some events afterward, then we have a form of indeterminism. Relative to our chosen substantial entities, space–time points considered as the elements of M, what happens at what space–time locations is radically undetermined. This can be

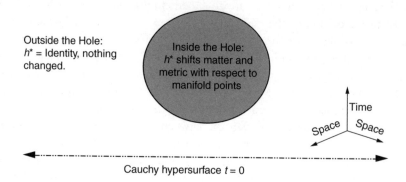

Outside the Hole:
h^* = Identity, nothing
changed.

Inside the Hole:
h^* shifts matter and
metric with respect to
manifold points

Time

Space Space

Cauchy hypersurface $t = 0$

$<M, h^*\mathbf{g}, h^*\mathbf{T}>$: identical to $<M, \mathbf{g}, \mathbf{T}>$ everywhere
outside the Hole (hence, at all times before $t=0$).

Figure 9.1 A hole diffeomorphism's effect

presented as an argument against the kind of substantivalism (*manifold* substantivalism) we started from.

Note however two points. First, this indeterminism is unobservable: M_1 and M_2 are qualitatively indistinguishable. Second, but relatedly, the hole problem assumes that the identities of the manifold points are given or specified, in some sense, independently of the material/observable processes occurring in spacetime (represented by g and T). In fact, one way of thinking of a hole automorphism is as a (continuous) *permutation* of the points underlying physical processes, or as a *re-labeling* of the points. Not surprisingly, most responses to the hole argument have departed from these points, arguing that substantivalism can be reinterpreted in ways that do not lead to the apparent indeterminism.

Metric and matter Derivation of the metric of space–time from a (somehow!) antecedently given specification of the relational distribution of matter is a characteristic *Machian* ambition. Mach's *Science of Mechanics* (1989), at least as Einstein read it, proposed that inertia should be considered an effect of relative acceleration of a body with respect to other bodies – most notably, the "fixed stars." Transplanting this idea to the context of GTR seems to indicate having the inertial structure (which g determines) determined by the relational matter distribution. GTR does not seem to fulfill this idea in general. In some models (notably Friedman–Robertson–Walker (FRW) Big Bang models), this idea seems intuitively fulfilled. But making the idea both precise and satisfiable in GTR has proven difficult if not impossible, despite the efforts of outstanding physicists such as Einstein, Sciama, Wheeler, and Dicke. And as we noted above, superficially at least the determination relation seems to go the other way (from metric to matter).

Despite these difficulties, a Machian program for extending (or restricting) GTR has appealed to many thinkers. In addition to anti-absolutist prejudices, there are a couple of reasons for this. First, GTR does yield some non-Newtonian inertial effects of the kind Mach speculated on: the so-called "frame-dragging" effects. Second, it is difficult to take as a mere coincidence the fact that the FRW models, which seem most Machian intuitively, are also those that seem to best describe our cosmos.

And there are difficulties with taking the metric "field" as a substantial entity that either subsumes ordinary matter or is in causal interaction with it. The former idea, which can be thought of as "super-substantivalism", would involve extending our notion of the metric in an attempt to derive fine-grained properties of matter in terms of fine-grained perturbations (knots, singularities?) of the former. Attempts by Einstein and others along these lines have not led to notable successes.

The less ambitious idea of the metric and (ordinary) matter as ontological peers in mutual interaction faces challenges too. The EFE give a regularity, but in order to view the regularity causally, we would ideally like to be able to quantify the strength of the interaction, in terms of energy or momentum exchange. But the metric field's energy, if it exists at all, is poorly understood and very different from ordinary energy (Hoefer, 2000). Attempts through the end of the twentieth century to detect the most intuitively causal-looking interactions – absorption of gravitational wave energy – were uniformly negative.

Current work

The relationism/substantivalism issue was dominated through the late 1980s and 1990s by responses to the Earman and Norton hole argument. The argument has pushed philosophers who have more sympathy with substantivalist views than relationist views to make more precise their ontological claims (Maudlin, 1990; Butterfield, 1989; Stachel, 1993). Depending on whether they view substantivalism as primarily attractive due to the Quinean indispensability argument or rather the metric-based considerations, philosophers re-work their views in different ways.

The hole argument inspired those with relationist leanings to revive the idea, advocated by Reichenbach earlier in the twentieth century but effectively killed by Earman (1970) and Friedman (1983), that GTR can be interpreted as fully compatible with relationism. Teller (1991), Belot (1999) and Huggett (1999) are examples of this approach. What makes this position possible is

(a) focusing on the point-manifold-indispensability argument for substantivalism primarily, and
(b) taking a liberal attitude toward the idea of *relations* between material things.

If the manifold is taken as only representing the continuity, dimensionality and topology of space–time (as some substantivalists would agree anyway), then what's really indispensable is the metric. Can it be interpreted relationally? The philosophers who argue that it can are *not* claiming a Machian reduction of metrical structure to material relations. Instead, they claim that the metric itself can be interpreted as giving the structure of actual *and possible* spatiotemporal relations between material things. g is not a thing or substance. Where matter is present, it is crucial to the definition of local standards of acceleration and non-acceleration; the EFE record just this relationship. In many ways, the desires of traditional relationists (especially Leibniz, Huygens and Mach) are – arguably – met by GTR when interpreted this way.

Current work has served to clarify the various types of substantivalist view that may be brought forth, and the strengths and weaknesses of them. To a lesser extent, relationist alternatives have also been clarified. Others feel unhappy about both alternatives, and their reasons stem from a conviction that the ontological categories of absolutism, substantivalism and relationism have no clear meanings in GTR and have thus outlived what usefulness they ever had.

Rynasiewicz has published two provocative papers on this topic. His 1996 paper argues that the categories of absolute and relational simply do not apply in GTR, so it is a mug's game trying to see which one "wins" in that theory. Tracing relational and absolutist ideas from Descartes through to Einstein and Lorentz, the core of his case is that the metric field of GTR is a bit like a Cartesian subtle matter and a bit like Newton's absolute space, but in the end not enough like either (though it is a lot like *ether*). In his 2000 paper, he does a similar historical/conceptual analysis of the notions of absolute and relative motion, concluding that the notions are impossible to define in GTR. While it is possible to mount counterarguments in defense of the traditional notions (Hoefer, 1998), it is impossible to deny that GTR is an awkward theory to comprehend using traditional concepts of space and time.

Robert Disalle (1994, pp. 278–9) argues along similar lines. He offers a positive way to understand space–time after we have freed ourselves from the outmoded categories. In his 1995, he argues that a chief mistake of the tradition is thinking of space–time structure as an entity that we postulate to *causally explain* phenomena of motion. It can't do the job of explaining motions because it is simply an expression of the *facts* about those motions – when certain coordinating definitions are chosen to relate spatio-temporal concepts with physical measurements and processes. The point is nicely made by analogy. When pre-nineteenth century thinkers asserted the Euclidean nature of space, they were claiming that observations of length, angle and distance will always conform to the rules of Euclid's geometry. But saying space is Euclidean is not giving a *causal explanation* of rulers and compasses behaving as they do, and it is not the postulation of a new, substantial "thing" in which rulers etc. are embedded. Nor is it, however, a claim that all spatial facts are reducible to observables or measurement outcomes. One can, says Disalle, be a realist about space (or space–time)'s structure, without making the mistake of inappropriate reification.

Future work

The notions of relational and substantival spacetime may have reached a sort of impasse when it comes to the interpretation of GTR's overall structure, as presented in entry-level textbooks. This hardly means that we have an adequate understanding of space–time's ontology, a comfortable resting place for philosophical curiosity. A search of the abstracts of recent work in the foundations of GTR and quantum gravity will show numerous occurrences of words like *relational* and *absolute, Leibniz* and *Mach*. This is because philosophers and physicists alike still want to deepen their understanding of the world's ontology. There is still important work that can be done on classical GTR. For example, what is the status of energy conservation laws? Does matter–energy really get exchanged between ordinary matter and empty space–time? How might relationalists understand parity nonconservation? Are there Machian replacements for or restrictions of GTR that are observationally equivalent over the standard range of current tests? (See Barbour and Bertotti (1982) and Barbour (1999).)

Conventionalism about Space–time

Some of the most basic principles of science (and perhaps mathematics) seem to be true as a matter of definitional choice. They are not *quite* purely analytic or trivial; they can not be demonstrated true simply on the basis of prior stipulative definitions and logical rules. Further, incompatible-looking alternative principles are conceivable, even though we may not be able to see how a useful framework could be built on them. Such principles are often held to be true by *convention*.

One example in mathematics is the famous parallel postulate of Euclidean geometry. Physical examples are less common and typically fraught with controversy. Perhaps Newton's famous 2nd law, $\mathbf{F} = m\mathbf{a}$, is an example. This may be thought a poor choice, for surely, as the center of his mechanics, the 2nd law is far from true by definition. But in the Newtonian paradigm, the 2nd law served as ultimate arbiter of the questions

(a) whether a force existed on a given object; and
(b) if so, what its magnitude was.

Any failure of the $\mathbf{a}$ of an object to conform to expectation was grounds for assuming that an unknown or unexpected force was at work, not grounds for questioning the 2nd law.

Of course, there is no guarantee that one can *always* maintain any putative conventional truth, come what may. Rather, one can usually imagine (or experimentally find, as in the example at hand) circumstances in which unbearable tensions arise in our conceptual frameworks from the insistence on retention of the conventional principle, and one is effectively forced to give it up. (Duhem (1954) gives a classic discussion of these matters.) If this is right, then the original claim of conventionality looks like something of an exaggeration. Are there in fact *any* choices in the creation of adequate physical theory that are genuinely free, conventional choices (as, e.g. choice of units is), without being completely trivial (as, again, choice of units is)? Many philosophers have thought that space–time structures give us true examples of such conventionality.

History

Before the eighteenth century all philosophers of nature assumed the Euclidean structure of space; it was thought that Euclid's axioms were true *a priori*. The work of Lobachevsky, Riemann and Gauss destroyed this belief; they demonstrated, first, that consistent non-Euclidean constant-curvature geometries were possible, and later that even variably curved space was possible. It was also apparent that our experience of the world could not rule out these new geometries, at least in the large. But what, exactly, does it mean to say that space is Euclidean or Riemannian? A naïve-realist interpretation can, of course, be given: there exists a *thing*, space, it has an intrinsic structure, and that structure conforms to Euclid's axioms. But some philosophers – especially empiricists such as Reichenbach – worried about how space is related to observable properties. These philosophers realized that our physical theories always contain assumptions or postulates that *coordinate* physical phenomena with spatial and temporal structures. Light rays in empty space travel in straight lines, for example; rigid bodies moved through empty space over a closed path have the same true length afterward as before; and so on. So-called *axioms of coordination* are needed to give meaning and testability to claims about the geometry of space.

The need for axioms of coordination seems to make space for conventionalism. For suppose that, under our old axioms of coordination, evidence starts to accumulate that points toward a non-Euclidean space (triangles made by light rays having angles summing to less than 180°, for example). We could change our view of the geometry of space; but equally well, say conventionalists, we could change the axioms of coordination. By eliminating the postulate that light rays in empty space travel in straight lines (perhaps positing some "universal force" that affects such rays), we could continue to hold that the structure of space itself is Euclidean. According to the strongest sorts of conventionalism, this preservation of a conventionally chosen geometry can *always* be done, come what may. Poincaré (1952) defended the conventionality of Euclidean geometry; but he also made an empirical conjecture, now regarded as false: that it would always be *simpler* to construct mechanics on assumption of Euclidean geometry.

Discussions of conventionalism took a dramatic turn because of the work of Einstein. With its variably curved space–time, GTR posed new challenges and opportunities for both sides on the conventionality of geometry. Cassirer, Schlick, Reichenbach, and Grünbaum are some notable figures of twentieth century philosophy who argued for the conventionality of space–time's geometry in the context of GTR. Recent scholars have tended to be skeptical that any non-trivial conventionalist thesis is tenable in GTR; Friedman (1983) and Nerlich (1994) are prominent examples here.

But it was in 1905, rather than 1915, that Einstein gave the greatest wind to conventionalist's sails. In the astounding first few pages of the paper that introduced the special theory of relativity (STR), Einstein overthrew the Newtonian view of space–time structure, and in passing, noted that *part* of the structure with which he intended to replace it had to be chosen by convention. That part was *simultaneity*. Einstein investigated the operational significance of a claim that two events at different locations happen simultaneously, and discovered that it must be defined in terms of some clock synchronization procedure. The obvious choice for such a procedure was to use light-signals: send a signal at event A from observer 1, have it be received and reflected back by observer 2 (at rest relative to 1) at event B, and then received by 1 again at event C. The event B is then simultaneous with an event E, temporally mid-way between A and C (Figure 9.2).

Or is it? To suppose so is to assume that the velocity of light on the trip from A to B is the same as its velocity from B to C (or, more generally, that light has

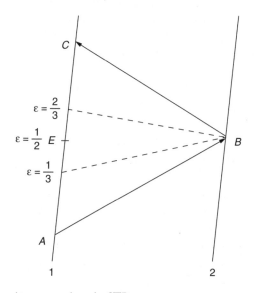

Figure 9.2   Simultaneity conventions in STR

the same velocity in a given frame in all directions). This seems like a very good thing to assume. But can it be verified? Einstein thought not. All ways of directly measuring the one-way velocity of light seemed to require *first* having synchronized clocks at separated locations. But if this is right, we are going in circles: we need to know light's one-way velocity to properly synchronize distant clocks, but to know that velocity, we need antecedently synchronized clocks.

To break the circle, Einstein thought we needed simply to adopt a conventional choice: we *decide* that event E is simultaneous with B (i.e. that light's velocity is uniform and direction-independent). Other choices are clearly possible, at least for the purposes of developing the dynamics and kinematics of STR. Following Reichenbach, these are synchronizations with $\varepsilon \neq \frac{1}{2}$ (ε being the proportion of the round-trip time taken on the outbound leg only). Adopting one of these choices is a recipe for calculational misery of a very pointless kind. But the Einstein of 1905, and many philosophers of an operationalist/verificationist bent since then, thought that such a choice cannot be criticized as *wrong*. Ultimately, they say, distant simultaneity is not only frame-relative, but partly conventional.

Taking up the challenge of establishing a one-way velocity for light, Ellis and Bowman (1967) argued that *slow clock transport* offers a means of synchronizing distant clocks that is independent of the velocity of light. In STR, when a clock is accelerated from rest in a given frame up to some constant velocity, then decelerated to rest again at a distant location, there are time-dilation effects that prevent us from regarding the clock as having remained in synch with clocks at its starting point. And calculation of the size of the effect depends on having established a distant-simultaneity convention (i.e. a choice of ε). So it looks as though carrying a clock from observer 1 to observer 2 will not let us break the circle. But Ellis and Bowman noted that the time dilation effect tends to zero as clock velocity goes to zero, and this is independent of ε-synchronization. Therefore, an "infinitely slowly" transported clock allows us to establish distant synchrony, and measure light's one-way velocity. Conventionalists were not persuaded, and the outcome of the fierce debate provoked by Ellis and Bowman's paper was not clear.

In 1977, David Malament took up the conventionalist challenge from a different perspective. One way of interpreting the claim of conventionalists such as Grünbaum is this: the observable *causal structure* of events in an STR-world does not suffice to determine a unique frame-dependent simultaneity choice. By "causal structure" we mean the network of causal connections between events; loosely speaking, any two events are causally connectable if they *could* be connected by a material process or light-signal. In STR, the "conformal structure" or light-cone structure at all points is the idealization of this causal structure. It determines, from a given event, what events could be causally connected to it (toward the past or toward the future). Grünbaum and others believed that the causal structure of space–time by no means singles out any preferred way of cutting up space–time into "simultaneity slices".

Malament showed that, in an important sense, they were wrong. The causal/conformal structure of Minkowski space–time *does* pick out a unique frame-relative foliation of events into simultaneity slices. Or rather, more precisely, the conformal structure suffices to determine a unique relation of *orthogonality*. If we think of an ε-choice as the choice of how to make simultaneity slices relative to an observer in a given frame, then Malament showed that the conformal structure is sufficient to define a unique, orthogonal foliation that corresponds to Einstein's $\varepsilon = \frac{1}{2}$ choice. For many philosophers, this result marked the end of the debate over conventionality of simultaneity. (But see Janis, 1983 and Redhead, 1993 for conventionalist responses.)

Current work

A recent paper by Sarkar and Stachel (1999) tries to re-open the issue of conformal structure and simultaneity relations. Stachel and Sarkar note that one of Malament's assumptions was that the causal connectability relation is taken as *time-symmetric*, i.e. that it does not distinguish past-future from future-past directions of connection. They argue that it is possible to distinguish the backward from forward light cones using only the causal-connectability relation Malament starts from. If this is granted, and we do not impose the condition that any causally-definable relation must be time-symmetric, then the uniqueness result Malament proved fails. Many different cone-shaped foliations become definable. Stachel and Sarkar advocate the backward-lightcone surface as an alternative simultaneity surface choice that could be made. It remains true, however, that only the genuine orthogonality relation ($\varepsilon = \frac{1}{2}$) is transitive and location-independent. These are two of the core features of classical simultaneity. To put forward Stachel and Sarkar's alternative relation as a genuine candidate for a distant-synchrony relation is therefore, at best, awkward and out of line with core intuitions about simultaneity.

Still, many philosophers of physics feel dissatisfied with even this much of a concession to conventionalism. They suspect that, even if it may have been in some sense possible to do physics with $\varepsilon \neq \frac{1}{2}$ in 1905, more recent quantum field theory has surely ruled that out. Zangari (1994) argued that the mathematics of spinor fields in Minkowski space–time – used in describing spin-$\frac{1}{2}$ particles, for example – is only consistent in frames with standard synchrony. Gunn and Vetharaniam (1995) claimed that Zangari was mistaken, and that using a different formalism, the Dirac equation could be derived in a framework including $\varepsilon \neq \frac{1}{2}$ frames. Karakostas (1997) has argued that both of the preceding authors' arguments are flawed, though Zangari's main claim is correct. And most recently, Bain (2000) argues that none of these authors has it exactly right. There is *always* a way to do physics using arbitrary coordinates (including those corresponding to non-standard simultaneity choices); but whether that amounts to the conventionality of simultaneity in an interesting sense remains a tricky question.

In trying to see one's way through the dense thicket of technical claims and counter-claims in these papers, it helps to fall back on the notion of *general covariance*. Kretschmann hypothesized in 1918 that *any* physical theory could be expressed in a generally covariant form, i.e. in a form that is valid in arbitrary coordinates. Nonstandard-synchrony frames do correspond to coordinate systems allowed under general covariance. Karakostas does not deny Kretschmann's claim. Instead, he notes that generally covariant treatments of spinor fields can be done, but they have to introduce a geometric structure (a "frame" or "*vierbein* field") that effectively picks out the orthogonal (= standard simultaneity) direction for a given observer in a given frame. This is a typical sort of move when theories with absolute space–time structures are given in a generally covariant form. Geometric objects or fields replace privileged coordinates or frames, but the "absoluteness" is only shifted, not removed. In the case of spinor fields, it seems that *something* that effectively encodes the Einstein-standard synchrony relation is mathematically necessary. Can the conventionalist respond by claiming that this necessary structure is, withal, not a *simultaneity* structure? Bain claims that she can; for spinor fields have nothing to do with rods and clocks, and the measurement of light's one-way velocity – i.e. with the original point conventionalists made.

Conventionalist claims – concerning both geometry and simultaneity – seem to be constantly in danger of collapsing into triviality: the trivial claim that, if we are mathematically clever and not afraid of pointless hard work, we can choose any perverse sort of coordinate system we like, and then claim that the coordinates reflect the geometric/simultaneity relations we have "chosen." Perhaps we can do this; but to suppose that this amounts to a genuine choice of spatio-temporal facts is to be somewhat disingenuous about the content of such facts. To be sure, axioms of coordination are needed to link pure geometric concepts to observable phenomena. But the axioms we choose are themselves constrained in many ways by the need to cohere with further practices and metaphysical assumptions. In practice, these constraints seem to fully determine, or even over-determine our "choices" regarding geometry. What keeps the debate concerning conventionality of *simultaneity* alive is the way in which our "conventional choices" play only a completely trivial role *qua* axioms of coordination. Just as one can do physics with any choice of ε, one can also do physics without *any* choice of clock synchronization.

Future work

Relativity theory (STR and GTR) provides the natural home for at least limited forms of conventionalism, though it remains a subject of dispute just how significant the conventionality is. The work of Karakostas, Bain and others points in the direction future work on these topics will take: toward new physics. One would also expect that advances in the general methodology of science will continue to bear on these issues.

Black Holes and Singularities

Our best theories tell us that stars eventually run out of nuclear fuel. When they do so, they leave equilibrium and undergo gravitational collapse, ending as white dwarves if the collapsing core's mass $M < 1.4$ solar masses, as neutron stars if $5 > M > 1.4$, or as black holes if $M > 5$. Black holes are regions of space–time into which matter can enter but from which matter can not escape. Their end states are singularities, which for now we might associate with a "hole" in space–time or a point where the space–time metric "blows up" and is ill-defined. There is some astronomical evidence for the existence of black holes, and they are relevant to a number of questions that interest philosophers, such as whether time travel is possible and whether the past and future are finite (Weingard, 1979). However, we here focus on singularities, as they are more general since they can exist without black holes, and they also pose several different philosophical questions that are the subject of active research.

History

Singularities are hardly novel to GTR. The classical Coulomb field when combined with STR goes to infinity at points. Collapsing spherical dust clouds and other highly symmetric solutions provide examples of singularities in Newtonian gravitational theory. But singularities in GTR are especially puzzling, as we will see.

The existence of singularities to EFE was known from the theory's inception. Hilbert, for instance, wrote about the notorious singularities in the Schwartzchild solution as early as 1917. The line element of this solution has singularities at $r = 0$ and $r = 2M$. Einstein in 1918 worried about them only because he took them as a threat to Machianism. Singularities in the solutions to the field equations didn't cause general alarm for many more decades because they were not very well understood (Earman, 1999; Earman and Eisenstaedt, 1999). They were viewed as unacceptable pathologies, but it was assumed that they were defects of only certain models. From 1918 until the mid-1950s, it was not realized that the singularities in these space–times were "essential" in some sense. There were two other options.

First, a singularity might be merely a "coordinate singularity" and not a feature of the space–time. To illustrate the distinction, consider coordinizing a sphere. It is a theorem that no single coordinate system can cover the sphere without singularity. This represents a problem for the coordinate system, not the sphere. The sphere is a perfectly well-defined geometric object; moreover, there are ways of covering the sphere without singularity using two different coordinate patches. The Schwartzchild solution caused particular mischief in this regard during the first half of the twentieth century; it famously emerged that only one ($r = 0$) of its two apparent singularities is genuine – the "Schwartzchild radius" ($r = 2M$) is a mere artifact of the coordinates.

Second, like the singularities in classical gravitational theory, relativistic singularities might be due to an artificial symmetry of the solution. The singular nature of a solution of Newton's equations representing a perfectly spherical collapse of dust is real enough. It is no artifact of the coordinates chosen. But the feeling is, what chance is there that this is our world? Our world does not have its matter arranged like dust formed in a *perfect* sphere. Change the distribution somewhat and the singularity disappears. Why worry? Similarly, when it became clear that (for example) the Schwarzchild and Friedman solutions contained genuine singularities, the hope was that these arose from the artificial symmetries invoked; after all, the Schwartzchild solution represents the geometry exterior to a spherically symmetric massive body and the Friedman solutions represents a homogeneous and isotropic matter distribution. The singular solutions were hoped to be in some sense "measure zero" in the space of all the solutions of EFE.

These hopes were dashed by singularity theorems in the 1950s by Raychaudhuri and Komar, and especially by theorems in the 1960s and early 1970s by Penrose, Geroch and Hawking. These theorems appear to demonstrate that singularities are generic in space–times like ours. They assume what seem to be plausible conditions on the stress-energy of matter to force geodesics to cross; they then employ global conditions on the geometry to show that these geodesics terminate in a singularity.

These advances in the 1960s and 1970s were made possible in part by the new, minimal definition of a singularity. Without going into the details, a space–time is said to be singular according to these theorems just in case it contains a maximally extended timelike geodesic that terminates after the lapse of finite proper time. Briefly put, a space–time is singular iff it is timelike geodesically incomplete. (This definition can be extended to cover null and spacelike curves, and can be extended in other ways too – to so-called 'b-incompleteness' – but we will not go into this here.) The idea behind this definition is that it must be a serious fault of the space–time, one worthy of the name singularity, if the life of a freely falling immortal observer nevertheless terminates in a finite time.

However fruitful this definition, it has proved to be controversial, as have the significance of the singularity theorems. The current work in philosophy on these topics, largely driven by Earman (1995), focuses on these two questions.

Current work

This section focuses on the analysis of singularities. We concentrate on this topic not because we feel that it is any more important than other questions – indeed, we feel the opposite, that (for instance) the question of the significance of singularities for GTR is far more important – rather, we so concentrate because it is a necessary point of entry into the literature. One cannot successfully evaluate the significance of singularities without first knowing what they are.

Naively, one has the idea that a singularity is a hole in space–time surrounded by increasing tidal forces that destroy any approaching object. This picture cannot be correct for general relativistic space–times. The reason is simple: the singularities here are singularities in the metric space itself, so there is literally no location for a hole. General relativity requires a manifold with a smooth Lorentz metric, so by definition there are no locations where the metric is singular. Fields *on* space–time can be singular at points; but space–time itself has nowhere to be singular.

Following Geroch (1968), commentators have identified several quite distinct meanings of singularity. To name a few, and sparing details, consider the following conditions proposed for making a space–time singular:

(a)   curvature blowup: a scalar curvature invariant, e.g. Ricci, tensor goes unbounded along a curve in space–time
(b)   geodesic incompleteness: see above
(c)   missing points: points are "missing" from a larger manifold, arising from the excision of singular points.

All three, we suppose, are involved in our intuitive idea of a space–time singularity. And for a Riemannian space, (b) and (c) are co-extensive. The Hopf–Rinow theorem states that, for connected surfaces, the conditions of being a complete metric space and being geodesically complete are equivalent. A metric space is complete if every Cauchy sequence of points in it converges to a point in that space. Intuitively, incompleteness is associated with missing points. For instance, the plane minus the origin, the surface $\mathbb{R}^2 - \{(0,0)\}$, is not complete because the Cauchy sequence $\{(1/n, 0)\}$ converges to a point excised from the plane. It is also not geodesically complete since there are no geodesics joining points $(-1,0)$ and $(1,0)$, so here we see a connection between geodesic incompleteness and missing points.

However, a relativistic space–time is not a Riemannian space, but a pseudo-Riemannian one, and the Hopf–Rinow theorem does not survive the change. None of the three definitions are co-extensive: the literature shows that while (c) implies (b), (b) does not imply (c); (a) implies (b), but (b) does not imply (a); and (a) seems to imply (c), but (c) does not imply (a). The official definition, (b), thus seems to act as a kind of symptom of the other two pathologies. Even here there are counterexamples. A curve might be incomplete even if the curvature is behaving normally, as happens in Curzon space–time; and as Misner shows, a curve might be incomplete even in a compact, and hence, complete and "hole-less", space–time.

It is of interest to see how hard it is to even make sense of definition (c). As mentioned above, a relativistic space–time has no room for singular points in the metric. Definition (c) would then have us look for the traces of an excised point, i.e. look for what is not there. How do you find points which are not on the space–time but which have been removed? Looking at the topology will not help since, in general, a variety of non-singular metrics can be put on any given topology (for instance, the Schwartzchild topology of $R^2 \times S^2$ is compatible with plenty of non-singular topologies). Although this way of understanding singularities is still active, it may be that the whole idea of a singularity as some localizable object is misleading.

Once we have an understanding of singularities in GTR, the next question to ask is about their significance. Do they "sow the seeds of GTR's demise" as is often alleged? Or are they harmless, perhaps even salutary, features of the theory? Earman (1996) provides an argument for tolerating singularities; but many physicists claim that they represent a genuine deficiency of the theory.

Future work

The topic of singularities is really a new one for philosophers of science. We can scarcely mention all the areas open to future endeavor. The majority of our focus below depends, perhaps naturally, on relatively recent ideas in physics.

Good arguments needed Earman (1996) pieces together and criticizes various arguments for the widespread belief that singularities sow the seeds of GTR's demise. A survey of the literature shows that there is a dearth of good argument supporting this belief. Can a good argument be articulated on behalf of this opinion that does not rely on misleading analogies with other pitfalls in the history of science?

Are there really singularities? The singularity theorems do not fall out as deductive consequences of the geometries of relativistic space–times. To say anything, the stress-energy tensor must be specified, and, in fact, all the theorems use one or another energy condition. The so-called weak energy condition, for instance, states that the energy density as measured by any observer is non-negative. But, is it reasonable to suppose these hold? The philosopher Mattingly (2000) sounds a note of skepticism, pointing out that various classical scalar fields and quantum fields violate all the conventional energy conditions. Even if Mattingly's skepticism is not vindicated, a better understanding of the relation between the energy conditions and real physical fields is certainly worth having.

Quantum singularities Philosophers may also wish to cast critical eyes over some of the methods suggested for escaping singularities with quantum mechanics. It is sometimes said that one should define a quantum singularity as the vanishing of the expectation values for operators associated with the classical quantities that vanish at the classical singularity. Then it is pointed out that the radius of the universe, for example, can vanish in what is presumed to be an infinite density and curvature singularity, even though the expectation value does not vanish (Lemos, 1987). This is sometimes taken as showing that quantum mechanics smoothes over the classical singularity. But, is this really so? Callender and Weingard (1995), for example, argue that this quantum criterion for singular status fits poorly with some interpretations of quantum mechanics.

Black hole thermodynamics Hawking's (1975) "discovery" that a black hole will radiate like a blackbody strengthened Beckenstein's work supporting an analogy between classical thermodynamics and black holes. The field known as black hole thermodynamics was spawned, and there are now thought to be black hole counterparts to most of the concepts and laws of classical thermodynamics. For instance, the black hole's surface gravity divided by 2π acts like the temperature and its area divided by 4 acts like the entropy. Physicists enticed by this analogy often claim that it is no analogy at all, that black hole thermodynamics *is* thermodynamics and that (for instance) the surface gravity really *is* the temperature. The signifiance of these startling claims and the analogy are certainly worthy of investigation by philosophers of science.

Information loss A related topic is the black hole "information loss paradox" that arises from Hawking's (1975) result. Take a system in a quantum pure state and throw it into a black hole. Wait for the black hole to evaporate back to the mass it had when you injected the quantum system. Now you have a system of a black hole with mass M plus a thermal mixed state, whereas you started with a black hole with mass M plus a pure state. Apparently, you have a process that converts pure states into mixed states, which is a non-unitary transformation prohibited by quantum mechanics (such a transformation allows the sum of the probabilities of all possible measurement outcomes to not equal 1). See Belot et al. (1999) and Bokulich (2000) for some philosophical commentary on this topic.

Cosmic censorship Perhaps the biggest open question relevant to singularities and many other topics in gravitational physics is the status of Penrose's cosmic censorship hypothesis; for a recent assessment, see Penrose (1999). This hypothesis is often glossed as the claim that naked singularities cannot exist; that is, that singularities are shielded from view by an event horizon, as happens in spherically symmetric gravitational collapse. Naked singularities are unpleasant because they signal a breakdown in determinism and predictability. If a naked singularity occurs to our future, then no amount of information on the space-like hypersurface we inhabit now will suffice to allow a determination of what happens at all future points. Singularities are, intuitively, holes out from which anything might pop. A singularity that we can see means we might see anything in the future, since the causal past will not sufficiently constrain the singularity.

Stated as the claim that naked singularities cannot exist, however, the hypothesis is clearly false, since there are plenty of relativistic space–times that violate it. Though formulated in a variety of non-equivalent ways (Earman, 1995, ch. 3), it is common to speak of weak and strong versions of the claim. Weak cosmic censorship holds that gravitational collapse from regular initial conditions never creates a space–time singularity visible to distant observers, i.e. any singularity that forms must be hidden within a black hole. Strong cosmic censorship holds that any such singularity is never visible to any observer at all, even someone close to it. By "regular initial data" we mean that the space–times are stable with respect to small changes in the initial data. Elaborating this definition further obviously requires some care.

The only consensus on the topic of cosmic censorship is that the hypothesis is both important and not yet proven true or false. Regarding the latter, there are plenty of counter-examples to both formulations of the hypothesis, though especially to strong cosmic censorship; see, for example, Singh (1998). Current and future work will dwell on whether these examples really count. In the background there is, you might say, the "moral" cosmic censorship hypothesis, which claims that the only naked singularities that occur are Good ones, not Bad ones. The exact formulations of Good and Bad depend, as one would expect, on the character of the particular investigator: prudish investigators hope GTR doesn't offer so much as a hint of nakedness, whereas the more permissive will lower their standards.

It is important to know whether some version of the hypothesis is true. If a cosmic censor operates, then many topics dear to philosophers will be affected. A cosmic censor will naturally affect what kinds of singularities we can expect, and therefore influence the question of their significance for GTR (Earman, 1996); it would mean the time-travel permitting solutions of EFE such as Gödel's will not be allowed; that the possibility of spatial topology change (Callender and Weingard, 2000) will not be possible, and so on. And a lack of a cosmic censor will also bear on much of the physics of potential interest to philosophers, e.g. black hole thermodynamics hangs crucially on the existence of a cosmic censor.

There are also philosophical topics about cosmic censorship that need further exploration. To name two, what is meant by "not being allowed" in the statement of cosmic censorship and how do white holes (the time reverses of black holes) square with the hypothesis and the time symmetry of EFE?

Horizons and Uniformity

History

The observed isotropy (or near isotropy) and presumed homogeneity of our universe suggest that we inhabit a world whose large scale properties are given by the well-known Friedman standard model. In this model, the world "began" in a hot dense fireball known as the Big Bang, and matter has since expanded and cooled ever since. The rate of expansion and cooling depend on the equation of state for the cosmological fluid, and the ultimate fate of the universe (closed or open) depends on the curvature. Part of the corroboration of this model comes from the observed uniformity of the cosmic microwave background radiation (CMBR). Neglecting some recently detected small inhomogeneities (which are themselves not defects but harmonic oscillations expected in some Big Bang models), these observations show that the temperature of this radiation is uniform to at least one part in 10,000 in every direction we look.

When coupled with the Friedman model, the uniformity of the CMBR produces a puzzle. To see this, we need to resolve an apparent contradiction between one's naïve view of the Big Bang singularity with the fact that in the Friedman model not all bodies can communicate with each other, even merely a fraction of a second after the Big Bang. Consider two nearby co-moving particles at the present time. The scaling factor, a, is the distance between the particles, say one light-second. Because the universe is expanding $da/dt > 0$. Now one would expect that, since $a \to 0$ as $t \to 0$ for all the particles, any particle could have been in causal contact with any other at the Big Bang. Since they are all "squashed together," a light pulse from one could always make it to any other particle in the universe. This is not so.

First off, there is *no* point on the manifold where $t = 0$ and $a = 0$; this point is not well-defined and it is not clear, anyway, that all the particles in the universe occupying the same point really makes sense. So this "point" does not count. But now it is a question of how fast the worldlines are accelerating away from each other and whether light signals from each can reach all the others. Will light emanating from a body "just after" the Big Bang singularity be able to reach an arbitrary body X by time t, where t is some significantly long time, possibly (in a closed universe) the end of time? In general, the answer is No, for there are some (realistic) values of the expansion parameter that do not allow the light signal to catch up to X by t. The space–time curvature is the key here. Imagine that you and a friend are traveling in opposite directions on a flat plane. Assuming nothing travels faster-than-light, can you evade a light pulse sent out in your direction from your friend? No: though you may give it a good run for its money, eventually it will catch you if the universe is open. Now imagine that you're moving on a balloon and the balloon is being quickly inflated. Then, depending on the speed of inflation and your velocity, you may well be able to escape the light signal, possibly for all time.

The curvature due to the expansion and deceleration causes the worldlines of galaxies to curve. In two spatial dimensions, our past lightcone becomes pear-shaped rather than triangular (Figure 9.3).

Note that due to this curvature we cannot "see" the entire Big Bang. A useful picture of the causal situation emerges if we "straighten out" the curvature, much as we do when we use a Mercator projection when we draw a flat picture of the earth (Figure 9.4).

Here the top of the large triangle is the point we are at right now, and the two shaded triangles are the past null cones of two points, separated by an angle A,

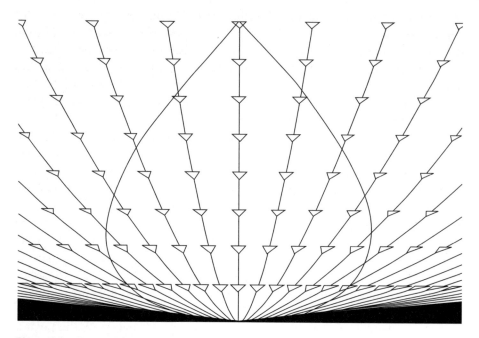

Figure 9.3   Our past light cone

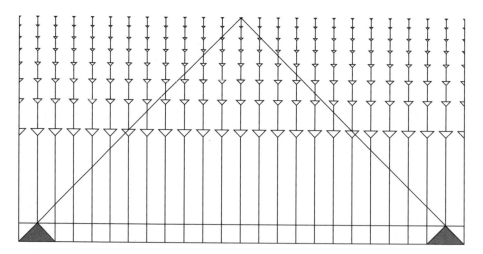

Figure 9.4   Straightening out the curvature

that we can see in our past. If *A* is sufficiently great the shaded regions do not intersect. But since the past null cone of a point represents all the points with which it might have had causal contact, this means that no point in the shaded regions could have had causal contact with each other (ignoring the possibility of faster-than-light travel). A *particle horizon* is defined as the maximum coordinate distance that one can see from a given point in space–time. From the diagram, one can see that the points at the top of the shaded regions have horizons that preclude them from seeing each other's pasts.

The puzzle about horizons arises from the fact that a Friedman model like that pictured can be said to fairly represent our universe, where the shaded regions are points in our past where matter decoupled from radiation. Since they share no common causal past, this means that they have no mechanism in common that will make the microwave radiation's temperature the same. How, then, did they arrive at the same temperature? It seems that, short of denying that the early universe can be approximately represented by a Friedman model, the only answer is that the universe was "born" in a highly isotropic and homogenous state. This necessary special initial state is the cause of the horizon problem.

Current work

In physics, the main response to this puzzle is to change the physics of expansion. Though there are other responses, the one known as inflation is almost universally maintained. In inflationary scenarios, the standard Friedman expansion is jettisoned in an early epoch in favor of a period of exponential expansion; the universe then undergoes a phase transition that slows it down back to the more moderate Friedman expansion. The details of this period vary with different proposals (there are more than fifty). Inflation does not remove particle horizons; instead, it increases the size of each point's past null cone so that pairs will overlap. The shaded past lightcones in the diagram would intersect while remaining proper subsets of each other. The hope is that the common causal past between two points will be large enough so that it accounts for their uniform temperature.

Work by Penrose (1989), Earman (1995) and Earman and Mosterin (1999) have severely criticized inflation for failing to deliver on its original promises. The theory, they say, does not rid cosmology of the need for special initial conditions to explain the apparent uniformity of the cosmic background radiation, nor does it enjoy much in the way of empirical success.

Future work

The horizon problem shares some general features with other well-known "problems" in physics. The problem of the direction of time (well, one of them) asks for an explanation of the thermodynamic arrow of time and ends up requiring the postulation of a very special initial condition of low entropy (Price, 1996). Philosophically, it is non-trivial whether requiring "special" boundary conditions is a genuine defect of a theory. For the situation with entropy and the direction of time, many do not see the special posit as a genuine failing of the theory to provide a scientific explanation (Callender, 1997); in the cosmological case with horizons, however, it is orthodoxy now that it is a genuine failing of the standard model that it cannot explain the uniformity of the cosmic background radiation. But is it? The failing is certainly not one of empirical disconfirmation, since given the "special" initial conditions the model is empirically adequate – we even get a deterministic explanation through time of why we see the features we do. Earman and Mosterin do much to criticize inflation as a solution to this problem, but the larger issue, common to this topic and others – whether there really is a problem here at all – is left open.

A related issue is whether the notion of "specialness" can be sharpened. In the cosmological case considered here, it is especially problematic to specify in exactly what sense the boundary conditions are "special," as Penrose (1989) emphasizes. By contrast, in the thermodynamic case this is somewhat clearer since one is talking about a statistical theory (statistical mechanics) equipped with a standard probability measure with respect to which the needed initial conditions do indeed occupy small measure. To be sure, there are problems in this case too in justifying this "natural" probability measure, but they appear to pale in comparison to the problems of defining a probability for cosmic inflation.

Finally, the question of whether there was inflation will probably ultimately be decided by observation and experiment rather than philosophical argument. Recent and future improvements in observational cosmology (e.g. CMB measurements, measurements of type 1a supernovae at high redshifts) have opened up the possibility of empirical support or disconfirmation of some inflation scenarios. The epistemology of this optimistic, burgeoning branch of physics is yet another field ripe for philosophical analysis.

Conclusion

The specter hanging over all future work in this field is quantum gravity. It is widely believed that general relativity is inconsistent with quantum field theory; "quantum gravity" is the research program that seeks a third theory that unifies, or at least makes consistent, these two theories. Though no such theory yet exists, there are some well-developed approaches such as string theory and canonical quantum gravity as well as some less developed theories such as topological quantum field theory and twister theory; for a philosophical slant, see Callender and Huggett (2001) and references therein. We believe it is fair to say that all of these theories are quite radical in their implications for space and time. If any of them, or remotely similar descendents, succeed, they may well have dramatic consequences for virtually all of the issues discussed above.

Acknowledgments

Figures 9.3 and 9.4 are reproduced, with permission, from Ned Wright's Cosmology Tutorial (http://www.astro.ucla.edu/%7Ewright/cosmolog.htm).

Notes

1   Whether this really spoils the relational ambitions of a hypothetical physics set in them is a difficult question. If space's structure is nothing more than what is implied by all the distance/angle relations between physical things – as one form of relationism holds – then the geometry of space must be an empirical matter, not something we can fix *a priori*.
2   Actually, there is a further element of absoluteness in GTR, namely the demand that the metric field have signature (1,3) and hence be "locally" like Minkowski space–time. See Brown (1997) for an illuminating discussion of this posit.
3   In fact, Hartry Field's (1980) uses the point-manifold, interpreted realistically, to eliminate platonic entities from the mathematics of physics. So, from his perspective, the manifold is not just indispensable in GTR, but in all of physical science.

References

Arntzenius, F. and Maudlin, T. (2000): "Time Travel and Modern Physics," *Stanford Online Encyclopedia of Philosophy*, http://plato.stanford.edu/entries/time-travel-phys.

Bain, J. (2000): "The Coordinate-Independent 2-Component Spinor Formalism and the Conventionality of Simultaneity," *Studies in History and Philosophy of Modern Physics*, 31B, 201–26.

Barbour, J. (1999): *The End of Time: The Next Revolution in our Understanding of the Universe*. London: Weidenfeld & Nicolson.

Barbour, J. and Bertotti, B. (1982): "Mach's Principle and the Structure of Dynamical Theories," *Proceedings of the Royal Society of London* A382, 295–306.

Belot, G. (1999): "Rehabilitating Relationalism," *International Studies in the Philosophy of Science*, 13, 35–52.

Belot, G. Earman, J. and Ruetsche, L. (1999): "The Hawking Information Loss Paradox: The Anatomy of a Controversy," *British Journal for the Philosophy of Science*, 50, 189–229.

Bokulich, P. (2000): "Black Hole Remnants and Classical vs. Quantum Gravity," available at the *PSA 2000* program page http://hypatia.ss.uci.edu/lps/psa2k/program.html.

Brown, H. (1997): "On the Role of Special Relativity in General Relativity," *International Studies in the Philosophy of Science*, 11, 67–80.

Brown, H. and Pooley, O. (2001): "The Origin of the Space–Time Metric: Bell's 'Lorentzian Pedagogy' and its Significance in General Relativity," in C. Callender and N. Huggett (eds.), *Physics Meets Philosophy at the Planck Scale*, Cambridge University Press, 256–74. Also available at gr-qc/9908048.

Butterfield, J. (1989): "The Hole Truth," *British Journal for the Philosophy of Science*, 40, 1–28.

Callender, C. (1997): "What is 'The Problem of the Direction of Time'?" *Philosophy of Science*, Supplement, 63(2), 223–34.

Callender, C. and Huggett, N. (eds.) (2001): *Physics Meets Philosophy at the Planck Scale*. New York: Cambridge University Press.

Callender, C. and Weingard, R. (1995): "Bohmian Cosmology and the Quantum Smearing of the Initial Singularity," *Physics Letters*, A208, 59–61.

Callender, C. and Weingard, R. (2000): "Topology Change and the Unity of Space," *Studies in History and Philosophy of Science*, 31B, 227–46.

Disalle, R. (1994): "On Dynamics, Indiscernibility and SpaceTime Ontology," *British Journal for the Philosophy of Science*, 45, 265–87.

Disalle, R. (1995): "SpaceTime Theory as Physical Geometry," *Erkenntnis*, 42, 317–37.

Duhem, P. (1954): *The Aim and Structure of Physical Theory*, trans. P. Wiener, Princeton, NJ: Princeton U.P.

Earman, J. (1970): "Who's Afraid of Absolute Space?" *Australasian Journal of Philosophy*, 48, 287–319.

Earman, J. (1989): *World enough and space–time: absolute versus relational theories of space and time*. Cambridge, Mass.: MIT Press.

Earman, J. (1995): *Bangs, Crunches, Whimpers and Shrieks: Singularities and Acausalities in Relativistic Spacetimes*. New York: Oxford University Press.

Earman, J. (1996): "Tolerance for Space–time Singularities," *Foundations of Physics*, 26(5), 623–40.

Earman, J. (1999): "The Penrose–Hawking Singularity Theorems: History and Implications," in H. Goemmer, J. Renn, J. Ritter and T. Sauer (eds.), *The Expanding Worlds of General Relativity*, Einstein Studies, vol. 7. Boston: Birkhauser.

Earman, J. and Eisenstaedt, J. (1999): "Einstein on Singularities," *Studies in History and Philosophy of Modern Physics*, 30, 185–235.

Earman, J. and Mosterin, J. (1999): "Inflation," *Philosophy of Science*, 66, 1–49.

Earman, J. and Norton, J. (1987): "What Price Space–Time Substantivalism? The Hole Story," *British Journal for the Philosophy of Science*, 38, 515–25.

Earman, J. and Norton, J. (1996): "Infinite Pains: The Trouble with Supertasks," in A. Morton and S. Stich (eds.), *Benacerraf and his Critics*, Cambridge, MA: Blackwells, 231–61.

Ellis, B. and Bowman, P. (1967): "Conventionality in Distant Simultaneity," *Philosophy of Science*, 34, 116–36.

Field, H. (1980): *Science without Numbers: A Defence of Nominalism*. Oxford: Blackwell.

Friedman, M. (1983): *Foundations of Space–Time Theories: Relativistic Physics and Philosophy of Science*. Princeton, N.J.: Princeton University Press.

Geroch. R. (1968): "What is a Singularity in General Relativity?" *Annals of Physics*, 48, 526–40.

Gunn, D. and Vetharaniam, I. (1995): "Relativistic Quantum Mechanics and the Conventionality of Simultaneity," *Philosophy of Science*, 62, 599–608.

Hawking, S. W. (1975): "Particle Creation by Black Holes," *Communications in Mathematical Physics*, 43, 199–220.

Hoefer, C. (1998): "Absolute versus Relation Space–Time: For Better or Worse, the Debate Goes On," *British Journal for the Philosophy of Science*, 49, 451–67.

Hoefer, C. (2000): "Energy Conservation in GTR," *Studies in the History and Philosophy of Modern Physics*, 31, 187–99.

Huggett, N. (1999): "Why Manifold Substantivalism is Probably Not a Consequence of Classical Mechanics," *International Studies in the Philosophy of Science*, 13, 17–34.

Janis, A. (1983): "Simultaneity and Conventionality", in R. Cohen (ed.), *Physics, Philosophy and Psychoanalysis*, Reidel: Dordrecht, 101–10.

Karakostas, V. (1997): "The Conventionality of Simultaneity in the Light of the Spinor Representation of the Lorentz Group," *Studies in the History and Philosophy of Modern Physics*, 28, 249–67.

Lemos, N. (1987): "Conservation of Probability and Quantum Cosmological Singularities," *Physical Review*, D36, 2364.

Mach, E. (1989): *The Science of Mechanics: A Critical and Historical Account of its Development*, trans. T. McCormack, LaSalle, Ill: Open Court.

Malament, D. (1977): "Causal Theories of Time and the Conventionality of Simultaneity," *Noûs*, 11, 293–300.

Mattingly, J. (2000): "Singularities and Scalar Fields: Matter Theory and General Relativity," available at the *PSA 2000* program homepage, http://hypatia.ss.uci.edu/lps/psa2k/program.html.

Maudlin, T. (1990): "Substance and Space–Time: What Aristotle Would Have Said to Einstein," *Studies in History and Philosophy of Science*, 21, 531–61.

Nerlich, G. (1994): *The Shape of Space*. Cambridge, New York: Cambridge University Press.

Penrose, R. (1989): "Difficulties with Inflationary Cosmology," *Annals of the New York Academy of Sciences*, 571, 249–64.

Penrose, R. (1999): "The Question of Cosmic Censorship," in *Journal of Astrophysics and Astronomy*, 20(3–4), 233–48.

Poincaré, H. (1952): *Science and Hypothesis*. New York: Dover Books.

Price, H. (1996): *Time's Arrow and Archimedes' Point: New Directions for the Physics of Time*. Oxford, Oxford University Press.

Redhead, M. (1993): "The Conventionality of Simultaneity," in J. Earman, A. Janis, G. Massey and N. Rescher (eds.), *Philosophical Problems of the Internal and External Worlds: Essays on the Philosophy of Adolf Grünbaum*, Pittsburgh: University of Pittsburgh Press, 103–28.

Rynasiewicz, R. (1996): "Absolute Versus Relation Space–Time: An Outmoded Debate?" *Journal of Philosophy*, 93, 279–306.

Rynasiewicz, R. (2000): "On the Distinction between Absolute and Relative Motion," *Philosophy of Science*, 67, 70–93.

Sarkar, S. and Stachel, J. (1999): "Did Malament Prove the Non-Conventionality of Simultaneity in the Special Theory of Relativity?" *Philosophy of Science*, 66, 208–20.

Savitt, S. (2000): "There is No Time Like the Present (in Minkowski Space–Time)," *Philosophy of Science*, 67, Proceedings, S563–S575.

Singh, T. P. (1998): "Gravitational Collapse, Black Holes and Naked Singularities," available at gr-qc/9805066.

Stachel, J. (1993): "The Meaning of General Covariance," in J. Earman, A. Janis, G. Massey and N. Rescher (eds.), *Philosophical Problems of the Internal and External Worlds: Essays on the Philosophy of Adolf Grünbaum*, Pittsburgh: University of Pittsburgh Press, 129–60.

Teller, P. (1991): "Substance, Relations, and Arguments About the Nature of Space–Time," *Philosophical Review*, 100, 363–97.

Weingard, R. (1979): "Some Philosophical Aspects of Black Holes," *Synthese*, 42, 191–219.

Zangari, M. (1994): "A New Twist in the Conventionality of Simultaneity Debate," *Philosophy of Science*, 61, 267–75.

Chapter 10

Interpreting Quantum Theories

Laura Ruetsche

Introduction: Interpretation

The foundational investigation of quantum theories is inevitably specialized, but it ought not be exclusively so. Continuities of theme and approach should link it to the foundational investigation of other physical theories, to the general philosophy of science, to metaphysics and epistemology more broadly construed. The interpretation of quantum theories is the furnace in which these links are forged. To interpret a physical theory is to say what the world would be like, if the theory were true. A realist about a theory believes that theory to be true. Interpretation gives the realist's belief content, tells the constructive empiricist what he does *not* believe, and makes available to all parties the understanding of a theory constituted by a grasp of its truth conditions. Interpretation can promote theory development: Howard Stein offers the example of interpretive questions about the ether's state of motion in Maxwell theory, questions whose answers "revolutionized the theory and deepened our understanding of nature very considerably" (Stein, 1972, p. 423).

Having issued this apology for interpretation, this chapter surveys the interpretation of quantum theories. It chronicles past highlights (pp. 200–9); covers current work (pp. 209–17); and presents future directions (pp. 217–21). The remainder of this section sets the stage.

The Heisenberg–Born–Jordan matrix mechanics and Schrödinger's wave mechanics were twin formulations of quantum theory so fraternal it took von Neumann to pinpoint their relation. He called the structure they shared "Hilbert Space."[1] Pure quantum states are normed Hilbert space vectors; quantum observables are self-adjoint Hilbert space operators; once the Hamiltonian operator $\hat{H}$ is provided, Schrödinger's equation determines dynamical trajectories through state space. Because classical observables are functions from state space elements to the reals, a system's classical state fixes the values of all classical observables pertaining to it. In quantum mechanics (QM), this is not so. A state $|\psi\rangle$ does not in general fix the value of an observable $\hat{A}$; rather $|\psi\rangle$, via the Born Rule, determines a probability distribution over $\hat{A}$'s possible values. In its standard Hilbert space formulation, QM lacks what I'll call a *semantics*, an account of which observables have determinate values on a quantum system, and of what those values are or might be.

The quartet {state space, observables, dynamics, semantics} characterizes what is or can be true for a theory over time, and so constitutes an interpretation of a theory. Correlatively, degrees of freedom available to those engaged in interpretive projects include freedoms to propose and modify members of the quartet. One way to see the venerable debate about the nature of space(time) is as a debate about how best to tune classical theory's state space, observable set, and dynamics to one another. But interpretations can be – many interpretations of QM are – efforts in creative physics.

Bohr and Complementarity

Interpretive efforts can also retard creative physics, as Einstein feared Bohr's philosophy of complementarity would. The doctrine is too intricate to explicate in a short space, so I will settle for listing a few of its key elements, some of which persist in influence, by seeming to those working at present either to deserve explication, or to constitute exculpation.

Bohr denies that position and momentum can be simultaneously determinate on a quantum system. Position and momentum are what he styles *complementary* modes of description, "complementary in the sense that any given application of classical concepts precludes the simultaneous use of other classical concepts which in a different guise are equally necessary for the elucidation of the phenomenon" (1934, p. 10). For Bohr, it is as though reality were a stereoscopic image we were constrained to view one eye at a time. The doctrine originates in an insistence on the use of classical concepts, which Bohr couples to an operationalism governing their use. He observes that the experimental circumstances warranting the use of the momentum concept are incompatible with those warranting the use of the position concept. The upshot is the complementarity of position and momentum concepts, representatives of the complementary classes of kinematic (that is, spatio-temporal) and dynamic (that is, subject to conservation laws) concepts. Bohr takes the position-momentum uncertainty relations to express – and be explained by – this deeper principle of complementarity (1934, p. 57); see also Murdoch (1987, ch. 3).[2]

Bohr repeatedly emphasizes that the quantum of action is central to the doctrine. But the quantum of action seems to have gone missing from the foregoing reconstruction. One place it might lurk is a loophole through which a sort of counterfactual discourse might sneak. Having bolted our diaphragm to the table, we may with Bohr's blessing speak of the position of an electron passing through our experimental arrangement. Could we also, and in defiance of complementarity, speak of its momentum, by appeal to experimental results we *would have* obtained, *had we* instead dangled our diaphragm from a spring balance? Not if the *uncontrollable* exchange of the quantum of action blocks such extrapolation. A disturbance theory of measurement fertilizes yet another root of complementarity.

Consider how Bohr's philosophy could interact with living physics. A physics community embracing the philosophy of complementarity would thereby abandon the project, declared inconceivable by the doctrine of complementarity, of "completing" QM by developing a theory which described the simultaneously determinate positions and momenta of systems. Einstein (Fine, 1986, p. 18) feared that "the Heisenberg-Bohr tranquilizing philosophy – or is it religion? – is so delicately contrived that, for the time being, it provides a gentle pillow for the true believer from which he cannot very easily be aroused." In 1935, with Podolsky and Rosen, he issued a wakeup call.

The Einstein-Podolsky-Rosen (EPR) Argument

Bohr denies that complementary magnitudes are simultaneously determinate. Einstein, Podolsky and Rosen (1935) argue that quantum statistics themselves imply that Bohr is wrong. Crucial to their argument is the "criterion of reality":

> If without in any way disturbing a system we can predict with certainty . . . the value of a physical quantity, then there exists an element of physical reality corresponding to this physical quantity (Einstein et al., 1935, p. 777).

(I take the consequent to be equivalent to "this physical quantity has a determinate value.") They argue that there are circumstances in which *complementary* observables satisfy the reality criterion. Their key move is to consider quantum states of *composite* systems instituting *correlations* between observables pertaining to component subsystems. Bohm (1951) reformulates the argument for a pair of electrons in the spin singlet state:

$$|\Psi\rangle_{singlet} = \frac{1}{\sqrt{2}}(|\rightarrow\rangle_I |\leftarrow\rangle_{II} - |\leftarrow\rangle_I |\rightarrow\rangle_{II}) \qquad (10.1)$$

Although (10.1) expresses $|\psi\rangle_{singlet}$ in terms of eigenstates $|\rightarrow\rangle$ and $|\leftarrow\rangle$ of the x-component of spin $\hat{\sigma}_x$, $|\psi\rangle_{singlet}$ assumes biorthogonal form for, and institutes perfect correlations between, $\hat{\sigma}_n$ eigenstates of the two systems for all n. Thus, $|\psi\rangle_{singlet}$ assigns Born rule probability 1 to the experimental result that the outcomes of $\hat{\sigma}_n$ measurements on systems I and II disagree. EPR consider a pair of electrons prepared in $|\psi\rangle_{singlet}$ and sent to laboratories remote from one another.

Measuring $\hat{\sigma}_x$ on system I affords the prediction, with certainty, that an $\hat{\sigma}_x$ measurement performed on system II will yield the opposite result. The remoteness of the laboratories ensures that a measurement on system I cannot in any way disturb system II – provided the universe is "local" in a way that renders distance an assurance of isolation. By the reality criterion, then, $\hat{\sigma}_x$ on system II is an element of reality. EPR could well have stopped here (Fine, 1986, ch. 3) musters evidence that Einstein wishes they had). They have shown that, for those who would withhold determinateness from the quantum realm, it is as though the spin measurement in first laboratory, instantaneously and at a distance, brings into being an element of reality in the second laboratory.

But EPR continue. We might rather have measured $\hat{\sigma}_y$ on system I. $|\psi\rangle_{singlet}$ anticorrelates $\hat{\sigma}_y$ eigenstates just as well as it anticorrelates $\hat{\sigma}_x$ eigenstates. By parity of reasoning, in this counterfactual situation, $\hat{\sigma}_y$ on system II would be an element of reality. EPR again appeal to locality to conclude from this that $\hat{\sigma}_y$ on system II *is* an element of reality – otherwise "the reality of $[\hat{\sigma}_x]$ and $[\hat{\sigma}_y]$ depend on the process of measurement carried out on the first system, which does not disturb the second system in any way. No reasonable definition of reality could be expected to permit this" (Einstein et al., 1935, p. 780). (Bohr's reply to EPR is to permit what they deem impermissible: the nonlocal dependence of system II's matters of fact on system I manipulations. There is "no question of a mechanical disturbance," Bohr writes, but there is "the question of an influence on the very conditions which define the possible types of predictions regarding the future behavior of the system" (Bohr, 1935, p. 699).) Because the correlations $|\psi\rangle_{singlet}$ institutes are thoroughgoing, if the EPR argument works, it works for every spin observable. Those convinced by the argument should undertake the project of "completing" QM, for instance, by devising a theory which attributes a determinate value to every element of reality established by the EPR gambit, a theory which moreover respects a "locality" requirement of the sort EPR exploit. (Those convinced *ab initio* that the project of completing QM is worth undertaking needn't be constrained by locality, or by reconstituting reality EPR element by EPR element.) One of John Bell's groundbreaking contributions to the foundations of QM was to bring "local" hidden variable theories (HVTs) in contact with empirical data.

Bell's Theorem and Other No-Go Results

Bell's theorem shows that local HVTs are committed to sets of statistical predictions known as Bell Inequalities. Insofar as there exist quantum states predicting the violation of the Inequalities, Bell's theorem sets up a crucial test of local HVTs vs. standard QM. Experiment upholds QM, violates the Inequalities, and falsifies local HVTs. The field is set for the game of experimental metaphysics. To play, show how to derive Bell Inequalities from a set of premises bearing philosophically fraught names ("determinism," "completeness," "locality"). Observe that the experimental violation of the Inequalities reveals at least one of these premises to be false. Invoking priors of various sorts, single out leading suspects. The literature is vast; see Cushing and McMullin (1989) for a sample. In this section, I'll review a few of its defining moments, express a concern that locality is a red herring, and touch upon questions the violation of the Bell Inequalities raises about the nature of explanation.

The Bell inequalities

Like the EPR argument, Bell's (1964) theorem concerns distant correlations established by $|\psi\rangle_{singlet}$. In Bell's version of the experimental setup, the distant devices need not measure the same component of spin. Thus the generic outcome of a Bell correlation measurement is $(x,y|a,b)$ where $x, y \in \{+,-\}$ are the outcomes of measurements of spin components $\hat{\sigma}_a, \hat{\sigma}_b$ on particles I and II respectively. The Born rule probability $|\psi\rangle_{singlet}$ assigns $(x,y|a,b)$ is $\frac{1}{2}sin^2\theta ab/2$, where θ_{ab} is the angle between orientations a and b. Consider how a HVT might handle such probabilities. Let λ denote a complete set of parameters by which such a theory characterizes the state of a physical system; let Λ denote the full set of such states. Let $Pr_\lambda (x,y|a,b)$ be the probability the hidden state λ assigns the experimental result $(x,y|a,b)$. So-called deterministic HVTs countenance only probabilities of 1 or 0; stochastic HVTs countenance non-trivial probabilities. A quantum system has a hidden state $\lambda \in \Lambda$, we know not which; a normalized probability density $\rho(\lambda)$ over Λ encodes our ignorance. To obtain the empirical probability for a Bell-type measurement outcome, a HVT integrates, over the set Λ, the probabilities each λ assigns this outcome, weighted by the density $\rho(\lambda)$:

$$Pr(x, y|a, b) = \int_\Lambda Pr_\lambda (x, y|a, b)\rho(\lambda)d\lambda \tag{10.2}$$

To derive the Bell Inequalities, one imposes additional constraints on the HVT's probability assignment. Appealing broadly to intuitions about locality, Bell required the joint probability to *factorize* into probabilities for outcomes on each wing, which probabilities conditionalize only on settings proper to that wing:

$$Pr_\lambda (x, y|a, b) = Pr_\lambda (x|a) \times Pr_\lambda (y|b) \tag{10.3}$$

HVTs obedient to the factorization condition (3) obey the Inequality[3]

$$-1 \leq Pr(+,+|a,b) + Pr(+,+|a,b') + Pr(+,+|a',b') - Pr(+,+|a',b)$$
$$- Pr(+|a) - Pr(+|b) \leq 0 \tag{10.4}$$

This is a Bell Inequality. There are quadruples of orientations (a,a',b,b') – for instance $(\pi/3,\pi,0,2\pi/3)$ – for which standard QM predicts its violation. Upholding standard QM, experiment falsifies local HVTs.

For the purposes of probing locality, the factorization condition is blunt. In his 1983 dissertation – Jarrett (1986) provides a précis – John Jarrett sharpened it, by demonstrating its equivalence to the pair of conditions:

$$\text{Pr}_\lambda(x|a,b) = \text{Pr}_\lambda(x|a) \qquad\qquad \text{(Jarrett Locality)}$$

$$\text{Pr}_\lambda(x|a,b,y) = \text{Pr}_\lambda(x|a,b) \qquad\qquad \text{(Jarrett Completeness)}$$

The first expresses the desideratum that the outcome of a particle I measurement be independent of detector II's setting (ergo Shimony's (1984b) label: "parameter independence"). Jarrett equates it to a prohibition on superluminal signaling, which prohibition he supposes the special theory of relativity (STR) to issue. If Jarrett Locality fails, by changing the setting of her detector, a physicist in laboratory I can send instantaneously to laboratory II a signal in the form of altered measurement statistics (Shimony calls this "controllable non-locality" or action-at-a-distance). Jarrett Completeness expresses the desideratum that the outcome of a particle I measurement be independent of the outcome of a particle II measurement (ergo Shimony's label: "outcome independence"). Because the laboratory I physicist has no control over laboratory II outcomes, she can not exploit breakdowns in Jarrett locality to signal (Shimony call this "uncontrollable non-locality" or "passion-at-a-distance").

The violation of the Bell Inequalities implies that one of the assumptions generating them must be false. Having furnished his factorization of (10.3), and supposing our commitment to the special theory of relativity theory to commit us, at least morally, to Jarrett Locality, Jarrett fingers Completeness as the culprit (1986, p. 27). Setting $\lambda = |\psi\rangle$, standard quantum mechanics itself can be cast as a stochastic hidden variable theory violating completeness: $|\psi\rangle_{\text{singlet}}$ makes particle I probabilities sensitive to particle II outcomes. It appears that the quantum domain is ruled by passion-at-a-distance. Enlisting a Lewis-style counterfactual analysis of causation, Butterfield (1992) has argued that this violation of Jarrett completeness signals a causal connection between distant wings of the aparatus. Much care has been lavished on articulating relativistic locality constraints suited to this stochastic setting, so that the question of whether QM and the STR can "peacefully coexist" (Redhead, 1983) might be settled once and for all.

No-Go results without locality

I would advocate postponing the question. STR does not issue bans on superluminal causation. It does not address causation at all. It rather requires of that class of space–time theories formulated in Minkowski space–time that they be Lorentz-covariant.[4] Non-relativistic QM, which is not a space–time theory, is not subject to STR's requirements. So the question of whether STR and QM can peacefully coexist is ill-posed. Another question – can there be Lorentz-covariant quantum theories? – is well-posed. Quantum field theory (QFT) associates observables

$\hat{A}(\mathcal{D})$ with regions of space–time $\mathcal{D}$. The inhomogeneous Lorentz group[5] Λ is represented on the Hilbert space which is the common domain of these observables by a group of unitary operators $\hat{U}(\Lambda)$. QFT so formulated is Lorentz covariant iff the observables associated with the Lorentz transform $\Lambda\mathcal{D}$ of a region $\mathcal{D}$ is the corresponding unitary transform of the observables associated with $\mathcal{D}$:

$$\hat{A}(\Lambda\mathcal{D}) = \hat{U}(\Lambda)\hat{A}(\mathcal{D}) \qquad\qquad \text{(LC-QFT)}$$

That there are QFTs satisfying (LC-QFT) should settle the peaceful coexistence question. In the QFT context, bans on superluminal signal propagation are expressed by the microcausality requirement that operators associated with space-like separated regions commute (intuitively, it does not matter what order they act in). That this microcausality requirement is independent of the requirement of Lorentz covariance suggests that the folkloric connection between STR and the prohibition on superluminal signal propagation is only that.

Bell's theorem may be profitably analyzed without recourse to locality notions tenuously linked to STR. Fine (1982a,b) showed the (Clauser–Horne form of) the Bell Inequalities to be equivalent to

1   the existence of a deterministic HVT
2   the existence of joint distributions for all pairs and triples of observables
3   the existence of a stochastic HVT satisfying (10.3).

Intuitions about locality might motivate (3), but they are not directly implicated in either (1) or (2), which simply offer ambitious patterns of determinate value assignment. Indeed, a family of arguments originating with Bell (1966 – which he wrote *before* the 1964 Bell Inequalities paper) but refined by Kochen and Specker reveals that the project of assigning determinate values to sufficiently rich sets of observables is untenable, if the value assignment is subject to prima facie reasonable constraints.

Here's an informal sketch of Bell's version of the No-Go result; see Redhead (1987, ch. 5) for more details and references. Consider a project of determinate value assignment satisfying

(Spectrum) $\hat{O}$'s determinate value $[\hat{O}]$ is one of its eigenvalues

and

(FUNC)   If $\hat{A} = f(\hat{B})$, then $[\hat{A}] = f([\hat{B}])$

In a Hilbert space of dimension three, any trio $\{\hat{P}_i\}$ of mutually orthogonal projection operators furnishes a resolution of the identity operator $\hat{I}$:

$$\hat{I} = \hat{P}_1 + \hat{P}_2 + \hat{P}_3 \qquad\qquad (10.5)$$

By the Spectrum rule $[\hat{I}] = 1$ and $[\hat{P_i}] \in \{0,1\}$. The $\{\hat{P_i}\}$ commute pairwise; there is therefore an operator of which each of them is a function. So the FUNC rule requires

$$[\hat{I}] = [\hat{P_1}] + [\hat{P_2}] + [\hat{P_3}]$$

(10.6)

Equations (10.5) and (10.6) together imply that for any trio of mutually orthogonal projectors, one of them will be assigned the value 1 while the other two will be assigned the value 0. This assignment induces a linear, normalized map from the set of projection operators on Hilbert space to the interval [0,1] – indeed to the set {0,1} containing only the endpoints of that interval. This map is also a probability measure over the closed subspaces of Hilbert space. According to Gleason's theorem, for Hilbert spaces of dimension three or greater, all such probability measures are continuous. But the map induced by the project of complete determinate value assignment is discontinuous – intuitively, as it sweeps through the set of projectors, it is going to have to leap from a projector it maps to 0 to a projector it maps to 1, without assigning intermediate projectors intermediate values. A HVT inducing such a map from Hilbert space operators to their determinate values is therefore inconsistent.

Bell needs infinitely many observables – the full set of projection operators on a three-dimensional Hilbert space – to generate the contradiction. Kochen and Specker showed that 117 projectors on a four-dimensional Hilbert space could not without contradiction be assigned determinate values obedient to the FUNC and Spectrum rules; Bell–Kochen–Specker type contradictions for ever smaller sets of observables have been emerging ever since.[6] Mermin's excellent presentation of Bell–Kochen–Specker results (1993) situates one version of the Bell Inequalities among them. The No-Go argument just sketched attributes $\hat{P_1}$ the same determinate value whether it's considered an element of the orthogonal triple $T = \{\hat{P_1}, \hat{P_2}, \hat{P_3}\}$ or an element of the *different* orthogonal triple $T' = \{\hat{P_1}, \hat{P_2'}, \hat{P_3'}\}$. It assigns a non-maximal observable a *non-contextual* value, that is, one not relativized to a particular eigenbasis of the observable. (The question of contextualizing does not arise for maximal observables, whose eigenbases are unique.) Contextualizing determinate value assignments, one can avert No-Go results, by without contradiction assigning $\hat{P_1}$ in the context of the basis T a value different from the one it's assigned in the context of the basis T'.

While such a move might seem shamefully *ad hoc*, it is precisely the move Bell makes after presenting his version of the No Go result. The argument, he writes, "tacitly assumed that the measurement of an observable must yield the same value independently of what other measurements must be made simultaneously" (Bell, 1966, p. 451). To see what this vaguely Bohrian pronouncement has to do with contextualism, and to anticipate its connection with the Bell inequalities, consider the dramatically non-maximal composite system observable $\hat{I} \otimes \hat{\sigma}_x$. One way to select an eigenbasis from the myriad available for this observable is to specify a spin observable for particle one: for instance $\hat{\sigma}_x \otimes \hat{\sigma}_x$ has a unique eigenbasis which is also an eigenbasis for $\hat{I} \otimes \hat{\sigma}_x$. To attribute $\hat{I} \otimes \hat{\sigma}_x$ a non-contextual value admitting faithful measurement is to assume that a $\hat{I} \otimes \hat{\sigma}_x$ measurement has the same outcome regardless of which particle I measurement is made. Seeing no reason to suppose that measurement outcomes are in general insensitive to measuring environments, Bell rejects the non-contextuality requirement.

Whether this "judo-like maneuver" (Shimony, 1984a) of invoking Bohr to protect ambitious plans of value assignment succeeds or not, it suggests a connection between Bell–Kochen–Specker arguments and the Bell Inequalities. The locality assumptions invoked in deriving the Inequalities are a species of a non-contextuality requirement. Mermin (1993) shows how to use locality-as-non-contextuality to convert an eight-dimensional Bell–Kochen–Specker result into one version of the Bell Inequalities. (What is lost in the translation is the state independence of the Bell–Kochen–Specker result; contradiction ensues in the converted case only for certain states.) I would regard this conversion as further evidence that to focus on locality is to distort the discussion. What precipitates No-Go results are overambitious plans of non-contextual determinate value assignment, whether the systems at issue are composite and spatially separated, or simple. Others (including Bell!) would say that it is only in the cases where locality motivates the requisite non-contextuality that the No-Go results have any bite.

Correlation and explanation

In articulating his principle of the common cause, Hans Reichenbach heeded twentieth-century revolutions in physics. Taking quantum mechanics to preclude deterministic causes and relativity to preclude non-local ones, he offered *common causes* as causes which act both locally and stochastically. Roughly, where A and B are events correlated in the sense that

$$\Pr(A\&B) \neq \Pr(A) \times \Pr(B)$$

their common cause C is an event in the overlap of their backwards lightcones rendering A and B probabilistically independent in the sense that

$$\Pr(A\&B|C) = \Pr(A|C) \times \Pr(B|C)$$

The principle of the common cause frames an influential and intuitively attractive account of explanation. Correlations – for instance, the correlations effected by $|\psi\rangle_{singlet}$ – are what require explanation; explanation proceeds by specifying a common cause for the correlated events. Straightforwardly applied to quantum correlations, the principle comes to grief. Articulated to regulate demands for explanation in the context of *statistical* theories, the principle, applied to the perfect (anti)correlations established by $|\psi\rangle_{singlet}$, is satisfied only by *deterministic* common causes, that is, Cs such that $\Pr(A|C)$, $\Pr(B|C) \in \{0,1\}$ (van Fraassen,

1989). What's more, the assumption that there are common causes for correlations observed in the Bell experiments implies the Bell Inequalities (van Fraassen, 1989). Thus any theory satisfying Reichenbachian demands for explanation will be empirically false.

Explanatory activity adheres to standards: not all demands for explanation are legitimate; not all putative *explanans* are satisfactory. Philosophers of science would like to tell the difference. One way to tell the difference is, as it were, ahead of time, by articulating a template to which scientific explanations always and everywhere conform. Taking the common causal account of explanation as just such a template, Fine (1989) and van Fraassen (1989) present its quantum travails as evidence that essentialism about explanation is misplaced, that explanatory strategies arise within the various sciences variously. But for many, the feeling persists that QM's capacity to predict correlations falls dramatically short of a capacity to explain those correlations.

The Measurement Problem

These No-Go results can be read as fables whose moral is that we ought not be too ambitious in ascribing quantum observables determinate values. One way to moderate our ambition is to adopt the semantics typically announced by textbooks:

> [I]t is strictly legitimate to say that $\hat{O}$ has a value in a state $|\psi\rangle$ if and only if a measurement of $\hat{O}$ on this state is certain to yield a definite result – i.e. if and only if $|\psi\rangle$ coincides with an eigenvector of $\hat{O}$ (Gillespie, 1973, p. 61).

Although this *eigenstate/eigenvalue link* averts No-Go results, there is another debacle in store for it. A measurement is an interaction between an object system S and an apparatus R prepared in its ready state $|p_0\rangle$, ideally one that establishes a perfect correlation between eigenstates of the object observable $\hat{O}$ and pointer observable $\hat{P}$. If measurement is a quantum mechanical process, this correlation-establishing evolution should be Schrödinger evolution, and so implemented by a unitary operator $\hat{U}_M$:

$$\hat{U}_M(|o_i\rangle|p_0\rangle) = |o_i\rangle|p_i\rangle \tag{10.7}$$

The right-hand side of (10.7) is the post-measurement state, a state in which both the object and pointer observables have determinate values, according to textbook semantics; a state in which the pointer value reflects the value of the object observable. This consolidates the status of evolution driven by $\hat{U}_M$ as *measurement* evolution. But consider what happens when an object system initially in a superposition $\Sigma_i c_i |o_i\rangle$ of $\hat{O}$ eigenstates is subject to a measurement of the sort just described. To obtain the post-measurement state of the composite system, apply $\hat{U}_M$ to the premeasurement state.

$$\hat{U}_M \left[\sum_i c_i |o_i\rangle |p_0\rangle \right] = \sum_i c_i \hat{U}_M (|o_i\rangle |p_0\rangle) = \sum_i c_i |o_i\rangle |p_i\rangle \tag{10.8}$$

(Use $\hat{U}_M$'s linearity to move from the first expression to the second, and (10.7) to move from the second to the third.) Unitary measurement leaves the object + apparatus system in the entangled state $\Sigma_i c_i |o_i\rangle |p_i\rangle$ which is *not* an eigenstate of the pointer observable $\hat{P}$. According to textbook semantics, then, the pointer observable has no determinate value, and the measurement has no outcome. (One version) of *the measurement problem* is that if measurement processes obey the laws of quantum dynamics, then measurements rarely have outcomes. Cautious enough to avoid No-Go results, textbook semantics are too cautious to accommodate the manifest and empirically central fact that experiments happen. If QM as interpreted by textbook semantics were true, we'd be unable to confirm it!

Recognizing this problem, von Neumann (1955 [1932]) responded by invoking the *deus ex machina* of measurement collapse, a sudden, irreversible, discontinuous change of the state of the measured system to an eigenstate of the observable measured. According to this (quite orthodox – many texts accord this "Collapse Postulate" axiomatic status) way of thinking, reconciling textbook semantics with the datum that there are empirical data requires suspending unitary dynamics in measurement contexts, and interpreting Born Rule probabilities as probabilities for collapse. Collapse is a Humean miracle, a violation of the law of nature expressed by the Schrödinger equation. If collapse and unitary evolution are to coexist in a single, consistent theory, situations subject to unitary evolution must be sharply and unambiguously distinguished from situations subject to collapse. And despite evocative appeals to such factors as the intrusion of consciousness or the necessarily macroscopic nature of the measuring apparatus, no one has managed to distinguish these situations clearly.

Contemporary Work

I now have on hand material sufficient to frame much recent philosophical work on QM. The challenge is to offer an interpretation of the theory which makes sense of measurement outcomes without running afoul of NoGo results. Such an interpretation will have to revise one or more of the following naive identifications, the set of which precipitates the measurement problem:

Quantum *states* are normed vectors $|\psi\rangle$ on a Hilbert space $\mathcal{H}$.
Quantum *observables* are self-adjoint operators on $\mathcal{H}$.
Quantum *dynamics* is unitary Schrödinger dynamics.
Quantum *semantics* are given by the eigenstate/eigenvalue link.

The revisions that require the least new physics, are semantic revisions; revisions which retain the standard state space but reconfigure its dynamical trajectories are more radical; most radical of all are revisions to the fundamental state space and observable set of QM. A recurrent feature of interpretations of QM is that their conservative exteriors hide radical hearts.

Changing the dynamics: The GRW model

The GRW model of quantum processes (Ghirardi, Rimini and Weber, 1986) – see also Pearle (1989) – would avoid having to reconcile Schrödinger and non-Schrödinger evolution by dispensing with Schrödinger evolution. GRW offers in its stead a more general form of state evolution, to which Schrödinger evolution is nearly approximate. The GRW equation of motion for an isolated quantum system supplements the usual unitary term with a non-unitary term. The effect of this extra term is, rarely and at random, but with a uniform probability per second (10^{-15}), to multiply the system's configuration space state $|\psi(x) >$ by a Gaussian (bell curve) of width 10^{-7} meters, then normalize. The result of a hit by a Gaussian centered at $x = q$ is a wave function $|\psi_q(x) >$ localized about q. Given that a particle in the state $|\psi(x) >$ is hit by a Gaussian, the GRW dynamics set the probability that it's hit by a Gaussian centered at $x = q$ equal to the Born Rule probability $|\psi(q)|^2$ that a position measurement performed on a system in the state $|\psi(x) >$ has the outcome q.

Generally, when systems interact, their composite state becomes entangled. For instance, a purely unitary $\hat{\sigma}_x$ measurement coupling a pointer system containing N particles to an electron in initial state $c_+| \rightarrow > + c_-| \leftrightarrow >$ generates the post measurement state

$$c_+| \rightarrow > \bigotimes_{i=1}^{n} |\chi_+(x)\rangle_i + c_-|\leftarrow > \bigotimes_{i=1}^{n} |\chi_-(x)\rangle_i \qquad (10.9)$$

where $|\chi_\pm(x) > i$ represents the ith particle in a pointer localized about $x = \pm L$. As the number of particles in the pointer grows, so too does the probability that one of them experiences a GRW collapse. The entanglement of (10.9) ensures that multiplying the state of any particle in the pointer by a Gaussian centered at $+L$ renders the second term on the right-hand side negligible, and so leaves the composite system localized about $+L$. Because our measuring apparatuses (generally) couple a macroscopic number of systems together, such a reduction is overwhelmingly likely to occur practically immediately upon the completion of measurement.

Thus, the GRW dynamics imply that the quantum states of individual systems will almost always Schrödinger evolve, while the quantum states of macroscopic measuring apparatuses are almost always highly localized. But this does not render GRW an unqualified success. It accounts only for measurement outcomes recorded in positions. However, it may not be that all measurement outcomes are so recorded (Albert (1992, ch. 5) presents one which, *prima facie*, is not). And in addition to modifying the dynamics of the naive interpretation, GRW must modify its semantics, and perhaps even its observable set. For GRW reductions are not reductions to strictly localized states (that is, states $|\varphi(x)\rangle$ such that for some finite interval Δ, $\int_\Delta \varphi^*(x)\varphi(x)dx = 1$ – recall that there are no point-valued position eigenstates). Rather, they are reductions to states with infinite tails in configuration space. The *problem of tails* is that adhering strictly to the orthodox semantics motivating their pursuit of reduction, GRW cannot attribute even interval-valued determinate positions to even systems in post-reduction states such as $|\psi_q(x)\rangle$. Relieving us of peculiar measurement dynamics, GRW does not supply our pointers with determinate positions.

A possible recourse is to liberalize eigenstate/eigenvalue semantics so that "System S in $|\psi(x)\rangle$ is localized in the interval Δ" is true iff

$$\int_\Delta \psi^*(x)\psi(x)dx > 1 - \varepsilon$$

where $0 < \varepsilon < \frac{1}{2}$ (Albert and Loewer, 1996). Setting $\varepsilon = 0$ reinstitutes the eigenstate/eigenvalue link; setting $\varepsilon = \frac{1}{2}$ allows incompatible propositions (for instance, those associated with the projectors P_Δ and $I - P_\Delta$) to be true at once. Setting ε somewhere in between implies that a system in the state $|\psi(x)\rangle$ can be *localized in* Δ while a system in the state $|\psi'(x)\rangle$ is *localized in* Δ', where Δ and Δ' are disjoint, even though $|\psi(x)\rangle$ and $|\psi'(x)\rangle$ are not orthogonal. If so, GRW's *localized* observable is not a standard quantum mechanical one. For quantum observables, self-adjoint operators, are projection-valued measures which associate distinct eigenvalues of the observable with orthogonal subspaces of Hilbert space. Though Δ and Δ' are distinct values of GRW's *localized* observable, $|\psi(x)\rangle$ and $|\psi'(x)\rangle$ are not orthogonal, and *localized* is not a self-adjoint operator. By liberalizing textbook semantics, GRW makes the more radical interpretive move of revising QM's observable set.[7]

Changing the state space: The Bohm theory

On Bohm's causal interpretation – originating in Bohm (1952); see Cushing et al. (1996) for subsequent developments – all particles have determinate positions. Thus Bohm attributes a system of N particles of mass m moving in three dimensions a determinate *configuration* $Q \in \mathbb{R}^{3N}$ in a configuration space of their possible joint positions. The quantum wave function $\psi(x_1, \ldots, x_{3N})$ for the system can be expressed as a function over this configuration space. Manipulating the Schrödinger equation, and reasoning by analogy with other bits of physics, Bohm offers a set of velocity functions

$$\dot{x}_i = \upsilon((x_1, \ldots, x_{3N})) = \frac{1}{m} \mathrm{Im}\left(\frac{\nabla_i \psi(x_1, \ldots, x_{3N})}{\psi(x_1, \ldots, x_{3N})}\right)$$

which make each component of each particle's velocity depend both on the quantum state and on the configuration of the composite system.[8] Possessing at all times "precisely definable and continuously varying values of position and momentum" (Bohm, 1952, p. 373), a Bohmian particle follows a deterministic trajectory.[9]

Bohmian Mechanics is the *guidance condition* (the velocity functions $\dot{x}_i$) along with the requirement that $\psi(x_i)$ evolve in accordance with the Schrödinger equation. Given the appropriate initial conditions, an ensemble of particles following their Bohmian trajectories can reconstitute quantum statistics. If at some initial time t_0, the distribution of determinate positions among particles in an ensemble assigned $\psi(x_i, t_0)$ is described by the probability density $|\psi(x_i, t_0)|^2$, then at all later times, probability densities are well-behaved, and described by the appropriate Schrödinger developments of $\psi(x, t_0)$. Bohm's *distribution postulate* is that $|\psi(x_i)|^2$ does give the probability density.[10]

Bohm's interpretation does not assign noncontextual determinate values to observables other than position. Non-position observables it relegates to the realm of dispositions manifested in the post-measurement positions of pointer systems. These dispositions are contextual: whether a particle described by some superposition of spin eigenstates will wind up in a position indicating spin up or spin down depends not only on the initial position of the particle but also on how the measuring device is configured; Albert (1992, ch. 7) gives a simple illustration. And positions themselves are subject to manifestly non-local influences: the velocities of individual particles are functions of the configurations of the composite systems they comprise, so that (reverting to the EPR case) changes in particle II's position instantaneously alter particle I's velocity. Bohmians deem this non-locality benign. Maintaining that we can not predict or control particle positions, they argue that we can not harness the non-locality for signaling purposes.

Stingy, contextual, non-local, the Bohm interpretation avoids No-Go results. It accounts for measurement outcomes recorded in particle positions. And it seems to its adherents "the most obvious," "most natural," and "simplest" (Dürr et al., 1996, pp. 21, 24) account of quantum phenomena – so much so that Cushing (1994) has suggested that had Bohm beaten Bohr to prominence, standard physics curricula would include Bohmian, rather than quantum, mechanics.

Pleas for the naturalness of Bohmian mechanics sometimes derive illicit support from glosses like the following (which Bell supplied for its ancestor, de Broglie's pilot wave theory):

> De Broglie showed in detail how the motion of a particle passing through just one of two holes in a screen could be influenced by waves propagating through both holes. And so influenced that the particle does not go to where the waves cancel out, but is attracted to where they cooperate. This idea seems to me so natural and simple, to resolve the wave-particle dilemma in such a clear and ordinary way, that it is a great mystery to me that it was so generally ignored. (Bell, 1987, p. 191)

On the picture enchanting Bell, the particle surfs the pilot wave through space. It is a picture that results when the Bohm theory's state space – the configuration space $\mathbb{R}^3$ on which the wave function of a single particle is defined – is identified with physical space. Such an identification threatens to confuse the representational with the concrete, and is anyway foiled once the theory considers systems composed of $N > 1$ particles, with wavefunctions $\psi(x_1, \ldots, x_{3N})$. Then surf's up not in $\mathbb{R}^3$ but in a $3N$-dimensional configuration space which it is not tempting to identify with physical space.

Natural or not, the Bohm theory is significant. By refusing to constitute matters of fact from determinate quantum observables, Bohm not only circumvents the usual No-Go results, but shows how they stack the deck against the non-quantum physicist by foisting upon her a quantum-theoretic space of possibilities.

Changing the semantics: Modal interpretations

Modal interpretations – see Kochen (1985), Healey (1989), Dieks (1989), van Fraassen (1991) and, for more recent work, Dieks and Vermaas (1998) – would resolve the Measurement Problem by maintaining the universality of Schrödinger evolution while revising the eigenvector/eigenvalue link. A stock example of a modal interpretation exploits the biorthogonal decomposition theorem, according to which any vector $|\psi\rangle^{SR}$ in the tensor product space $\mathcal{H}_S \otimes \mathcal{H}_R$ admits a decomposition of the form

$$|\Psi\rangle^{SR} = \sum_i c_i |a_i\rangle |b_i\rangle$$

where $\{c_i\}$ are complex coefficients, $\{|a_i\rangle\}$ and $\{|b_i\rangle\}$ are sets of orthogonal vectors on $\mathcal{H}_S$ and $\mathcal{H}_R$ respectively, and the summation index i does not exceed the dimensionality of the smaller factor space. If the set $\{|c_i|^2\}$ is non-degenerate, then this biorthogonal decomposition of $|\psi\rangle^{SR}$ is unique. Modal interpretations replace the orthodox eigenvector/eigenvalue link with the following semantic rule:

If $|\Psi\rangle^{SR} = \sum_i c_i |a_i\rangle |b_i\rangle$ is the unique biorthogonal decomposition of the state of a composite $S + R$ system, then subsystem S has a determinate value for each $\mathcal{H}_S$ observable with eigenbasis $\{|a_i\rangle\}$, and subsystem R has a determinate value for each $\mathcal{H}_R$ observable with eigenbasis $\{|b_i\rangle\}$. $|c_i|^2$ gives the probability that these observables' actual values are the eigenvalues associated with $|a_i\rangle |b_i\rangle$.

Consider the unitarily evolved post-measurement state

$$|\Psi\rangle^{S+R} = \sum_i c_i |o_i\rangle |p_i\rangle$$

The eigenbasis of the pointer observable $\hat{P}$ conspires in its biorthogonal decomposition. By modal semantics, then, the pointer observable $\hat{P}$ is determinate on the apparatus system after measurement. Moreover, the probability that $\hat{P}$'s actual value is p_n is just the Born Rule probability. Thus would modal interpretations explain what textbook interpretations can not: how measurement interactions obedient to the laws of quantum dynamics issue determinate outcomes corroborating quantum statistical predictions.

Four problems for this stock modal interpretation are listed here:

(i) What to say when the biorthogonal decomposition is degenerate
     In the extreme case where $\mathcal{H}_S$ and $\mathcal{H}_R$ are each of dimension $N > 2$ and

$$|\Psi\rangle^{SR} = \sum_i \frac{1}{\sqrt{N}} |a_i\rangle |b_i\rangle$$

the eigenbasis of every observable on the component systems conspires in some biorthogonal decomposition, and Kochen–Specker contradictions threaten.

(ii) What to say about the dynamics of possessed values
     A viable option, one preserving the status of the modal interpretation as an interpretation that succeeds not by developing new physics but by adjusting semantics to existing physics is: nothing. Dickson (1998a) describes modal dynamics which are dramatically underdetermined by the requirement that they return single time probabilities conforming to the Born Rule, and discusses that underdetermination.

(iii) What to say about state preparation, the laboratory processes whereby we assign states to quantum systems
     Modal interpretations cannot avail themselves of the standard account that measurement collapse leaves the prepared system in the eigenstate of the measured observable corresponding to the eigenvalue obtained. Perhaps modal interpretations can account for preparation by appeal to conditional probabilities: the "prepared" state is the one mimicking the post-preparation composite state's predictions for the prepared system, conditional on the "outcome" of the preparation – Wessels (1997) treats preparation along these lines. Adopting standard quantum expressions for conditional probabilities, modal interpretations can take this way with preparation at the cost, in certain settings, of violating the Markov consistency requirement that

$$\Pr(a|b) = \sum_i \Pr(a|c_i) \times \Pr(c_i|b)$$

where $\{c_i\}$ is an exhaustive set of mutually exclusive events intermediate between a and b. Using non-standard conditional probabilities, modal interpretations embark on value state dynamics, with the class of candidate dynamics narrowed to those that make sense of preparation.

(iv) What to make of non-ideal measurements (Albert, 1992, appendix)
     These are measurements which fail to correlate eigenstates of the designated pointer observable with orthogonal states of the object system, so that the

pointer eigenbasis fails to furnish a biorthogonal decomposition of the post-measurement composite state. By the biorthogonal decomposition theorem, some apparatus eigenbasis will furnish a biorthogonal decomposition, and observables with this eigenbasis, not the pointer observable, are determinate after measurement, according to modal semantics. Perfectly error-free measurements confront modal intepretations with this problem, and there is a class of observables whose only error-free measurements are of this sort (Ruetsche, 1995).

Responses to (iv) (and also (i) and (iii)) appeal to *decoherence* processes – interactions between the pointer and its environment that tend to correlate distinct pointer eigenstates with nearly orthogonal states of the environment.[11] The suggestion is that decoherence carries post non-ideal measurement systems into states biorthogonally decomposed by apparatus observables close enough to the designated pointer observables that one needn't fret (Bacciagaluppi and Hemmo, 1996). Because decoherence is not perfect, this response leaves the modal interpretation with its own version of the problem of tails, a problem whose resolution might lie in the now-familiar maneuver of constituting matters of fact from something other than determinate quantum observables.

Relative state formulations

"Postulat[ing] that a wave function that obeys a linear wave equation everywhere and at all times supplies a complete mathematical model for every isolated physical system without exception" (Everett, 1983 [1957], p. 316), Hugh Everett's Relative State Formulation promises an interpretation according to which the quantum state description is complete and the quantum dynamics are universal. Although the entangled post-measurement state

$$|\Psi>^{S+R} = \sum_i c_i |o_i> |p_i>$$

associates no $\hat{O}(\hat{P})$ eigenstates with the object (apparatus) simpliciter, it correlates $\hat{O}$ and $\hat{P}$ eigenstates with one another. This illustrates Everett's moral that "the state of one subsystem does not have an independent existence, but is fixed only by the state of the remaining subsystem" (1983 [1957], p. 316), so that "it is meaningless to ask the absolute state of a subsystem – one can only ask the state relative to a given state of the remainder of the system" (1983 [1957], p. 317). (Relatively speaking) when system S has determinate $\hat{O}$ value o_n, system R has determinate $\hat{P}$ value p_n, and "this correlation is what allows one to maintain the interpretation that a measurement has been performed" (1983 [1957], p. 320). Thus Everett purports to reconcile the uncollapsed composite state $|\psi>^{S+R}$ with determinate measurement outcomes.

But the terms of reconciliation are notoriously unclear. An option proposed by physicists but embraced by the science fiction community is that "the universe is constantly splitting into a stupendous number of branches, all resulting from the measurement-like interactions between its myriads of components" (DeWitt, 1970, p. 161); within each branch, the relative state of the pointer registers a determinate outcome. Criticisms of this version of Everett (Albert and Loewer, 1988) include that its profligate creation of new universes violates the conservation of mass/energy required by unitary evolution, and that it makes hash of quantum probabilities by rendering every outcome *certain* to occur along some branch. What's more, to disambiguate this version of Everett, its proponents must furnish an account of when splitting occurs, and into what branches. Such an account would serve also on the von Neumann collapse interpretation to distinguish systems subject to collapse from systems evolving unitarily, rendering that interpretation consistent, unambiguous, and free of suspect metaphysics.

More recent Everett-style interpretations have responded to the disambiguation problem in one of two broad ways. The more fanciful notes that it is, after all, only our determinate experiences which must be reconciled with universal unitary evolution, and so postulates "eigenstates of mentality" – brain states to which correspond mental states whose contents are determinate beliefs – as a preferred basis of relative states. Perhaps the most astonishing variation of this approach is the Many Minds interpretation, a radical dualism which invites us to

> Suppose that every sentient physical system there is is associated not with a single mind but rather with a *continuous infinity* of minds; and suppose (this is part of the proposal too) that the measure of the infinite subset of those minds which happen to be in some particular mental state at any particular time is equal to the square of the absolute value of the coefficient of the brain state associated with that mental state, in the wave function of the world at that particular time. (Albert, 1992, p. 130)

A more prosaic response (Griffiths, 1993; Hartle, 1990) to the disambiguation problem offers *consistent* (or *decoherent*) histories as the preferred basis of relative states. A time-indexed set of determinate observables generates a family of *histories* for a system; an individual history in the family assigns observables in the time-indexed set determinate values. Given the initial state of the system and the unitary operator governing its evolution, a generalized Born Rule assigns probabilities to such histories. A family of histories is said to be consistent if the probabilities so assigned do not "interfere" – roughly, they are Markov consistent. Thus histories in a consistent family admit multi-time probability assignments that constitute a tractable dynamics.

The rub is that while the initial state and the system Hamiltonian constrain which families of histories are consistent, they don't determine a unique family of consistent histories. So, while there may be a consistent family of histories declaring the pointer observable determinate at measurement's completion, there will also be other consistent families which do not. What assures that a consistent family containing the pointer observable, and not one excluding it, corresponds to what actually occurs in the laboratory? Branding families of consistent histories which foil merger into a family satisfying the non-interference condition "complementary," Griffiths rejects this yen for reassurance on broadly Bohrian grounds: "A question of the form, 'Which of these really took place?' asked in terms of comparing two mutually incompatible histories, makes no sense quantum mechanically" (Griffiths 1993, p. 2204).

Like the perspectival metaphysics of the many worlds approach, this response is philosophically suspect. Yet Everett-style approaches are the preferred quantum framework for many working physicists. Rovelli (1997) sees in "relational QM" the seeds of a solution to the problem of time in quantum gravity; Hartle (1990) puts the consistent histories approach, and the tractable (if perspectival) dynamics it underwrites, to cosmological use. Meanwhile, interpretations of QM more philosophically respectable languish relatively unloved.[12] Saunders offers a stark diagnosis: "The disturbing feature of both the Bohm and GRW approaches is that they seem to require that we *redo* high energy physics" (Saunders, 1996, pp. 125–6). Requiring a preferred time foliation, both approaches fundamentally (if not phenomenologically) violate Lorentz and general covariance, and thus deprive physicists of a powerful criterion for winnowing down the set of acceptable theories. This should remind us at least that non-relativistic QM is not the only game in town – a lesson those working on the foundations of quantum theories have increasingly taken to heart.

Future Directions: Interpreting QFT

With apologies to those who have been working in the field for years – for a very recent review, see Huggett (2000) – I offer QFT – and quantum gravity, a theory about whose eventual shape QFT on curved spactimes might hold a clue – as one future direction for the philosophy of quantum theories. Moving from the least to the most exotic space–time settings, this section sketches some issues that are kicked up by the pursuit of quantum theories in such settings.

Minkowski space–time

The proper setting for questions about "locality," QFT is also a provocative one. A striking example is the Reeh–Schlieder theorem, which states that where $\{A(O)\}$ is the set of observables the theory associates with an open bounded region of space–time O and $|0\rangle$ is the Minkowski vacuum state, $\{A(O)\,|0\rangle\}$ is *dense* in the theory's state space – that is, any state the theory recognizes can be approximated arbitrarily closely by acting on the vacuum by polynomial combinations of observables in $\{A(O)\}$. If it were appropriate to model events in the region O as applications of elements of $\{A(O)\}$ to the global vacuum state, this would mean that machinations in local regions could produce arbitrary approximations of arbitrary global states! The model is not apt, but its whiff of non-locality is. The Reeh–Schleider theorem implies that $|0\rangle$ is an eigenstate of no observable associated with a finite space–time region, which in turn implies that the vacuum spreads correlations far and wide. Redhead (1995) illuminates the Reeh–Schleider theorem by explicating analogies between how the vacuum stands to local algebras of observables and how the spin-singlet state stands to algebras of spin observables pertaining to the component systems. Clifton et al. (1998) show that states with $|0\rangle$'s feature that given any pair of space–time regions, any observable from one is correlated with some observable from the other, are dense; Butterfield (1994) discusses the capacity of such correlations to violate Bell-type inequalities (they can, even maximally). The nature and extent of such non-local features of QFT, as well as the theory's hospitability to causal talk, are topics of ongoing research.

To see how questions about the ontology of QFT, as well as its state space, arise, we need to go into a bit more detail. The canonical approach to quantization casts a classical theory in Hamiltonian form, then promotes its canonical observables q_k, p_k to symmetric operators $\hat{q}_k, \hat{p}_k$ obeying *canonical commutation relations* arising from the Poisson brackets of the classical theory. A classical *field* theory assigns a field configuration $\varphi(x)$ and a conjugate momentum density $\pi(x) \equiv \partial L/\partial \dot{\phi}$ (where L is the theory's Lagrangian density) to every point x of space–time; its quantization proceeds by finding operators $\hat{\phi}(x)$ and $\hat{\pi}(x)$ obeying the relevant canonical commutation relations.[13] I will refer in what follows to a mathematically well-behaved exponential form of these commutation relations known as the Weyl relations, and call sets of operators satisfying them *representations* of the Weyl relations.

A simple classical field is the Klein–Gordon field $\varphi(x)$, which satisfies

$$(g^{ab}\nabla^a\nabla_a - m^2)\varphi(x) = 0$$

Its solutions can be Fourier-decomposed into uncoupled normal modes with angular frequency ω_k, and so the classical field can be modeled as an infinite collection of independent oscillators. The textbook route to quantization exploits this analogy by introducing creation and annihilation operators $\hat{a}_k^\dagger$ and $\hat{a}_k$ for field modes obeying

$$[\hat{a}_k,\hat{a}_{k'}] = 0 = [\hat{a}_k^\dagger,\hat{a}_{k'}^\dagger], [\hat{a}_k,\hat{a}_{k'}^\dagger] = \delta_{kk'}\hat{I} \qquad (10.10)$$

Formal expressions for operators $\hat{\phi}(x)$ and $\hat{\pi}(x)$ satisfying the canonical commutation relations can be constructed from these. The resulting quantization is the free boson field; imposing anti-commutation relations in lieu of (10.10) yields the free fermion field.

The state $|0>$ such that $\hat{a}_k|0>=0$ for all k is the lowest energy eigenstate of the quantum Hamiltonian for the free boson field. The state $(\hat{a}_k^\dagger)^n |0>$ is an eigenstate of the Hamiltonian with the same energy a system of n particles each with energy $\hbar\omega_k$ would have – provided the momenta and rest masses of these particles are given by standard relativistic expressions. Thus, the theory tempts a particle interpretation:

- the vacuum state $|0>$ is the no particle state
- the state $\hat{a}_k^\dagger|0>$ describes one particle of energy $\hbar\omega_k$. . . .
- $\hat{N}_k = \hat{a}_k^\dagger\hat{a}_k$ is the number operator for particles of type k
- $\hat{N} = \sum_j \hat{a}_j^\dagger\hat{a}_j$ is the total number operator

and so on. Countering this temptation in the first instance are some distinctly unparticulate features of the theory so interpreted (Teller, 1995, ch. 2). For one thing, the theory hosts states with indeterminate particle numbers. For another, even states which are eigenstates of the total number operator are constrained by (10.10) to be symmetric – that is to be unchanged under permutations of particle labels. Whether this, and their ensuing obedience to Bose-Einstein statistics, deprives bosons of the genidentity one might expect from particles has been a topic of lively debate, well-represented in Castellani (1998).

A prior challenge to the viability of particle interpretations has excited somewhat less interest among philosophers. Consider two quantum theories, each taking the form of a Hilbert space $\mathcal{H}$, and a collection of operators $\{\hat{O}_i\}$. When are these theories physically equivalent? A natural criterion of equivalence is that the theories recognize the same set of states, that is, probability distributions over eigenprojections of their observables. And a sufficient condition for this is that the theories be *unitarily equivalent* in the sense that there exists a unitary map $\hat{U}: \mathcal{H} \rightarrow \mathcal{H}'$ s.t. $\hat{U}^{-1}\hat{O}_i'\hat{U} = \hat{O}_i$ for all values of i, in which case the expectation values assigned observables $\{\hat{O}_i\}$ by any state $|\psi\rangle$ in the first theory are duplicated by those assigned $\{\hat{O}_i'\}$ by the state $\hat{U}|\psi\rangle$ in the second. If the observable set is rich enough, unitary equivalence is necessary as well. If physical equivalence is unitary equivalence, the quantization of a classical theory yields a unique quantum theory if and only if all representations of the relevant Weyl relations are unitarily equivalent. The Stone–Von Neumann theorem ensures that representations of Weyl relations expressing the quantization of a classical theory with a finite dimensional state space are unique upto unitary equivalence. But classical fields have infinitely many degrees of freedom. The Stone–Von Neumann theorem does not apply. Indeed, the Weyl relations encapsulating the quantization of classical Klein–Gordon theory admit continuously many inequivalent representations.

Let $\{\hat{a}_k, \hat{a}_k^\dagger\}$ be one quantization of some classical field theory, and $\{\hat{a}_k', \hat{a}_k'^\dagger\}$ be another, unitarily inequivalent to the first. In general, the primed vacuum state will not be a state in the unprimed representation, nor will the primed total number operator be an operator there, and *mutatis mutandis*. One might say that associated with the unitarily inequivalent quantizations are *incommensurable* particle notions. Even granting that it is appropriate to run a particle interpreta-tion of a quantization $\{\hat{a}_k, \hat{a}_k^\dagger\}$, one cannot run a sensible particle interpretation of QFT unless one can privilege as *physical* a unitary equivalence class of representa-tions admitting particle interpretations – as Saunders (1988) discusses, not all do.

The default setting for a QFT is Minkowski space–time. And this furnishes a de facto criterion of privilege: physical representations respect the symmetries of the space–time, in the sense that their vacua are invariant under its full isometry group. Coupled with the requirement that physical representations admit only states of non-negative energy, this singles out a unitary equivalence class of rep-resentations. But this strategy for privilege breaks down in generic curved space–time settings, which do not supply the symmetries it requires.[14]

The *algebraic approach* to quantum theories grounds an entirely different response to unitarily inequivalent representations. The algebraic approach articu-lates the physical content of a theory in terms of an abstract algebra $\mathcal{A}$. Observ-ables are elements of $\mathcal{A}$, and states are normed, positive linear functionals $\omega: \mathcal{A} \rightarrow \mathbb{C}$. The expectation value of an observable $A \in \mathcal{A}$ in state ω is simply $\omega(A)$. Abstract algebras can be realized in concrete Hilbert spaces. A map π from ele-ments of the algebra to the set of bounded linear operators on a Hilbert space $\mathcal{H}$ furnishes a Hilbert space representation of the algebra. In particular, all Hilbert spaces carrying a representation of the Weyl relations are also representations of the abstract Weyl algebra. For the algebraist, "[t]he important thing here is that the observables form some algebra, and not the representation Hilbert space on which they act" (Segal, 1967, p. 128). Inequivalent representations need not puzzle him, for conceiving the state space of a quantum theory as the space of algebraic states, he has rendered unitary equivalence an inappropriate criterion of physical equivalence. (Early proponents of the algebraic approach concocted baldly operationalist motivations for alternative glosses on physical equivalence; see Summers (1998) for a review.) Nor need he trouble with particles, for particle notions are (at least prima facie) the parochial residues of concrete representations.

Standard quantum states are probability measures over closed subspaces of Hilbert space. The class of algebraic states is broader than the class of such probabil-ity measures. There are, for instance, algebraic states which can accomplish what no Hilbert space state can: the assignment of precise and punctal values to continuous observables (Clifton, 1999). Some would advocate restricting admissable algebraic states. A restriction that looks down the road to quantum gravity is the Hadamard condition, which requires admissable state to be states for which a prescription assigning the stress-energy tensor an expectation value succeeds. (Provocatively, in closed space–times such states form a unitary equivalence class (Wald, 1994, §4.6). Both mathematical and physical features of algebraic states merit, and are receiving, further attention, attention which should inform discussion about the state space, and maybe even the ontology, appropriate to QFT.

Curved space–time

Different notions of state demand different dynamical pictures. Hilbert space dynamics are implemented by unitary Hilbert space operators. Having jettisoned Hilbert spaces as essential to QFT, the algebraist has jettisoned as well this account of the theory's dynamics. In its stead, he implements quantum field dynamics by means of automorphisms of the abstract algebra $\mathcal{A}$ of observables (that is, structure-preserving maps from $\mathcal{A}$ to itself). A question of equipollence arises: is it the case that every dynamical evolution implementable by an automorphism on the abstract algebra is also implementable as a unitary evolution in some fixed Hilbert space? Algebraic evolution between Cauchy slices related by isometries can be implemented unitarily, but more general algebraic evolution can not be; see Arageorgis et al. (2001) for details. One moral we may draw from these results is that in exactly those space–times whose symmetries furnish principles by appeal to which a unitary equivalence class of representations might be privileged, dynamical automorphisms are unitarily implementable. In more general settings, the algebraic formulation is better suited to capturing the theory's dynamics.

Unitarity breaks down even more dramatically in the exotic space–time setting of an evaporating black hole. Hawking has argued that a pure to mixed state transition – the sort of transition von Neumann's collapse postulate asserts to happen on measurement – occurs in the course of black hole evaporation. Not only unitarity but also symmetries of time and pre/retrodiction are lost if Hawking is right. Belot et al. (1999) review reactions to the Hawking Information Loss Paradox; not the least of the many questions the Hawking paradox raises is how to pursue QFT in non-globally hyperbolic space–times.

Quantum gravity

The QFTs discussed so far are free field theories, whereas the QFTs brought into collision with data from particle accelerators are *interacting* field theories, whose empirical quantitites are calculated by perturbative expansions of the free field. The divergence of these expansions calls for the art and craft of renormalization, chronicled in Teller (1995, chs 6 and 7). Cushing (1988) argues that this (and every other!) feature of QFT raises not "foundational" but "methodological" issues. Insofar as methodological predilections are affected by foundational commitments and affect the shape of future theories, the two domains might not be so cleanly separable as Cushing suggests; ongoing work on Quantum Gravity is one place to look for their entanglement.

Notes

1 See Hughes (1989, pt. I) for an introduction. As space limitations prohibit even a rudimentary review, I attempt in what follows to minimize technical apparatus.

2 Such explanation has its limits. Consider $\hat{\sigma}_x$ and $\hat{\sigma}_y$, perpendicular components of

intrinsic angular momentum or spin. They obey uncertainty relations, and their measurement requires incompatible experimental apparatus. Yet spin is explicitly and infamously a quantum phenomenon, and $\hat{\sigma}_x$ and $\hat{\sigma}_y$ are individually conserved. They are not *classical* concepts precluding one another's application, nor occupants different sides of the kinematic/dynamic divide.

3 For a derivation, which is simply a matter of bringing a home truth about sums of 1s and −1s to bear upon probabilities the HVT assigns, see Redhead (1987, pp. 97–8). This form of Bell Inequality applies to both deterministic and stochastic HVTs. A particularly simple derivation of a Bell Inequality, due to Wigner, applies to deterministic HVTs requiring perfect (anti-) correlation; see Hughes (1989, pp. 170–2). One can derive Bell-type inequalities without the intermediary of hidden variables, provided one assumes that joint probability distributions for non-commuting observables are well-defined; see Redhead (1987, pp. 81–3) for this Stapp-Eberhard form of the inequalities.

4 Maudlin (1994) discusses STR's real and imagined implications, and the constraints they place on the interpretation of QM.

5 Conventionally denoted by the same letter as, but not to be confused with, the space of hidden variable states.

6 Bub (1997) gives a thorough review, and presents Bub and Clifton's trend-bucking "Go" theorem, which characterizes the *largest* set of observables that can without contradiction be attributed determinate values obedient to the Spectrum and FUNC rules.

7 So setting ε also has the repercussions that our discourse about localization features oddities reminiscent of discourse involving vague predicates. For instance, each of a pair of predicates ("is localized in Δ" and "is localized in Δ'") can be true of some system S without their conjunction ("is localized in $\Delta \cap \Delta'$") being true of S. Whether we can live with this is a topic of ongoing debate; see, for instance, Clifton and Monton (1999).

8 The analogies plied in Bohm's original presentation invoke a quantum potential with disquieting features; Dürr et al. (1996) attempt to eliminate this invocation by showing how the velocity function is suggested (if not implied) by symmetry considerations alone.

9 Vink (1993) extends Bohm's approach to assign every observable a determinate value (albeit a contextual one), and offer for those possessed values a generalization of Bohmian dynamics which is stochastic when the observables are discrete.

10 Valentini (1991) would like to unifiy the role of $\psi(x_i)$ in the Bohm theory by proving a "quantum H-theorem" according to which arbitrary initial distributions evolve under the influence of the Bohmian equations of motion to the distribution $|\psi(x_i)|^2$. This would render the distribution postulate otiose. Dickson (1998b, pp. 123–5) offers criticisms of Valentini's approach.

11 Zurek (1982) offers toy models of decoherence processes, as well as the claim that decoherence solves the measurement problem. An apparatus entangled not only with the object system but also with its environment is still entangled, and not a system to which eigenstate/eigenvalue semantics attribute determinate values. To respond to the measurement problem, decoherence proposals need to be accompanied by non-standard semantics. Modal semantics work admirably.

12 But see Huggett and Weingard (1994) for Bohmian approaches to QFTs, and Pearle (1992) for Lorentz-invariant quantum field version of continuous spontaneous localization.

13 Mathematical nicety demands that the quantum field be cast not (as the foregoing suggests) as a map from space–time points to operators, but as operator-valued distributions over space–time regions. Wald (1994) is an excellent introduction to this and other issues discussed in this section.

14 Notoriously, it even breaks down in a subset of Minkowski space–time. Positive energy states correspond to solutions to the Klein–Gordon equation that oscillate with purely positive frequency. States in the standard Minkowski representation are positive frequency with respect to time as measured by families of inertial observers. But restricting our attention to the right Rindler wedge of two-dimensional Minkowski space–time setting $c = 1$, this is the region where x is positive and $|x| < t$ – we can quantize the Klein-Gordon field by admitting solutions that oscillate with positive frequency with respect to time as measured by observers whose accelerations are constant, for Lorentz boost isometries generate a global time function for the Rindler wedge. The Rindler representation we thereby obtain has a natural particle interpretation – but the Rindler representation is unitarily inequivalent to the Minkowski representation! This is sometimes, and loosely, expressed as the Unruh effect: observers accelerating through the Minkowski vacuum "see" a thermal flux of particles (Wald, 1994, ch. 5).

References

Albert, D. (1992): *Quantum Mechanics and Experience*. Cambridge, Mass: Harvard.

Albert, D. and Loewer, B. (1988): "Interpreting the Many Worlds Interpretation," *Synthese*, 77, 195–213.

Albert, D. and Loewer, B. (1996): "Tails of Schrödinger's Cat," in R. Clifton (1996, pp. 81–92).

Arageorgis, A., Earman, J. and Ruetsche, L. (2001): "Weyling the Time Away: The Non-Unitary Implementability of Quantum Field Dynamics on Curved Space–time and the Algebraic Approach to Quantum Field Theory," *Studies in History and Philosophy of Modern Physics*, forthcoming.

Bacciagaluppi, G. and Hemmo, M. (1996): "Modal Interpretations, Decoherence and Measurements," *Studies in the History and Philosophy of Modern Physics*, 27B, 239–77.

Bell, J. (1964): "On the Einstein–Podolsky–Rosen Paradox," *Physics*, 1, 195–200. Reprinted in Wheeler and Zurek (1983, pp. 403–8).

Bell, J. (1966): "On the Problem of Hidden Variables in Quantum Mechanics," *Reviews of Modern Physics*, 38, 447–52. Reprinted in Wheeler and Zurek (1983, pp. 397–402).

Bell, J. (1987): *Speakable and Unspeakable in Quantum Mechanics*. Cambridge: Cambridge University Press.

Belot, G., Earman, J. and Ruetsche, L. (1999): "The Hawking Information Loss Paradox: The Anatomy of a Controversy," *British Journal for the Philosophy of Science*, 50, 189–229.

Bohm, D. (1951): *Quantum Theory*. Englewood Cliffs, NJ: Prentice-Hall.

Bohm, D. (1952): "A Suggested Interpretation of the Quantum Theory in Terms of 'Hidden' Variables, I and II," *Physical Review*, 85, 166–93. Reprinted in Wheeler and Zurek (1983, pp. 367–96).

Bohr, N. (1934): *Atomic Theory and the Description of Nature*. Cambridge: Cambridge University Press.

Bohr, N. (1935): "Can Quantum-Mechanical Description of Reality be Considered Complete?" *Physical Review*, 48, 696–702. Reprinted in Wheeler and Zurek (1983, pp. 145–51).

Brown, H. and Harre, R. (eds.) (1988): *Philosophical Foundations of Quantum Field Theory*. Oxford: Clarendon Press.

Bub, J. (1997): *Interpreting the Quantum World*. Cambridge: Cambridge University Press.

Butterfield, J. (1992): "Bell's Theorem: What it Takes," *British Journal for the Philosophy of Science*, 58, 41–83.

Butterfield, J. (1994): "Vacuum Correlations and Outcome Dependence in Algebraic Quantum Field Theory," in D. M. Greenberger and A. Zeilinger (eds.), *Fundamental Problems in Quantum Theory, Annals of the New York Academy of Sciences*, 755, 768–85.

Castellani, E. (1998): *Interpreting Bodies: Classical and Quantum Objects in Modern Physics*. Princeton: Princeton University Press.

Clifton, R. (ed.) (1996): *Perspectives on Quantum Reality: Non-Relativistic, Relativistic, and Field-Theoretic*. Dordrecht: Kluwer.

Clifton, R. (1999): "Beables for Algebraic Quantum Field Theory," in J. Butterfield and C. Pagonis (eds.), *From Physics to Philosophy*, Cambridge: Cambridge University Press, 12–44.

Clifton, R. and Monton, B. (1999): "Losing Your Marbles in Wavefunction Collapse Theories," *British Journal for the Philosophy of Science*, 50, 697–717.

Clifton, R., Feldman, D., Halvorson, H. and Wilce, A. (1998): "Superentangled States," *Physical Review*, A58, 135–45.

Cushing, J. (1988): "Foundational Problems in and Methodological Lessons from Quantum Field Theory," in Brown and Harre (1988, pp. 25–39).

Cushing, J. T. (1994): *Quantum Mechanics: Historical Contingency and the Copenhagen Hegemony*. Chicago: University of Chicago Press.

Cushing, J. T. and McMullin, E. (eds.) (1989): *Philosophical Consequences of Quantum Theory*. Dordrecht: University of Notre Dame Press.

Cushing, J. T., Fine, A. and Goldstein, S. (1996): *Bohmian Mechanics: An Appraisal*. Dordrecht: Kluwer.

DeWitt, B. (1970): "Quantum Mechanics and Reality," *Physics Today*, 23(9); reprinted in B. DeWitt and N. Graham (eds.), *The Many-Worlds Interpretation of Quantum Mechanics*, Princeton: Princeton University Press, 155–67.

Dickson, W. M. (1998a): "On the Plurality of Dynamics: Transition Probabilities and Modal Interpretations," in R. Healey and G. Hellman (eds.), *Quantum Measurement: Beyond Paradox*, Minneapolis: University of Minnesota Press, 160–82.

Dickson, W. M. (1998b): *Quantum Chance and Nonlocality*. Cambridge: Cambridge University Press.

Dieks, D. (1989): "Quantum Mechanics without the Projection Postulate and its Realistic Interpretation," *Foundations of Physics*, 19, 1397–1423.

Dieks, D. and Vermaas, P. (eds.) (1998): *The Modal Interpretation of Quantum Mechanics*. Dordrecht: Kluwer.

Dürr, D., Goldstein, S. and Zanghi, N. (1996): "Bohmian Mechanics as the Foundation of Quantum Mechanics," in Cushing et al. (1996, pp. 21–44).

Einstein, A., Podolsky, B. and Rosen, N. (1935): "Can Quantum Mechanical Description of Physical Reality be Considered Complete?" *Physical Review*, 47, 777–80. Reprinted in Wheeler and Zurek (1983, pp. 138–41).

Everett, H. (1957): "'Relative State' Formulation of Quantum Mechanics," *Review of Modern Physics*, 29, 454–62. Reprinted in Wheeler and Zurek (1983, pp. 315–23).

Fine, A. (1982a): "Joint Distributions, Quantum Correlations, and Commuting Observables," *Journal of Mathematical Physics*, 23, 1306–10.

Fine, A. (1982b): "Hidden Variables, Joint Probabilities, and the Bell Theorems," *Physical Review Letters*, 48, 291–95.

Fine, A. (1986), *The Shaky Game: Einstein, Realism and The Quantum Theory*. Chicago: University of Chicago Press.

Fine, A. (1989): "Do Correlations Need to be Explained?" in Cushing and McMullin (1989, pp. 175–94).

Ghirardi, G., Rimini, A. and Weber, T. (1986): "Unified Dynamics for Microscopic and Macroscopic Systems," *Physical Review*, D34, 470–84.

Gillespie, D. T. (1973): *A Quantum Mechanics Primer*. New York: International Textbook Company.

Griffiths, R. B. (1993): "The Consistency of Consistent Histories: A Reply to D'Espagnat," *Foundations of Physics*, 23, 2201–10.

Hartle, J. B. (1990): "The Quantum Mechanics of Cosmology," in S. Coleman, T. Piran and S. Weinberg (eds.), *Quantum Cosmology and Baby Universes*, Singapore: World Scientific, 65–151.

Healey, R. (1989): *The Philosophy of Quantum Mechanics: An Interactive Interpretation*. Cambridge: Cambridge University Press.

Huggett, N. (2000): "Philosophical Foundations of Quantum Field Theory," *British Journal for the Philosophy of Science*, 51, Supplement, 617–37.

Huggett, N. and Weingard, R. (1994): "Interpretations of Quantum Field Theory," *Philosophy of Science*, 61, 370–88.

Hughes, R. I. G. (1989): *The Structure and Interpretation of Quantum Mechanics*. Cambridge, MA: Harvard University Press.

Jarrett, J. (1986): "An Analysis of the Locality Assumption in the Bell Arguments," in L. M. Roth and A. Inomata (eds.), *Fundamental Questions in Quantum Mechanics*, London: Gordon and Breach, 21–7.

Kochen, S. (1985): "A New Interpretation of Quantum Mechanics," in P. Lahti and P. Mittelstaedt (eds.), *Symposium on the Foundations of Modern Physics*, Singapore: World Scientific, 151–69.

Maudlin, T. (1994): *Quantum Nonlocality and Relativity*. Oxford: Blackwell.

Mermin, D. (1993): "Hidden Variables and the Two Theorems of John Bell," *Reviews of Modern Physics*, 65, 803–15.

Murdoch, D. (1987): *Niels Bohr's Philosophy of Physics*. Cambridge: Cambridge University Press.

Pearle, P. (1989): "Combining Stochastic Dynamical State Vector Reduction with Spontaneously Localization," *Physical Review*, A39, 2277–89.

Pearle, P. (1992): "Relativistic Models of State Vector Reduction," in P. Cvitanovic, I. Percival and W. Wirzba (eds.), *Quantum Chaos–Quantum Measurement*, Dordrecht: Kluwer, 283–97.

Redhead, M. (1983): "Nonlocality and Peaceful Coexistence," in R. Swinburne (ed.), *Space, Time and Causality*, Dordrecht: Reidel, 151–89.

Redhead, M. (1987): *Incompleteness, Nonlocality, and Realism*. Oxford: Clarendon Press.

Redhead, M. (1995): "More Ado About Nothing," *Foundations of Physics*, 25, 123–37.

Rovelli, C. (1997): "Relational Quantum Mechanics," revised and updated as Los Alamos archive paper quant-ph/9609002. Originally published as "Relational Quantum Mechanics," *International Journal of Theoretical Physics*, 35, 1637–78.

Ruetsche, L. (1995): "Measurement Error and the Albert–Loewer Problem," *Foundations of Physics Letters*, 8, 331–48.

Saunders, S. (1988): "The Algebraic Approach to Quantum Field Theory," in Brown and Harre (1988, pp. 149–86).

Saunders, S. (1996): "Relativism," in Clifton (1996, pp. 125–42).

Segal, I. (1967): "Representations of the Canonical Commutation Relations," in F. Lurcat (ed.), *Cargese Lectures in Theoretical Physics*, New York: Gordon and Breach, 107–70.

Shimony, A. (1984a): "Controllable and Uncontrollable Non-Locality," in S. Kamefuchi, P. Fusaichi, and D. Drarmigan (eds.), *Foundations of Quantum Mechanics in Light of New Technology*, Tokyo: Physical Society of Japan, 225–30.

Shimony, A. (1984b): "Contextual Hidden Variable Theories and Bell's Inequalities," *British Journal for the Philosophy of Science*, 35, 25–45.

Stein, H. (1972): "On the Conceptual Structure of Quantum Mechanics," in R. A. Colodny (ed.), *Paradigms and Paradoxes: The Philosophical Challenge of the Quantum Domain*, Pittsburgh: University of Pittsburgh Press, 367–438.

Summers, S. J. (1998): "On the Stone–von Neumann Uniqueness Theorem and its Ramifications," to appear in M. Redei and M. Stolzner (eds.), *John von Neumann and the Foundations of Quantum Mechanics*, forthcoming.

Teller, P. (1995): *An Interpretive Introduction to Quantum Field Theory*. Princeton: Princeton University Press.

Valentini, A. (1991): "Signal Locality, Uncertainty and the Subquantum H-Theorem. I and II," *Physics Letters*, A156, 5–11 and 158, 1–8.

van Fraassen, B. (1989): "The Charybdis of Realism," in Cushing and McMullin (1989, pp. 97–113).

van Fraassen, B. (1991): *Quantum Mechanics*. Oxford: Oxford University Press.

Vink, J. (1993): "Quantum Mechanics in Terms of Discrete Beables," *Physical Review*, A48, 1808–18.

von Neumann, J. (1955 [1932]): *Mathematical Foundations of Quantum Mechanics*. Princeton: Princeton University Press. Translated from *Mathematische Grundlagen der Quantenmechanik* Berlin: Springer.

Wald, R. M. (1994): *Quantum Field Theory in Curved Space–Time and Black Hole Thermodynamics*. Chicago: University of Chicago Press.

Wessels, L. (1997): "The Preparation Problem in Quantum Mechanics," in J. Earman and J. D. Norton (eds.), *The Cosmos of Science*, Pittsburgh: University of Pittsburgh Press, 243–73.

Wheeler, J. and Zurek, W. (eds.) (1983): *Quantum Theory and Measurement*. Princeton: Princeton University Press.

Zurek, W. J. (1982): "Environment Induced Superselection Rules," *Physical Review*, D26: 1862–80.

Chapter 11

Evolution

Roberta L. Millstein

Introduction

It has become almost standard practice for philosophers of biology to bracket their writings with a pair of manifestoes. The first manifesto proclaims that the philosophy of science is not just about physics anymore. This is usually accompanied by an argument suggesting that a myopic focus on physics has led the philosophy of science to misrepresent the true nature of science. As we face a new millennium, the time has come to dispense with such proclamations. The philosophy of biology has come into its own and no longer needs to justify its existence. One look at the most recent Philosophy of Science Association conference should be enough to convince anyone of that fact: approximately one fourth of the presentations are in the philosophy of biology. The first manifesto has become manifest.

The second manifesto, on the other hand, often takes the form of a "call to arms" for philosophers to venture into fields of biology outside of evolutionary theory, such as ecology and molecular biology. That this call to arms has been at least partially successful is reflected in the inclusion of an essay in this volume on Developmental and Molecular Biology, distinct from the present essay on Evolution. Thus, I need not apologize, as many have done before me, for focusing exclusively on evolutionary theory. Yet, since many of the debates in the philosophy of evolution overlap and intertwine with those in the philosophy of developmental and molecular biology, it is not entirely possible to separate the issues. In fact, philosophers seldom use the phrase "philosophy of evolution." Philosophy of biology has often *meant* philosophy of evolution. However, perhaps it is time to be more explicit.

The philosophy of evolution considers issues that are both conceptual and empirical, and, consequently, it is practiced by philosophers and scientists alike. The following discussion will reflect that diversity. Some philosophy of evolution has looked to evolutionary theory to answer broader questions in the philosophy of science. For example, a recent volume explores epistemological issues through the lenses of evolution and other areas of biology (Creath and Maienschein, 2000). While I applaud such work – in fact, I think there ought to be more of it – it does not make up the bulk of the philosophy of evolution, most of which focuses on issues specific to the discipline (although often with broader implications for issues such as causality and explanation). So, my focus will be on the specific rather than the general. Even so, philosophy of evolution at the dawn of the twenty-first century suffers from an embarrassment of riches, both in terms of the number of interesting topics and in terms of the number of insightful analyses exploring these topics. I cannot hope to do them all justice in an essay of this length. Thus, some topics will be touched on only briefly, some in more depth, and perhaps some not at all. My intention is to provide a guide to what I take to be the most important and interesting issues in the philosophy of evolution today, and to point the reader to classic as well as more recent sources.

Mechanisms of Evolution

Arguably, and perhaps uncontroversially, Charles Darwin's greatest contribution to biology was his theory of natural selection. Others had proposed theories of evolution, but it was Darwin's theory of natural selection that made evolution plausible by providing a mechanism through which evolution occurs.[1] According to this theory, if three conditions obtain:

1   organisms within a species vary from one another
2   if some of those variations are heritable and advantageous to the organism
3   there are more organisms that can survive, leading to a "struggle for life"

then organisms with advantageous variations will tend to survive better and reproduce more, leading to an increase in organisms with advantageous variations (and a decrease in organisms with harmful variations) over the course of generations.

   According to Gould and Lewontin (1979), Darwin is often portrayed as having put forth natural selection as the sole mechanism of evolution in the first edition of the *Origin of Species*, only (misguidedly and under critical pressure) including other mechanisms in later editions. Against this view, Gould and Lewontin argue that Darwin was a pluralist concerning mechanisms of evolution (1979). Indeed, the last sentence of the Introduction to the first edition of the *Origin of Species* states: "I am convinced that Natural Selection has been the main but not exclusive means of modification" (Darwin, [1859] 1964). Darwin endorsed such evolutionary mechanisms as Lamarckian use and disuse[2] and sexual selection (which Darwin considered a different mechanism from natural selection). Thus, even for Darwin, natural selection and evolution were not the same thing; natural selection was the primary, but not the only, mechanism of evolution. In contrast, Alfred Russel Wallace, who independently arrived at the theory of natural selection, is usually interpreted as being much more of a selectionist than Darwin, at least for non-human animals.

The neutralist/selectionist debate

During the twentieth century, there was much debate over mechanisms of evolution – is natural selection the sole or primary mechanism of evolution, or do other mechanisms play a greater role? (Beatty, 1984; Provine, 1985). One debate in particular (in its earlier versions, between Sewall Wright and R. A. Fisher) concerned the relative roles of natural selection and the phenomenon of *random drift*.[3] Random drift occurs when physical differences between organisms are causally *irrelevant* to differences in their reproductive success, unlike natural selection, where physical differences between organisms *are* causally relevant to differences in their reproductive success (physical differences which confer an advantage tend to lead to greater reproductive success). For example, if a population of green and red moths is preyed on by a colorblind predator, any differences in reproductive success between the two types will not be due to the physical differences between the moths. As a result, particularly if the population is small, there may be a change in the distribution of types in the population over the course of generations. For example, the population of moths might become entirely red. This would be evolution, but not adaptive evolution. Thus, random drift is another possible mechanism of evolution, but it is a mechanism that tends to lead to an increase in nonadaptive (read: neutral) traits rather than adaptive ones. With the development of the neutral theory of molecular evolution (Kimura, 1969, 1983), a theory that claims that the majority of evolutionary changes at the molecular level are the result of random drift acting on neutral mutations, the scientific debate (often called the neutralist/selectionist debate) over the relative roles of natural selection and random drift intensified. Although some would proclaim the death of the neutralist/selectionist debate (Hey, 1999), the debate continues today. In one recent version of the debate, disputants differ over whether a "nearly neutral" theory makes conceptual and empirical sense (Ohta and Kreitman, 1996; Dover, 1997b, 1997a; Ohta, 1997). It is also interesting that even those who maintain the neutral model is not empirically adequate argue for its usefulness as a null model.

The debate over the relative roles of natural selection and random drift raises a number of philosophical issues. First, there are conceptual issues. Beatty (1984) argues that in some cases, random drift cannot be distinguished conceptually from natural selection, an argument that has been widely accepted. Beatty's argument rests on the fact that natural selection and random drift are both probabilistic concepts, which creates a conceptual overlap. However, if Beatty is right, the very foundations of the neutralist/selectionist debate are called into question. Shanahan (1992) suggests that random drift and natural selection are on a continuum. Millstein (2001) argues that when the two concepts are conceived as processes rather than outcomes, they can be distinguished from one another.

Second, if random drift plays even a small role in evolution, does that imply that evolution is indeterministic? Rosenberg (1988; 1994) claims that an omniscient account of evolution would have no need for the concept of random drift – that all instances of random drift can be explained in terms of natural selection. Rosenberg uses this claim to argue that although evolutionary *theory* is statistical, the evolutionary *process* is a deterministic one. According to Rosenberg, evolutionary theory is statistical purely for instrumental reasons; random drift serves merely as a useful fiction. A similar claim for the determinism of the evolutionary process is made by Horan (1994). Contra Rosenberg, Millstein (1996) argues that any evolutionary theory, omniscient or otherwise, must take random drift into account – that random drift is not eliminable from evolutionary theory. Brandon and Carson (1996) further challenge Rosenberg's and Horan's claims; they maintain that it is more reasonable for a scientific realist to conclude that the evolutionary process is fundamentally indeterministic. Most recently, Horan and Rosenberg join with Graves in an attempt to counter the arguments of Brandon and Carson (Graves et al., 1999). Weber (2001) and Millstein (1997) provide analyses of the positions of both camps.

However, without settling the debate between the determinist and the indeterminist, we can still ask whether the probabilities in evolutionary theory are in some sense objective, or whether they are purely epistemic, only appearing in the theory, because we find probabilities useful and tractable in evolutionary contexts. If the evolutionary process is *indeterministic*, then the answer to this question is clear; evolutionary theory is probabilistic in an objective sense. On the other hand, if the evolutionary process is *deterministic*, there may still be a sense in which the probabalities employed by evolutionary theory are objective (Sober, 1984; Brandon and Carson, 1996; Millstein, 1997). These issues bear further exploration.

The adaptationist programme and its challenges

In 1979, a debate related to, yet broader than the neutralist/selectionist debate was sparked by the publication of what would become a well-read and controversial essay entitled, "The Spandrels of San Marco and the Panglossian Paradigm: A Critique of the Adaptationist Programme" (Gould and Lewontin, 1979). As discussed above, natural selection leads to the accumulation of advantageous variations. According to Darwin, over long periods of time, the accumulation of advantageous variations in a population would lead organisms to become adapted to their physical environments as well as to become adapted to one another. Thus, natural selection can provide explanations for adaptations that we observe in nature, such as the long, thick, chisel-like beaks of woodpeckers that are adapted for drilling wood and chipping away tree bark, enabling woodpeckers to feed on insects and tree sap. For Darwin, however, these adaptations were not necessarily perfect; for example, a bee sting will cause a bee's own death ([1859] 1964, p. 472).

Gould and Lewontin charge that evolutionary biologists in England and the USA failed to heed Darwin's lessons – that so-called "adaptationists" not only focus almost exclusively on natural selection as the mechanism of evolution, but that they see the natural world as being as well adapted as it could possibly be ("Panglossian"). Furthermore, according to Gould and Lewontin, when adaptationists fail to explain the adaptive value of a particular trait, they either create another adaptationist story, simply assume there is another adaptationist story, or attribute the failure to an imperfect understanding of the circumstances of the organism. Instead of continuing to tell one adaptationist story after another, Gould and Lewontin maintain, evolutionists should consider other possible mechanisms, including, but not limited to, random drift. For example, the human chin may be a byproduct of developmental constraint, rather than a separate trait on which selection acts. Alternatively, a trait may have been selected for a specific function, but the trait (often in a modified form) is no longer serving that function. For example, it is believed that a bird's feathers evolved to assist in thermoregulation, but they are now used for a different function: flight. Much of evolution, Gould and Lewontin claim, may not be adaptive after all.

Gould and Lewontin's charges raise the question of what an *adaptation* is. On the standard picture of adaptation – Brandon (1990) calls it the "received view" – adaptation is a *historical* concept; a trait is an adaptation if and only if it is the product of selection. This received view of adaptation rejects the *ahistorical* conception of adaptation, which would consider a trait an adaptation if it were of current benefit to its possessor, providing a good fit with the environment. Proponents of the received view would reserve the term "adaptation" for its historical meaning, using the term "adaptive" or "adaptedness" or "aptation" to refer to the ahistorical meaning. Sterelny and Griffiths (1999) provide useful examples of these two terms from the perspective of the received view. The human appendix, as a product of natural selection, would be considered to be an adaptation, but not adaptive, since it no longer benefits us in our present environment. However, the ability to read is adaptive (beneficial to us in our present environment) without being the direct result of natural selection (it is more likely a side-effect of selection for other cognitive abilities). Some proponents of the historical view would restrict the term "adaptation" to traits that are currently serving the function for which they were selected, and many of these proponents have adopted the term *exaptation* (Gould and Vrba, 1982) to refer to a trait that is serving a function other than the one for which it was selected. The example of a bird's wings discussed above is an example of an exaptation.[4]

Recently, the received view of adaptation has come under some criticism. Elisabeth Lloyd (1992) argues that confusion between the historical and ahistorical senses of adaptation has contributed to some of the confusion in the units of selection debate. She further points out that the ahistorical conception of adaptation is at work in debates over the relationship between natural selection and sexual selection (should the product of sexual selection, for example, a peacock's tail, be considered an adaptation?). Grene (1997) also questions the historical conception of adaptation, calling it "a move to the a priori that almost makes the theory of natural selection a mere tautology." In making this criticism, she quotes Sterelny, who states that "Natural selection is no explanation of adaptation, if adaptation is by definition whatever selection undisturbed produces" (Sterelny, 1996, p. 197). This reexamination of the historical question is worthwhile, both for the reasons that Grene and Lloyd state, as well as to make sense of Gould and Lewontin's claim that there can be selection without adaptation and adaptation without selection. It would seem that we are more tied to the pre-Darwinian, ahistorical notion of adaptation than we would like to think; perhaps another term should be used to refer to a product of selection – for further discussion of the concept of adaptation, see Burian (1983), Rose and Lauder (1996) and Hull and Ruse (1998).

Many philosophers and biologists think that Gould and Lewontin sounded the death knell for the adaptationist program (or at least exposed its weaknesses), but others have leapt to its defense, notably Cronin (1991), Dawkins (1976), Dennett (1995), Futuyma (1988), and Mayr (1983). Some have argued that the Gould and Lewontin's critique relies on Karl Popper's falsificationism, and that since this view has been discredited, the charges against the adaptationist programme fall flat. Others have argued that falsifiability is an inappropriate criterion for a research programme. Still others point to the successes of the adaptationist programme, and defend its methodology. (Recent philosophical debate on adaptationism can be found in Dupré (1987), Brandon and Rausher (1996), Orzack and Sober (1996) and Godfrey-Smith (1999).)[5]

Far too much of the literature, in my view, mischaracterizes the adaptationist position (or its critique) as claiming that all traits are adaptive or mischaracterizes the anti-adaptationist position (or its critique) as claiming that all traits are non-adaptive. (Of course, it is generally one's opponent who is mischaracterized.) Rather, the debate is over the *degree* to which adaptation is found in nature, with both sides generally accepting that some traits are adaptive and some are non-adaptive. At least, that is the empirical, biological debate. But should philosophers of biology be taking sides on this empirical question? It certainly does seem that philosophers can clarify (and have clarified) these kinds of empirical debates, both in terms of the concepts and the arguments involved. However, in the end, the question is an empirical one, and philosophers should not take sides. Furthermore, as Beatty (1997) suggests, these kinds of "relative significance" debates may not even be resolvable. After all, if one is to argue that evolution has been significantly adaptive, what does that mean? Ninety percent? Greater than fifty percent? At least ten percent? The terms of the debate are vague, making resolution difficult. Furthermore, how would we even answer such questions? Certainly, we cannot examine all living populations and it is unclear what a representative sample would amount to in this situation. It seems more likely that there is a resolution to the *methodological* debate over whether it is better to pursue an adaptationist research programme or an anti-adaptationist research program. As Beatty (1987) argues, how we distribute the resources of the evolutionary community has to do with the questions we pose for ourselves. With regard to the question concerning the relative importance of selection versus drift, Beatty suggests that if this is a question we really want to answer, "then we really must give serious thought to distributing the resources of the evolutionary community between the pursuit of selection and drift hypotheses" (1987, p. 72). The same argument holds for adaptationist and nonadaptationist hypotheses in general.

Legacies of the adaptationist debate: Sociobiology, contingency, and laws of biology

The adaptationist debate has given rise to several other philosophical debates. For example, one outgrowth of the adaptationist programme is sociobiology – the application of the theory of natural selection to animal behavior, and more controversially, human behavior in particular; the classic text is Wilson (1975). Thus, sociobiology maintains that much of human behavior (and not just human physical characteristics) can be explained as adaptations that enhanced our ancestors' ability to survive. E. O. Wilson (1975) identified *altruism* as the central problem of sociobiology; a theory proposing to explain behavior via natural selection must explain how it is that behaviors which seem to promote the fitness of others at the expense of one's own fitness could have evolved (see the section on Units of Selection for further discussion of altruism).

Some of the debate surrounding sociobiology echoes the general debate over adaptationism – critics argue that it engages in Panglossian, untestable story telling, e.g., Gould (1980), while defenders question the adequacy of the Popperian criterion in this context, e.g. Caplan (1982). Critics also accuse sociobiology of being committed to genetic determinism – the view that we are completely determined by our genes, with little or no role for environmental or cultural factors (Lewontin et al., 1984). Extreme versions of sociobiology bite this bullet, while defenses that are more reasonable acknowledge a considerable role for non-genetic as well as genetic factors. Of course, such a response still leaves open the difficult-to-answer question of the extent to which we are affected by each of these factors. Perhaps more damning are the accusations that sociobiology simply entrenches our existing stereotypes and prejudices by providing purportedly scientific explanations for behaviors such as male aggression and female "coyness." There have been numerous feminist critiques of sociobiological explanations on grounds such as these – see, for example, Harding (1986) and Longino (1990) – Hrdy (1999) argues that a reexamination of the evidence overturns many of these sexist explanations.

Further criticisms are offered by Kitcher (1985), but in a subsequent paper Kitcher (1990) points to a new sociobiology that might avoid many of the problems of the past. This recent work (and philosophical controversy) in sociobiology focuses not on explaining specific behaviors, but on explaining broader human capacities as adaptations to an ancestral environment, a view known as "evolutionary psychology."[6] Other recent work attempts to draw ethical or epistemological implications from sociobiology – "evolutionary ethics" and "evolutionary epistemology," respectively; see, for example, Ruse (1998). Debate on these issues will likely continue into the future.

Controversies over adaptationism have also sparked a revival of the debate over whether there are laws in biology, a debate that has divorced itself from its previous context – the debate over whether biology is a legitimate science. The present debate takes the legitimacy of biology as a science for granted, and rightly so. Gould's anti-adaptationist *Wonderful Life* suggests that the broad-scale of the history of life is contingent (Gould, 1989). Beatty (1995) usefully distinguishes between a weak sense of contingency – evolutionary generalizations may hold for the present, although there are often exceptions, and those that do hold may change in the future course of evolution – and a strong sense of contingency that Gould seems to be endorsing – "replay the tape" of life from the same starting point, and we might arrive at a different outcome (one without human beings, for example). Since all biological generalizations are contingent, evolutionary outcomes, Beatty argues, there are no laws in biology. Waters, on the other hand, argues that causal regularities (as distinguished from distributions) are ubiquitous in biology, and that these regularities exhibit the most important features traditionally attributed to scientific laws (1998). Beatty's paper also led to a Philosophy of Science Association symposium featuring several responses to Beatty (and to the argument against laws given in Rosenberg 1994), as well as an additional paper from Beatty himself (Beatty, 1997). These papers provide a fruitful reexamination of the nature of scientific law and show that old philosophical problems can often yield new insights.

The Species Problems

Charles Darwin entitled his magnum opus *On the Origin of Species*. But what is a species? Ironically, Darwin denies the reality of species, leading one to wonder: Is it a book about nothing? Ghiselin (1969) and Beatty (1985) persuasively argue that the confusion is only apparent. In saying that species are "arbitrarily given for the sake of convenience" (Darwin, [1859] 1964, p. 52), Darwin is referring to the species *category*, pointing out that it cannot be clearly distinguished from "subspecies," which in turn blend into "varieties" and then "individual differences" (Darwin, [1859] 1964, p. 51). However, Darwin does accept the reality of species *taxa* such as the cabbage, the radish, and the onion, and the evolution of these and other species taxa by natural selection form the primary subject matter of the *Origin*. Yet Darwin leaves "species" undefined. Species concepts are used in many different areas of biology, including taxonomy, macroevolutionary biology, and ecology, making the defining of "species" an important issue in biology.

Each of the ways of talking about species – as a category or as a taxon – generates a conceptual problem. The species category problem concerns the ontological nature of the species category itself – what sort of thing is the category *species*? Is it a natural kind, an individual, a set, or something else? The species taxon problem, on the other hand, has to do with the criterion used to determine which organism are assigned to which species. Is it based on their morphological characteristics, their evolutionary history, whether they interbreed, or something else? We will examine each of these problems in turn.

Species category problem

For Carolus Linnaeus, originator of the modern system of biological classification, species were seen as natural kinds, a view that dates back to Aristotle (also called an "essentialist" or "typological" view of species). On this view, species have *essences*, in the Aristotelian sense. That is, one could specify the necessary and sufficient conditions for belonging to a kind – for example, the necessary and sufficient conditions for being a tiger. Furthermore, species were seen as static and unchanging.

Evolutionary theory presents a challenge to the view of species as unchanging essences. As Hull states:

> The only basis for a natural classification is evolutionary theory, but according to evolutionary theory, species developed gradually, changing one into another. If species evolved so gradually, they cannot be delimited by means of a single property or set of properties. If species can't be so delineated, then species names cannot be defined in the classic manner. If species names cannot be defined in the classic manner, then they can't be defined at all. If they can't be defined at all, then species can't be real. If species aren't real, then 'species' has no reference and classification is completely arbitrary (1965, p. 320).

In this way, evolutionary theory is sometimes seen as implying the unreality of species taxa, and thus the unreality of the species category. And yet organisms do, at least to some extent, sort into non-arbitrary groups. For example, both Bengal tigers and Siberian tigers are considered to be subspecies of the same species, tiger. On the other hand, Bengal tigers and Grant's zebras are not classified as the same species, although we could place them both into the category of "organisms that are striped." The latter category seems arbitrary, in a way that the category of tigers is not. So, perhaps the proper conclusion to draw is not that the species category is unreal, but that the essentialist view of species is false.

An alternative account that characterizes species as historical entities (often referred to as the Hull-Ghiselin view) has become prominent; see the classic papers of Ghiselin (1974) and Hull (1976, 1978) or the more recent paper by Ghiselin (1997). On this view, species are not kinds (or classes of any sort); they are *indi-*

viduals. This means that they are integrated and cohesive entities with a restricted spatiotemporal location (as opposed to spatiotemporally unrestricted classes). It is argued that species must be considered to be individuals because they are the entities that evolve. Species come into existence, and go out of existence, similar to the way that organisms (also individuals) are born and die. Thus, for example, if a population of animals resembling tigers in every way were to evolve independently on another planet, they would not *be* tigers; they would be spaciotemporally separated from our own tigers and would therefore be of a different species. *Panthera tigris* is a proper noun, not the name of a class or a type. The Hull-Ghiselin view is often connected with Mayr's biological species concept (to be discussed below).

According to Sterelny (1995), there is a consensus forming in favor of the Hull-Ghiselin view. Nonetheless, there are (as Sterelny admits) dissenters to this consensus. Kitcher (1984; 1989), for example, argues that the species category can be construed as a set of organisms sharing a common property, and that – contra, e.g., Hull (1978) – this view is *also* consistent with the evolutionary changes of a population over time. One could also be a pluralist concerning the question, viewing species categories "either individuals or sets, depending on the biological problem at issue" (Dupré, 1992). Yet another alternative views the species category as being somewhere intermediate between an individual and a class; Mayr prefers the term "population" (Mayr, 1987; Mishler and Brandon, 1987).

Species taxon problem

There are an amazing number – perhaps as many as twenty or more – of different conceptions that spell out the way in which organisms are to be placed into species taxa (Mayden, 1997). Here I will sketch the three primary alternatives.

Hull and Ghiselin argue that their "species as individuals" thesis, discussed above, is the natural interpretation of Ernst Mayr's influential *biological species concept.* According to Mayr, "*Species are groups of interbreeding natural populations that are reproductively isolated from other such groups*" (Mayr, 1996, p. 264; italics in original). This view coincides with Mayr's view of the speciation process, whereby the geographical isolation of populations leads to the evolutionary divergence of those populations, with the result that the populations are no longer able to interbreed (reproductive isolation). There are two common criticisms of this view. The first is that it is inapplicable to the classification of asexual organisms, a point that Mayr now acknowledges, suggesting that a different definition of species for asexual organisms should be developed (1996). This is a relaxation of his earlier position (Mayr, 1987) that only sexually reproducing organisms qualify as species. The second criticism comes primarily from botanists who point out that many plants are considered distinct species, yet they do sometimes interbreed. In response, Mayr now allows that organisms of different species may occasionally interbreed, so long as their biological properties prevent the "complete fusion" of the populations (Mayr, 1996, p. 265). These changes seem to be a substantial weakening of the biological species concept, with the latter change making the concept much less precise.

Sokal and Crovello (1970) offer additional criticisms of the biological species concept. They contend that the concept is not operational, arguing that extensive field observations of interbreeding are "impractical," leaving the biologist with "partial" or "circumstantial" evidence, and of course interbreeding cannot be observed in fossils at all. The biologist is forced to rely on the disimilarities between organisms, which they maintain are "an imperfect reflection of infertility between organisms" (Sokal and Crovello, 1970, p. 133). Furthermore, they charge that Mayr's biological species concept ties the variation in nature to a biased, "abstract ideal" that prevents the discovery of new insights into the mechanisms of evolution (Sokal and Crovello, 1970, p. 149). Their alternative, the *phenetic species concept*, places organisms into species on the basis of their overall similarities, thus allowing for a definition of species that remains constant with changes in evolutionary theory and which (purportedly) allows for an independent test of the theory. The idea of sorting organisms into species based on their similarities is not new; it pre-dates Darwin. Generally speaking, phenetic species concepts may categorize organisms into species based on their structural, behavioral, or genetic similarities.[7] One version of the phenetic species concept is called numerical taxonomy (Sneath and Sokal, 1973). This concept relies on computer analysis to determine the similarities between a large number of features. Phenetic species concepts have stirred up quite a bit of controversy, but currently are not all that popular, due in part to the difficulty (if not impossibility) of objectively specifying the nature of "similarity."

Phylogenetic (evolutionary) species concepts[8] are an alternative to biological species concepts and phenetic species concepts. Whereas biological species concepts focus on a *process* of evolution (specifically, speciation), phylogenetic species concepts focus on the *pattern* of evolution (Sterelny and Griffiths, 1999). That is, phylogenetic species concepts seek to reflect genealogical evolutionary history by characterizing species as lineages – "segments of the evolutionary tree" between two speciation events, or between a speciation event and an extinction event. Consider a branch of the evolutionary tree – a lineage. If that branch splits into two new branches, then two new species are formed, replacing the old species (the *stem species*). If the branch terminates, then the species has gone extinct. By focusing on the pattern of evolution, phylogenetic species concepts leave open the question of which particular process produced the pattern – the very issue that led the biological species concept into trouble. However, ultimately phylogenetic species concepts must face the same issue: when can we say that a lineage has divided into two new lineages? Additional criteria are needed, either reproductive isolation (which brings back all the problems of the biological species concept) or some other criteria. Sterelny and Griffiths (1999) suggest the ecological species concept (Van Valen, 1976) or the cohesion species concept (Templeton, 1989) as sources of possible criteria.

Recently, the plethora of species taxon concepts has led to various calls for pluralism (Kitcher, 1984; Mishler and Brandon, 1987; Ereshefsky, 1998; Dupré, 1999), as well as responses to these calls (Ghiselin, 1997; Mayden, 1997; Hull, 1999). In 1969, Hull remarked that, "the biological literature on the species concept is overwhelmingly large" (1969, p. 180, n. 10). If it was overwhelmingly large in 1969, you could fairly well drown in the philosophical and biological literature on species today. It is hard to predict how these issues will resolve themselves; the only certain prediction seems to be that we have not heard the last of the species taxa problem.

Tautology and the Nature of Fitness

No summary of the philosophy of evolution would be complete without a discussion of the "tautology problem," given the amount of space that has been devoted to it. Yet given a proper understanding of tautology and evolutionary theory, there is neither a *prima facie* tautology, nor is there a problem. Nonetheless, much interesting philosophical discussion about the nature of fitness has arisen as a result of the misunderstanding.

The "problem" isn't new, either. According to Hull (1969), evolutionists as far back as Darwin have been defending the theory of natural selection against the criticism that it is tautologous. Nonetheless, the criticism refuses to die, kept alive in large part by creationists who love to quote Popper's claim that "Darwinism is not a testable scientific theory but a *metaphysical research programme*" (Popper, 1974, p. 134; italics in original), but who ignore his subsequent recantation (Popper, 1978).

The standard criticism goes as follows: the principle of natural selection is "the survival of the fittest," but who are the "fittest"? Those that survive. The principle then becomes "the survival of those that survive." Thus, the critics charge, the theory of natural selection is a tautology, and is therefore circular and empty; it says nothing about the way the world is, since it is true regardless of the empirical reality. This claim is often conjoined with the claim that the theory of natural selection is unfalsifiable – a tautology cannot be proven false.

A few technical points regarding the standard criticism – if there is anything wrong with the phrase "the survival of the fittest," it is that it is an analytic statement, not that it is a tautology, as Sober (1984) points out. A tautology is a statement that is true by virtue of its logical form alone, such as "Either it is raining or it is not raining." An analytic statement, on the other hand, is a statement that is true by virtue of the meaning of its constituent words (i.e., true by definition), with the classic example being "a bachelor is an unmarried man." If the phrase "the survival of the fittest" were to be worded as a statement that could be true or false – it is not a statement in its current form – then it would be characterized as an analytic statement rather than a tautology. Still, even as an analytic statement, the critics' charge that the phrase is circular, empty, and/or unfalsifiable lingers.

One possible line of response to the standard criticism involves a reexamination of the concept of "fittest." If the only thing that makes one group of organisms fitter than another is that the first group in fact survived when the second did not, then this seems to be the source of the circularity. In response to this concern, Mills and Beatty (1979) and Brandon (1978) independently developed the propensity interpretation of fitness (although Brandon prefers the term "adaptedness"). On this view, fitness is not defined in terms of an organism's *actual* survival or reproductive success. Instead, fitness is an organism's *propensity*, or ability, to survive and reproduce in a particular environment. (Fitness is never defined absolutely, but always relative to a given environment; what enhances survival or reproductive success in one environment may not do so in a different environment.) Thus, "the survival of the fittest" is not "the survival of the survivors," but rather "the survival of those who have the greatest propensity to survive." The organisms that have the greatest propensity to survive may not in fact survive; consider, for example, two identical twins, one of which is struck by lightening and dies, the other which survives and leaves offspring. Both are equally fit (have the same propensity to survive and reproduce), yet one has greater actual reproductive success. In this manner, the propensity interpretation of fitness attempts to break the purported circularity of the theory of natural selection.

The propensity interpretation of fitness is not without its critics; see, for example Rosenberg (1982) and Rosenberg and Williams (1986). Even Beatty and Finsen (née Mills) return to point out some technical difficulties with their own position (Beatty and Finsen, 1989). Nonetheless, the view enjoys widespread acceptance among philosophers of biology (see, for example, Burian, 1983; Brandon and Beatty, 1984; Sober, 1984; Richardson and Burian, 1992). Sober (2000) responds to Beatty and Finsen's self-criticisms and points out that whereas the criticisms apply to the particular mathematical implementation of the propensity interpretation, they do not challenge the nonmathematical heart of the propensity interpretation.

In spite of the popularity of the propensity interpretation as an account of the concept of fitness, some philosophers – including Beatty (1992), who has changed his position on this issue – have argued that it does not actually solve the tautology problem. Rather, Waters suggests, if we spell out the principle of the survival of the fittest as "Organisms with greater higher fitnesses in (environment) E will probably have greater reproductive success in E than (conspecific) organisms with lower fitnesses" (1986, p. 211), there are two basic ways of interpreting the term "probably": the propensity interpretation and the frequentist interpretation. Waters argues that if one chooses the propensity interpretation, the principle is true by definition; if one chooses the frequentist interpretation, the principle is not analytic, but it is untestable.

If this argument is correct, does that mean that the theory of evolution is circular and unfalsifiable? It might, if the phrase "the survival of the fittest" actually described the theory – but it does not. The real problem with the standard criticism is that it misrepresents evolutionary theory, as Hull (1969) and Waters (1986) note. As discussed above, the present-day theory of evolution includes not only natural selection as a possible mechanism leading to the differential survival and reproduction of types; random drift is a possible mechanism, as are migration and mutation. In other words, in any particular case survival may not be "the survival of the fittest."

Even Darwin's theory of natural selection alone is not captured by this phrase;[9] as previously mentioned, Darwin described natural selection as a process requiring

1   a struggle for existence where not all organisms that are born can survive
2   heritable variations between organisms in the population, and
3   variations that confer a differential ability to survive and reproduce.

Whether any or all of these conditions obtain in a particular population is an empirical question, not a matter of definition, and thus we can test the population for the presence or absence of the three conditions. The theory of natural selection is neither circular nor vacuous.

The tautology problem ought to be a dead issue, even if there are those who refuse to let it go. However, its offspring, the proper conception of fitness, remains a fruitful area of research (Recent discussions of the concept of fitness in evolutionary theory appear in *Biology and Philosophy*, volume 6; see also Weber (1996), Stout (1998) and Abrams (1999).

Units of Selection

Life can be viewed as a hierarchy of levels, from genes, to cells, organs, and organisms, to populations and species, to yet higher levels. This raises the following questions: which of these levels (or units) does natural selection act upon? Or does it act on one level in some case, but act at different levels in other cases? Is there one level that it *usually* acts on? These questions form the core of what is known as the "units of selection"[10] problem (Lloyd, 1992 identifies four units of selection problems). The answers to these questions might seem to be arbitrary, if one assumes, for example, that what benefits the group also benefits the individual organism. Consider, however (as Darwin did), the case of sterile worker ants. Clearly it is not advantageous to the individual ants to be sterile (an organism that cannot reproduce has zero fitness), so how could such ants be the product of natural selection? Darwin's answer is that sterile worker ants could have been selected because they were "profitable to the community" (Darwin, [1859] 1964, p. 236). If this analysis is correct, sterile worker ants are an example of group selection, but not individual selection; sterility benefits the group, but not the individual. Nonetheless, most of Darwin's examples of natural selection are of individual rather than group selection, and many consider individual, organismic selection to be the "received view" of natural selection.[11] However, there is a great deal of controversy surrounding the received view; see, for example, Brandon and Burian (1984).

The case of the worker ant appears to be an extreme case of altruism, which Rosenberg usefully characterizes as behavior that "increases the reproductive fitness of another at the expense of one's own reproductive fitness" (Rosenberg, 1992) – an evolutionary sense of altruism which does not require conscious intention on the part of the altruist. Less extreme examples of group selection explanations for apparent altruism are offered by Wynne-Edwards (1962), who claims that organisms will limit their reproduction to preserve their food supply and prevent extinction of the population. Many consider the idea of group selection to have been dealt a deathblow in 1966 by G. C. Williams's classic *Adaptation and Natural Selection*. G. C. Williams argues that group selection, although not impossible, is unlikely, and that most purported cases of group selection can be explained more simply. However, the alternative explanations that he provides are not individual (organismic) selection explanations; they are explanations citing selection at the level of the gene, a view that is endorsed and popularized by Dawkins in *The Selfish Gene*.

What exactly is genic selection? To help clarify his position, Dawkins introduces the terms "replicator" (for which the gene is the primary, but not exclusive example) and "vehicle" (for which the organism is the primary, but not exclusive example). Hull modifies these terms and renames them to "replicator" and "interactor." Hull defines a replicator as "an entity that passes on its structure directly in replication" (1980, p. 318). A gene replicates itself (relatively) directly, whereas the traits of an organism are not passed directly to its offspring (genes are transmitted, and development must occur). An interactor, on the other hand, is defined as "an entity that directly interacts as a cohesive whole with its environment in such a way that replication is differential" (Hull, 1980, p. 318). Selection is then defined as "a process in which the differential extinction and proliferation of interactors cause the differential perpetuation of the replicators that produced them" (Hull, 1980, p. 318).[12] These definitions leave us with *two* levels of selection questions – we can ask, "at which levels does replication occur?" and "at which levels does interaction occur?"

So, in one sense, Dawkins's (and G. C. Williams's) genic selectionism does not go against the received view – Dawkins accepts organisms as interactors in the selection process, acknowledging that genes are "*[o]bviously* . . . selected by virtue of their phenotypic effects (Dawkins, 1982, p. 117). This might lead one to conclude, as Reeve and L. Keller do, that "the debate is resolved" since "the unit of replication is the gene (or, more precisely, the information contained in a gene), and the organism is one kind of vehicle for such genes, a vehicle being the entity on which selection acts directly"; participants on different sides of the debate simply have chosen to focus on one aspect rather than the other (1999, p. 5).

However, this conclusion is too hasty, and misses the larger units of selection debate over whether replicators or interactors are "the" causal agents in the selection process. As Sober explains Dawkins's position: "Those who argue that the single gene is the unit of selection often seem to think of genes as the deeper cause of evolution by natural selection . . . genes cause phenotypes, and phenotypes then determine survival and reproductive success" (Sober, 1984, p. 228); see also Lloyd (1988) for a thorough analysis of Dawkins's views. Furthermore, it is the replicators, not the interactors, which actually survive or fail to survive in the process of natural selection. On the other side of the debate are those who attribute the causal agency to the interactor, e.g., Sober (1984), Lloyd (1988) and Brandon (1990). Defenders of the interactor view often trace their views to Mayr's assertion that "natural selection favors (or discriminates against) phenotypes, not genes or genotypes" (1963, p. 184) or to Hull's definition of natural selection (quoted above), both of which attribute causal agency to the interactor. Proponents of the interactor view often acknowledges that replicator views are good for "bookkeeping" – that is, they are empirically adequate – but argue that they fail to capture the true causal picture. Since the primary interactor is usually taken to be the organism, whereas the primary replicator is usually taken to be the gene, Dawkins's focus on replicators *is* (in this sense) a challenge to the received view.

Thus, the question of which unit is "the" unit of selection has focused on two sub-questions:

- Are groups only rarely units of selection, if at all?
- Are genes or organisms best seen as the true (causal) units of selection?

Concerning the former question, Sober and D. S. Wilson argue that, properly defined, group selection is more prevalent than is usually supposed (Wilson and Sober, 1994; Sober and Wilson, 1998). Sober and D. S. Wilson also argue that attempts to explain apparent altruism as organismic selection – either by claiming that organisms that help their offspring are favored by selection (kin selection) or that organisms that help each other are favored by selection (reciprocal altruism) – can be better understood through the lens of group selection.

With regard to the latter question, some have argued against Dawkins's genic selectionism on the grounds it is reductionistic and that genes are context-sensitive; the same gene can enhance fitness in one context and reduce it in another (see, e.g., Wimsatt, 1980; Sober and Lewontin, 1982). Sterelny and Kitcher, on the other hand, maintain that this argument does not weaken the case for gene selection, since the fitness of any unit of selection is necessarily relative to context (1988). An alternative account is provided by Brandon (1990), who argues that conceptually, we can describe a dual hierarchy of interactors and replicators, but that it is an empirical question as to whether selection actually occurs at any of the interactor levels (e.g., chromosome, gene, organism, group, species). Brandon maintains that we have ample evidences for selection at level of the organism; the existence of selection at the other levels is suggestive, but not conclusive.

Brandon's account relies on Wesley Salmon's "screening off" account of causation; Sober (1984) uses an alternative conception of causation (causes increase the chances of their effects in all causally relevant background contexts). There has been much recent debate between these two accounts (Brandon et al., 1994; Sober and Wilson, 1994; van der Steen, 1996; Brandon, 1997; Hitchcock, 1997).

Waters (1991) points out that proponents of gene selection and proponents of individual selection share a common assumption of realism concerning the level of selection. (Presumably, the same can be said about proponents of group selection.) That is, they assume that "there is a uniquely correct identification of the operative selective forces and the level at which each impinges" (Waters, 1991, p. 554). Waters argues that our realism about the levels of selection must be *tempered*; we must acknowledge "that the causes of one and the same selection process can be correctly described by accounts which model selection at different levels" (Waters, 1991, p. 572). Sober and D. S. Wilson (1994) defend the realist position; they argue that different hypotheses about the units of selection will produce different predictions. Others claim neutrality on this realism issue (Lloyd, 1989) or make instrumentalist claims (Sterelny and Kitcher, 1988).

If there is anything like consensus on these contentious issues, it would probably be surrounding the idea of a hierarchical view of selection (the origins of which are in Lewontin, 1970). That is, most philosophers and biologists accept the idea that it is possible for selection to occur at more than one level – even G. C. Williams and Dawkins now accept the *possibility* of group selection. Accepting a hierarchical level of selection, however, leaves open the question of which level of selection predominates in nature – again, is it primarily the organism or the gene? Does group selection occur often, or very rarely if at all? Others – e.g., Eldredge and Gould (1972), Stanley (1979) and Lloyd and Gould (1993) – argue for the prevalence of species selection.[13] Gould and Lloyd (1999, p. 11904) declare "emerging consensus in favor of the interactor approach," but consensus on that point seems less clear.

Evolving Out of the Past and Into the Future

Many of the issues discussed above are longstanding issues in the philosophy of evolution. And yet, that doesn't mean the field is standing still – far from it. As I hope I have shown, progress has been made in many of these areas. It's just that, as tends to happen with philosophical analyses, the settling of some issues only raises further questions. I have tried to indicate the direction in which each of the debates seems to be heading, or ought to be heading. Here I will consider broader issues for the future.

In 1969, David Hull chastized the fledgling field of the philosophy of biology for being misinformed about biology and the issues and distinctions that biologists find important. Some three decades later, philosophers have taken Hull's admonition to heart, and yet there is still always more to do. For example, philosophers have explored issues surrounding macroevolution (large-scale evolutionary changes at or above the species level) such as the theory of punctuated equilibria (which challenges the Darwinian thesis of gradual evolution) and the related idea of species selection (discussed briefly above). They have examined the question of progress – does evolutionary theory overturn the idea that the history of life is progressive, or is there still some sense in which there is progress? If so, what is it? See, for example, Nitecki (1988), McShea (1991) and Ruse (1993). And yet, philosophers of evolution have only begun to scratch the surface on macroevolutionary issues such as contingency (discussed briefly above), the mass extinction debates – Grantham (1999b) calls for a philosophical examination of these debates – and stochastic macroevolutionary models (Grantham, 1999a; Millstein, 2000). What does it mean to say that there are "autonomous" theories of macroevolution, distinct from microevolutionary theories? Philosophy of evolution should continue to be vigilant about exploring topical issues in biology such as these.

Perhaps more controversially, I think that philosophers of evolution ought to pay more attention to philosophical issues that arise outside of academia. For example, the action that the Kansas Board of Education took in July 1999 that had the effect of minimizing the teaching of evolution highlights the importance of philosophers being involved in the creationist/evolutionist debates. Of course, there are some notable explorations in this area (Kitcher, 1982; Ruse, 1996; Pennock, 1999), but in general philosophers of evolution seem to shy away from such engagements, perhaps because of philosophical worries concerning general debates about the nature of science, or perhaps because of a general distaste for taking a stand on controversial issues of the day. Whatever the reason, I think we neglect our duties when we shy away from such controversies.

Notes

1 Plausible, that is, to present-day biologists and some of Darwin's contemporaries. Not everyone was initially persuaded by Darwin's arguments.

2 Lamarckian evolution, long considered to be a refuted theory, has recently experienced a resurgence with the publication of Cairns, Overbaugh, and Miller's controversial essay claiming to have demonstrated the occurrence of directed mutation in bacteria (Cairns et al., 1988). This, in turn, has led to a philosophical reexamination of Lamarckism in general and the concepts of directed and random mutation in particular; see, for example, Sarkar (1991), Keller (1992), Jablonka and Lamb (1995) and Millstein (1997).

3 Migration and mutation are also considered mechanisms of evolution, when evolution is construed as a change in gene frequencies (not an uncontroversial definition in itself).

4 Similar issues arise in discussions of the related notions of function and teleology, both of which are longstanding issues in the philosophy of biology. The classic works are by Cummins (1975) and Wright (1973). For an excellent collection containing these

and other germinal works as well as more recent work, see Allen, Bekoff, and Lauder, eds. (1998). Here, as in other areas of the philosophy of evolution, there have been calls for pluralism.

5   Other recent debates center on the connections between development and adaptationism. See Amundson (1994), Kaufman (1993), Griffiths (1996), and this volume.

6   The interface between philosophy of biology and philosophy of psychology in general is a recent and burgeoning area; see, for example, Hardcastle (1999).

7   Sokal and Crovello's version of the phenetic species concept construes "similarity" very liberally, including morphological, physiological, biochemical, behavioral, genetic, protein, and ecological similarities (1970, p. 150).

8   Here I speak of phylogenetic species concepts in general, rather than the particular phylogenetic species concept defended by Mishler and Brandon (1987), or the (different) phylogenetic species concept defended by Cracraft (1983).

9   In fact, the phrase "survival of the fittest" was coined by Herbert Spencer in 1864, not Darwin, and was not included in the *Origin of Species* until the fifth edition, at the urging of Alfred Russel Wallace. Philosophers and biologists have been sorry ever since.

10 In what follows, I will use "units of selection" and "levels of selection" interchangeably, although there are those who would distinguish them (Brandon, 1982).

11 In what follows, I will use "individual selection" and "organismic selection" interchangeably, even though, for example, some would argue that species are individuals as well (see Species section on page 234).

12 See Cain and Darden (1988) for an analysis of Hull's model of natural selection and an alternative characterization.

13 These debates appear to be examples of "relative significance" debates, *sensu* Beatty (see above discussion).

References

Abrams, M. (1999): "Propensities in the Propensity Interpretation of Fitness," *Southwest Philosophy Review*, 15(1), 27–35.

Allen, C., Bekoff, M. and Lauder, G. (eds.) (1998): *Nature's Purposes: Analyses of Function and Design in Biology*. Cambridge, MA: MIT Press.

Amundson, R. (1994): "Two Concepts of Constraint: Adaptationism and the Challenge from Developmental Biology," *Philosophy of Science*, 61(4), 556–78.

Beatty, J. (1984): "Chance and Natural Selection," *Philosophy of Science*, 51, 183–211.

Beatty, J. (1985): "Speaking of Species: Darwin's Strategy," in D. Kohn (ed.), *The Darwinian Heritage*, Princeton: Princeton University Press, 265–81.

Beatty, J. (1987): "Natural Selection and the Null Hypothesis," in J. Dupré (ed.), *The Latest on the Best: Essays on Evolution and Optimality*, Cambridge, MA: MIT Press, 53–75.

Beatty, J. (1992): "Fitness," in E. F. Keller and E. A. Lloyd (eds.), *Keywords in Evolutionary Biology*, Cambridge, MA: Harvard University Press, 115–19.

Beatty, J. (1995): "The Evolutionary Contingency Thesis," in G. Wolters and J. Lennox (eds.), *Concepts, Theories, and Rationality in the Biological Sciences: The Second*

Pittsburgh–Konstanz Colloquium in the Philosophy of Science, University of Pittsburgh, October 1–4, 1993, Pittsburgh, PA: University of Pittsburgh Press, 45–81.

Beatty, J. (1997): "Why Do Biologists Argue Like They Do?" *Philosophy of Science*, 64(4), Supplement, S432–S443.

Beatty, J. and Finsen, S. (1989): "Rethinking the Propensity Interpretation: A Peek Inside Pandora's Box," in M. Ruse (ed.), *What the Philosophy of Biology Is: Essays for David Hull*, Dordrecht: Kluwer, 17–31.

Brandon, R. N. (1978): "Adaptation and Evolutionary Theory," *Studies in History and Philosophy of Science*, 9, 181–206.

Brandon, R. N. (1982): "The Levels of Selection," in P. Asquith and T. Nickles (eds.), *PSA 1982, vol. I*, East Lansing, MI: Philosophy of Science Association, 315–23.

Brandon, R. N. (1990): *Adaptation and Environment*. Princeton, NJ: Princeton University Press.

Brandon, R. N. (1997): "Discussion: Reply to Hitchcock," *Biology and Philosophy*, 12(4), 531–8.

Brandon, R. N. and Beatty, J. (1984): "Discussion: The Propensity Interpretation of 'Fitness' – No Interpretation Is No Substitute," *Philosophy of Science*, 51, 342–7.

Brandon, R. N. and Burian, R. M. (eds.) (1984): *Genes, Organisms, Populations: Controversies Over the Units of Selection*. Cambridge, MA: MIT Press.

Brandon, R. N. and Carson, S. (1996): "The Indeterministic Character of Evolutionary Theory: No 'No Hidden Variables Proof' but No Room for Determinism Either," *Philosophy of Science*, 63(3), 315–37.

Brandon, R. N. and Rausher, M. D. (1996): "Testing Adaptationism: A Comment on Orzack and Sober," *American Naturalist*, 148(1), 189–201.

Brandon, R. N., Antonovics, J., Burian, R. M., Carson, S., Cooper, G., Davies, P. S., Horvath, C. D., Mishler, B. D., Richardson, R. C., Smith, K. C. and Thrall, P. (1994): "Discussion: Sober on Brandon on Screening-Off and the Levels of Selection," *Philosophy of Science*, 61(3), 475–86.

Burian, R. M. (1983): "Adaptation," in M. Grene (ed.), *Dimensions of Darwinism*, New York: Cambridge University Press, 287–314.

Cain, J. and Darden, L. (1988): "Hull and Selection," *Biology and Philosophy*, 3, 165–71.

Cairns, J., Overbaugh, J. and Miller, S. (1988): "The Origin of Mutations," *Nature*, 335, 142–5.

Caplan, A. (1982): "Say It Just Ain't So: Adaptational Stories and Sociobiological Explanations of Social Behavior," *The Philosophical Forum*, 13, 144–60.

Cracraft, J. (1983): "Species Concepts and Speciation Analysis," *Current Ornithology*, 1, 159–87.

Creath, R. and Maienschein, J. (eds.) (2000): *Biology and Epistemology*. Cambridge: Cambridge University Press.

Cronin, H. (1991): *The Ant and the Peacock*. Cambridge: Cambridge University Press.

Cummins, R. (1975): "Functional Analysis," *Journal of Philosophy*, 72, 741–65.

Darwin, C. ([1859] 1964): *On the Origin of Species: A Facsimile of the First Edition*. Cambridge: Harvard University Press.

Dawkins, R. (1976): *The Selfish Gene*. Oxford: Oxford University Press.

Dawkins, R. (1982): *The Extended Phenotype: The Gene as the Unit of Selection*. Oxford: W. H. Freeman and Company.

Dennett, D. C. (1995): *Darwin's Dangerous Idea*. New York: Simon & Schuster.

Dover, G. (1997a): "Neutralist–Selectionist–Molecular Drive Debate," *BioEssays*, 19(9), 836–7.

Dover, G. (1997b): "There's More to Life Than Selection and Neutrality," *BioEssays*, 19, 91–2.

Dupré, J. (ed.) (1987): *The Latest on the Best: Essays on Evolution and Optimality*. Cambridge, MA: MIT Press.

Dupré, J. (1992): "Species: Theoretical Contexts," in E. F. Keller and E. A. Lloyd (eds.), *Keywords in Evolutionary Biology*, Cambridge, MA: Harvard University Press, 312–17.

Dupré, J. (1999): "On the Impossibility of a Monistic Account of Species," in R. A. Wilson (ed.), *Species: New Interdisciplinary Essays*, Cambridge, MA: MIT Press, 3–22.

Eldredge, N. and Gould, S. J. (1972): "Punctuated Equilibria: An Alternative to Phyletic Gradualism," in T. Schopf (ed.), *Models in Paleobiology*, San Francisco: Freeman, 82–115.

Ereshefsky, M. (1998): "Species Pluralism and Anti-Realism," *Philosophy of Science*, 65(1), 103–20.

Futuyma, D. (1988): "Sturm Und Drang and the Evolutionary Synthesis," *Evolution*, 42(2), 217–26.

Ghiselin, M. T. (1969): *The Triumph of the Darwinian Method*. Chicago: University of Chicago Press.

Ghiselin, M. T. (1974): "A Radical Solution to the Species Problem," *Systematic Zoology*, 23, 536–44.

Ghiselin, M. T. (1997): *Metaphysics and the Origin of Species*. Albany: SUNY Press.

Godfrey-Smith, P. (1999): "Adaptationism and the Power of Selection," *Biology and Philosophy*, 14, 181–94.

Gould, S. J. (1980): "Sociobiology and the Theory of Natural Selection," in G. W. Barlow and J. Silverberg (eds.), *Sociobiology: Beyond Nature/Nurture?*, Boulder: Westview Press, 257–69.

Gould, S. J. (1989): *Wonderful Life: The Burgess Shale and the Nature of History*. New York: W. W. Norton & Company.

Gould, S. J. and Lewontin, R. C. (1979): "The Spandrels of San Marco and the Panglossian Paradigm: A Critique of the Adaptationist Programme," *Proceedings of the Royal Society of London, B*, 205, 581–98.

Gould, S. J. and Lloyd, E. A. (1999): "Individuality and Adaptation Across Levels of Selection: How Shall We Name and Generalize the Unit of Darwinism?" *Proceedings of the National Academy of Sciences*, 96(21), 11904–9.

Gould, S. J. and Vrba, E. S. (1982): "Exaptation: A Missing Term in the Science of Form," *Paleobiology*, 8, 4–15.

Grantham, T. (1999a): "Explanatory Pluralism in Paleobiology," *Philosophy of Science*, 66(3), Supplement, S223–S36.

Grantham, T. (1999b): "Philosophical Perspectives on the Mass Extinction Debates," *Biology and Philosophy*, 14, 143–50.

Graves, L., Horan, B. L. and Rosenberg, A. (1999): "Is Indeterminism the Source of the Statistical Character of Evolutionary Theory?" *Philosophy of Science*, 66(1), 140–57.

Grene, M. (1997): "Current Issues in the Philosophy of Biology," *Perspectives on Science*, 5(2), 255–82.

Griffiths, P. E. (1996): "The Historical Turn in the Study of Adaptation," *British Journal for the Philosophy of Science*, 47(4), 511–32.

Hardcastle, V. G. (ed.) (1999): *Where Biology Meets Psychology*. Cambridge, MA: MIT Press.

Harding, S. (1986): *The Science Question in Feminism*. Ithaca, NY: Cornell University Press.

Hey, J. (1999): "The Neutralist, the Fly and the Selectionist," *Trends in Ecology and Evolution*, 14(1), 35–7.

Hitchcock, C. R. (1997): "Discussion: Screening-Off and Visibility to Selection," *Biology and Philosophy*, 12(4), 521–9.

Horan, B. L. (1994): "The Statistical Character of Evolutionary Theory," *Philosophy of Science*, 61(1), 76–95.

Hrdy, S. B. (1999): *Mother Nature: A History of Mothers, Infants, and Natural Selection*. New York: Pantheon Books.

Hull, D. L. (1965): "The Effect of Essentialism on Taxonomy – Two Thousand Years of Stasis (II)," *British Journal for the Philosophy of Science*, 16, 1–18.

Hull, D. L. (1969): "What Philosophy of Biology Is Not," *Synthese*, 20, 157–84.

Hull, D. L. (1976): "Are Species Really Individuals?" *Systematic Zoology*, 25, 174–91.

Hull, D. L. (1978): "A Matter of Individuality," *Philosophy of Science*, 45, 335–60.

Hull, D. L. (1980): "Individuality and Selection," *Annual Review of Ecology and Systematics*, 11, 311–32.

Hull, D. L. (1999): "On the Plurality of Species: Questioning the Party Line," in R. A. Wilson (ed.), *Species: New Interdisciplinary Essays*, Cambridge, MA: MIT Press, 23–48.

Hull, D. L. and Ruse, M. (eds.) (1998): *The Philosophy of Biology*. Oxford: Oxford University Press.

Jablonka, E. and Lamb, M. J. (1995): *Epigenetic Inheritance and Evolution: The Lamarckian Dimension*. Oxford: Oxford University Press.

Kaufman, S. (1993): *Origins of Order*. Oxford: Oxford University Press.

Keller, E. F. (1992): "Between Language and Science: The Question of Directed Mutation in Molecular Genetics," *Perspectives in Biology and Medicine*, 35, 292–306.

Kimura, M. (1969): "Evolutionary Rate at the Molecular Level," *Nature*, 217, 624–6.

Kimura, M. (1983): *The Neutral Theory of Molecular Evolution*. Cambridge: Cambridge University Press.

Kitcher, P. (1982): *Abusing Science: The Case Against Creationism*. Cambridge, MA: MIT Press.

Kitcher, P. (1984): "Species," *Philosophy of Science*, 51, 308–33.

Kitcher, P. (1985): *Vaulting Ambition: Sociobiology and the Quest for Human Nature*. Cambridge, MA: MIT Press.

Kitcher, P. (1989): "Some Puzzles About Species," in M. Ruse (ed.), *What the Philosophy of Biology is*, Dordrecht: Kluwer, 183–208.

Kitcher, P. (1990): "Developmental Decomposition and the Future of Human Behavioral Ecology," *Philosophy of Science*, 57(1), 96–117.

Lewontin, R. C. (1970): "The Units of Selection," *Annual Review of Ecology and Systematics*, 1, 1–18.

Lewontin, R. C., Rose, S. and Kamin, L. (1984): *Not in Our Genes: Biology, Ideology, and Human Nature*. Cambridge, MA: Harvard University Press.

Lloyd, E. A. (1988): *The Structure and Confirmation of Evolutionary Theory*. New York: Greenwood Press.

Lloyd, E. A. (1989): "A Structural Approach to Defining Units of Selection," *Philosophy of Science*, 56, 395–418.

Lloyd, E. A. (1992): "Units of Selection," in E. F. Keller and E. A. Lloyd (eds.), *Keywords in Evolutionary Biology*, Cambridge, MA: Harvard University Press, 334–40.

Lloyd, E. A. and Gould, S. J. (1993): "Species Selection on Variability," *Proceedings of the National Academy of Sciences*, 90, 595–9.

Longino, H. (1990): *Science as Social Knowledge*. Princeton: Princeton University Press.

Mayden, R. L. (1997): "A Hierarchy of Species: The Denouement in the Saga of the Species Problem," in M. F. Claridge, H. A. Dawah and M. R. Wilson (eds.), *Species: The Units of Biodiversity*, New York: Chapman and Hall, 381–424.

Mayr, E. (1963): *Animal Species and Evolution*. Cambridge, MA: Harvard University Press.

Mayr, E. (1983): "How to Carry Out the Adaptationist Program?" *American Naturalist*, 121(3), 324–34.

Mayr, E. (1987): "The Ontological Status of Species: Scientific Progress and Philosophical Terminology," *Biology and Philosophy*, 2, 145–66.

Mayr, E. (1996): "What is a Species, and What Is Not?" *Philosophy of Science*, 63(2), 262–77.

McShea, D. (1991): "Complexity and Evolution: What Everybody Knows," *Biology and Philosophy*, 6, 303–24.

Mills, S. K. and Beatty, J. (1979): "A Propensity Interpretation of Fitness," *Philosophy of Science*, 46, 263–86.

Millstein, R. L. (1996): "Random Drift and the Omniscient Viewpoint," *Philosophy of Science*, 63(3), Supplement, S10–S18.

Millstein, R. L. (1997): "*The Chances of Evolution: An Analysis of the Roles of Chance in Microevolution and Macroevolution*," PhD dissertation, University of Minnesota.

Millstein, R. L. (2000): "Chance and Macroevolution," *Philosophy of Science*, 67(4), 603–24.

Millstein, R. L. (2001): "Are Random Drift and Natural Selection Conceptually Distinct?" *Biology and Philosophy*, forthcoming.

Mishler, B. D. and Brandon, R. N. (1987): "Individuality, Pluralism, and the Phylogenetic Species Concept," *Biology and Philosophy*, 2, 397–414.

Nitecki, M. H. (ed.) (1988): *Evolutionary Progress*. Chicago: University of Chicago Press.

Ohta, T. (1997): "More on the Neutralist–Selectionist Debate," *BioEssays*, 19, 359.

Ohta, T. and Kreitman, M. (1996): "The Neutralist–Selectionist Debate," *BioEssays*, 18(8), 673–84.

Orzack, S. H. and Sober, E. (1996): "How to Formulate and Test Adaptationism," *American Naturalist*, 148(1), 202–10.

Pennock, R. T. (1999): *Tower of Babel: The Evidence against the New Creationism*. Cambridge, MA: MIT Press.

Popper, K. R. (1974): "Intellectual Autobiography," in P. A. Schilpp (ed.), *The Philosophy of Karl Popper*, La Salle, IL: Open Court Press, 3–181.

Popper, K. R. (1978): "Natural Selection and the Emergence of Mind," *Dialectica*, 32, 339–55.

Provine, W. B. (1985): "Adaptation and Mechanisms of Evolution After Darwin: A Study in Persistent Controversies," in D. Kohn (ed.), *The Darwinian Heritage*, Princeton, NJ: Princeton University Press, 825–66.

Reeve, H. K. and Keller, L. (1999): "Levels of Selection: Burying the Units-of-Selection Debate and Unearthing the Crucial New Issues," in L. Keller (ed.), *Levels of Selection in Evolution*, Princeton: Princeton University Press, 3–14.

Richardson, R. C. and Burian, R. M. (1992): "A Defense of the Propensity Interpretation of Fitness," in D. Hull, M. Forbes and K. Okruhlik (eds.), *PSA 1992, vol. 2.* East Lansing: Philosophy of Science Association, 349–62.

Rose, M. R. and Lauder, G. V. (eds.) (1996): *Adaptation.* San Diego: Academic Press.

Rosenberg, A. (1982): "On the Propensity Definition of Fitness," *Philosophy of Science*, 49, 268–73.

Rosenberg, A. (1988): "Is the Theory of Natural Selection a Statistical Theory?" *Canadian Journal of Philosophy*, 14, Supplement, 187–207.

Rosenberg, A. (1992): "Altruism: Theoretical Contexts," in E. F. Keller and E. A. Lloyd (eds.), *Keywords in Evolutionary Biology*, Cambridge, MA: Harvard University Press, 19–28.

Rosenberg, A. (1994): *Instrumental Biology or the Disunity of Science.* Chicago: University of Chicago Press.

Rosenberg, A. and Williams, M. B. (1986): "Fitness as Primitive and Propensity," *Philosophy of Science*, 53, 412–18.

Ruse, M. (1993): "Evolution and Progress," *Trends in Ecology and Evolution*, 8(2), 55–9.

Ruse, M. (ed.) (1996): *But Is It Science? The Philosophical Question in the Creation/ Evolution Controversy.* Amherst, NY: Prometheus Books.

Ruse, M. (1998): *Taking Darwin Seriously: A Naturalistic Approach to Philosophy.* Amherst NY: Prometheus Books.

Sarkar, S. (1991): "Lamarck *Contra* Darwin, Reduction Versus Statistics: Conceptual Issues in the Controversy Over Directed Mutagenesis in Bacteria," in A. I. Tauber (ed.), *Organism and the Origins of Self*, Dordrecht: Kluwer, 235–71.

Shanahan, T. (1992): "Selection, Drift, and the Aims of Evolutionary Theory," in P. Griffiths (ed.), *Trees of Life: Essays in Philosophy of Biology*, Dordrecht: Kluwer, 131–61.

Sneath, P. H. A. and Sokal, R. R. (1973): *Numerical Taxonomy.* San Francisco: W. H. Freeman.

Sober, E. (1984): *The Nature of Selection: Evolutionary Theory in Philosophical Focus.* Cambridge, MA: MIT Press.

Sober, E. (2000): "The Two Faces of Fitness," in R. S. Singh, C. Krimbas, D. Paul and J. Beatty (eds.), *Thinking About Evolution*, Cambridge, MA: Cambridge University Press, forthcoming.

Sober, E. and Lewontin, R. C. (1982): "Artifact, Cause and Genic Selection," *Philosophy of Science*, 49, 157–80.

Sober, E. and Wilson, D. S. (1994): "A Critical Review of Philosophical Work on the Units of Selection Problem," *Philosophy of Science*, 61(4), 534–55.

Sober, E. and Wilson, D. S. (1998): *Unto Others: The Evoluton and Psychology of Unselfish Behavior.* Cambridge, MA: Harvard University Press.

Sokal, R. R. and Crovello, T. J. (1970): "The Biological Species Concept – A Critical Evaluation," *American Naturalist*, 104(936), 127–53.

Stanley, S. (1979): *Macroevolution: Pattern and Process.* San Francisco: Freeman.

Sterelny, K. (1995): "Understanding Life: Recent Work in Philosophy of Biology," *British Journal for the Philosophy of Science*, 46(2), 155–83.

Sterelny, K. (1996): "Explanatory Pluralism in Evolutionary Biology," *Biology and Philosophy*, 11, 193–214.

Sterelny, K. and Griffiths, P. E. (1999): *Sex and Death: An Introduction to the Philosophy of Biology.* Chicago: University of Chicago Press.

Sterelny, K. and Kitcher, P. (1988): "The Return of the Gene," *Journal of Philosophy*, 85, 339–61.

Stout, R. (1998): "The Evolution of Theoretically Useful Traits," *Biology and Philosophy*, 13(4), 529–40.

Templeton, A. (1989): "The Meaning of Species and Speciation: A Genetic Perspective," in D. Otte and J. Endler (eds.), *Speciation and Its Consequences*, Sunderland, MA: Sinauer Associates, 3–27.

van der Steen, W. J. (1996): "Screening-Off and Natural Selection," *Philosophy of Science*, 63(1), 115–21.

Van Valen, L. (1976): "Ecological Species, Multispecies, and Oaks," *Taxon*, 25, 233–9.

Waters, C. K. (1986): "Natural Selection without Survival of the Fittest," *Biology and Philosophy*, 1, 207–25.

Waters, C. K. (1991): "Tempered Realism about the Force of Selection," *Philosophy of Science*, 58, 553–73.

Waters, C. K. (1998): "Causal Regularities in the Biological World of Contingent Distributions," *Biology and Philosophy*, 13(1), 5–36.

Weber, M. (1996): "Fitness Made Physical: The Supervenience of Biological Concepts Revisited," *Philosophy of Science*, 63(3), 411–31.

Weber, M. (2001): "Determinism, Realism, and Probability in Evolutionary Theory: The Pitfalls, and How to Avoid Them," *Philosophy of Science (Proceedings)*, PSA 2000-Partl, forthcoming.

Wilson, D. S. and Sober, E. (1994): "Reintroducing Group Selection to the Human Behavioral Sciences," *Behavior and Brain Sciences*, 17, 585–608.

Wilson, E. O. (1975): *Sociobiology: The New Synthesis*. Cambridge, MA: Harvard University Press.

Wimsatt, W. C. (1980): "Reductionistic Research Strategies and their Biases in the Units of Selection Controversy," in T. Nickles (ed.), *Scientific Discovery: Case Studies, vol. ii*, Dordrecht: Reidel, 213–59.

Wright, L. (1973): "Functions," *Philosophical Review*, 82, 139–68.

Wynne-Edwards, V. C. (1962): *Animal Dispersion in Relation to Social Behavior*. Edinburgh, UK: Oliver and Boyd.

Chapter 12

Molecular and Developmental Biology

Paul Griffiths

Introduction

Philosophical discussion of molecular and developmental biology began in the late 1960s with the use of genetics as a test case for models of theory reduction. With this exception, the theory of natural selection remained the main focus of philosophy of biology until the late 1970s. It was controversies in evolutionary theory over punctuated equilibrium and adaptationism that first led philosophers to examine the concept of developmental constraint. Developmental biology also gained in prominence in the 1980s, as part of a broader interest in the new sciences of self-organization and complexity. The current literature in the philosophy of molecular and developmental biology has grown out of these earlier discussions under the influence of twenty years of rapid and exciting growth of empirical knowledge. Philosophers have examined the concepts of genetic information and genetic program, competing definitions of the gene itself, and competing accounts of the role of the gene as a developmental cause. The debate over the relationship between development and evolution has been enriched by theories and results from the new field of "evolutionary developmental biology." Future developments seem likely to include an exchange of ideas with the philosophy of psychology, where debates over the concept of innateness have created an interest in genetics and development.

Review of Past Literature

Reduction of Mendelian to molecular genetics

According to the classical account of theory reduction, one theory reduces to another when the laws and generalizations of the first theory can be deduced from those of the second theory with the help of bridge principles relating the vocabularies of the two theories (Nagel, 1961). In 1967, Kenneth Schaffner suggested that classical Mendelian genetics could be reduced to the new, molecular genetics in something like this way. In a series of papers, Schaffner (1967, 1969) outlined his "general reduction model" and argued for its applicability to the case of genetics. Despite the fact that Schaffner's model of reduction was less demanding than the classical model and allowed substantial correction of the reduced theory to facilitate its deduction from the reducing theory, his proposal elicited considerable skepticism. David Hull (1974) argued that key terms in the vocabulary of Mendelian genetics – gene, locus, allele, dominance and so forth – have no unique correlate in molecular biology. There is, for example, no single molecular mechanism corresponding to dominance. The phenotypic resemblance between heterozygote and dominant homozygote might be explained by the nature of the products of the two alleles, by gene regulation that compensates for the loss of one copy of an allele or by the existence of alternative pathways to the same outcome in morphogenesis. Definitions of dominance and other key Mendelian terms at the molecular level will be open-ended disjunctions of ways in which the Mendelian phenomena might be produced. Therefore, Hull and others argued, the generalizations of classical genetics cannot be captured by statements at a similar level of generality in molecular biology. So the theory of classical genetics is irreducible to theories in molecular biology.

The same fundamental issues were still under discussion ten years later, when Philip Kitcher (1984) put forward his "gory details" argument. Kitcher argued that classical Mendelian genetics offers explanations of many important biological phenomena which are complete in their own terms and are not improved by adding the "gory details" at the molecular level. The Mendelian ratios, for example, are explained by the segregation and independent assortment of chromosomes. Any mechanism that obeyed these two laws would produce Mendelian ratios and so, the details of how segregation and assortment are achieved, however important they are in their own right and as explanations of other facts, do not add anything to the explanation of Mendelian ratios. Kenneth C. Waters has rebutted this argument, arguing that classical genetic phenomena such as crossing-over in meiosis immediately raise questions that can only be addressed in a molecular framework, such as why recombination is more likely at certain points on the chromosomes. It is simply not plausible, Waters (1994,a,b) argues, to treat the relatively small number of exception-ridden generalizations identified by classical genetics as an explanatory framework that is complete in its own terms. Waters also proposes a definition of "gene" designed to rebut the charge that Mendelian genes do not display a unity at the molecular level. A gene is any relatively short segment of DNA that functions as a biochemical unit (Waters, 1994a, p. 407). Waters admits that this definition makes the gene a unit of indeterminate length and that it is the specific research context that determines whether a particular utterance of "gene" refers to a series of exons, an entire reading frame including both exons and introns, the reading frame plus adjacent regulatory regions or that complex plus other regions involved in regulating splicing and editing the transcript. Nevertheless, he argues, at the core of all these definitions of "gene" is the basic concept of a sequence that is transcribed to produce a gene product. Other authors have argued that Waters's definition creates a merely verbal unity between "genes" with different structures, different functions and different theoretical roles in molecular biology (Neumann-Held, 1998). The empirical facts that underlie this dispute are that reading sequences – the structural basis of the classical molecular conception of the gene – can be used to make a variety of products depending on the cellular context which regulates their expression and cuts, splices and edits the gene transcript. Reading sequences can also overlap one another. All these phenomena were unanticipated by early molecular biologists, let alone by premolecular Mendelian geneticists. The magnitude of these theoretical developments in genetics makes it highly plausible that there have been changes in the concept of the gene, which is the central theoretical construct of that discipline. Whether such conceptual change would make reduction impossible is less clear.

The thirty-year debate between reductionists and anti-reductionists has been complex and wide-ranging and numerous authors not mentioned here have made important contributions. A more adequate, but still brief, survey can be found in Sterelny and Griffiths (1999, chs 6–7) and an extended treatment in Sarkar (1998). For many philosophers, the main lesson of the debate is that traditional models of reduction do not capture the important role played in scientific progress by successful explanations of larger systems in terms of their smaller constituents (Wimsatt, 1976). Even committed reductionists, such as Waters, have adopted models of reduction very different from those with which the debate began. Schaffner himself has continued to make some of the most sophisticated contributions to the development of adequate models of the relationship between molecular biology and theories of larger units of biological organization. His work (Schaffner, 1993) has increasingly focused on the role of model systems and results of limited generality derived from the analysis of these systems.

Developmental constraints and evolution

It is generally accepted that the "modern synthesis" of Mendelian genetics and natural selection that put so many of the biological sciences on a common theoretical basis failed to include the science of developmental biology (Hamburger, 1980). The synthetic theory bypassed what were at the time intractable questions of the actual relationship between stretches of chromosome and phenotypic traits. Although it was accepted that genes must, in reality, generate phenotypic differences through interaction with other genes and other factors in development, genes were treated as "black boxes" that could be relied on to produce the phenotypic variation with which they were known to correlate. The black-boxing strategy allowed the two tractable projects – theoretical population genetics and the study of selection at the phenotypic level – to proceed. Selection could be studied at the phenotypic level on the assumption that variant phenotypes were generated in some unknown way by the genes and that phenotypic change would be tracked by change in gene frequencies. Population genetics, the mathematical core of the modern synthesis, could postulate genes corresponding to phenotypic differences and track the effect of selection on these phenotypic variants at the genetic level. One effect of this strategy was to direct attention away from ideas that would obstruct these research practices. Among these inconvenient ideas was the view that development does not always permit the phenotypes that selection would favor. This idea was revived in the "punctuated equilibrium" theory of Niles Eldredge and Stephen J. Gould (Eldredge and Gould, 1972; Gould and Eldredge, 1977). Traditional neo-Darwinian gradualism suggests that species evolve more or less continuously in response to local selection pressures. The fossil record, on the other hand, suggests that species remain largely unaltered for long periods of time and occasionally undergo dramatic periods of rapid evolutionary change. The punctuated equilibrium theory proposed that the fossil record be read at face value, rather than in the light of the gradualist model of evolution. The new theory needed an evolutionary explanation of this pattern and sought it in "developmental constraints." The range of variant phenotypes produced by genetic changes is constrained by the nature of the organism's developmental system so that selection is usually unable to produce dramatic reorganization of the phenotype. Conversely, a relatively small genetic change might, in the context of development as a whole, result in large phenotypic changes and very rapid evolution. Both possibilities can be understood using C. H. Waddington's metaphor of developmental canalization (Figure 12.1). Most small perturbations to the course of development are compensated so that the organism arrives at the same destination. Some, however, send development down an entirely new "channel."

A second source of the renewed interest in developmental constraint was the debate over the limits of adaptive explanation. Stephen J. Gould and Richard Lewontin (1979) strongly criticized "adaptationism" – the practice of seeking adaptive explanations for every feature of organisms. They suggested developmental constraint as one alternative explanation of biological form. There are, for example, many viviparous snakes, but no viviparous turtles. Perhaps this is to be explained adaptively: any transitional form of turtle would be less fit that its fully oviparous competitors. The ease with which other groups, such as snakes and sharks, have evolved viviparous and quasi-viviparous species suggests an alternative explanation. Perhaps the developmental biology of turtles means that no mutation produces the transitional forms. Gould and Lewontin also revived the traditional idea of the *bauplan* (body plan) or "unity of type" of a whole group of organisms. Crustaceans, for example, have the segmented body of other arthropods but are distinguished from other clades by the fusion of the first five segments to form a head. It seems unlikely that this character has been a critical component of the fitness of every crustacean, from lobster to barnacle, but it has remained stable through long periods of evolution. Perhaps this is to be explained

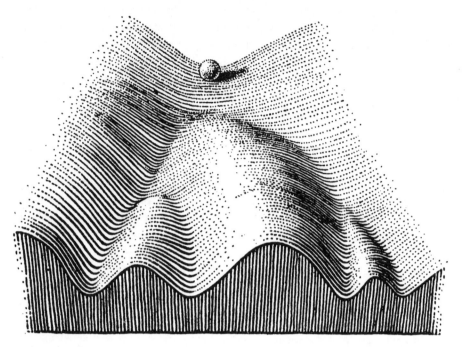

Figure 12.1   Canalization of development

by developmental constraint – the head is part of the basic body plan of this kind of organism – a highly canalized outcome of crustacean development.

There is no doubt that developmental constraints exist (Maynard Smith et al., 1985). A constraint can be defined fairly uncontentiously as a bias in the production of variation in a population. But there is little agreement about the evolutionary importance of constraints. Even more importantly for philosophy of biology, there is little agreement about how their importance might be measured. At one extreme, "process structuralists" like Brian Goodwin have argued that explanation in terms of natural selection have only a marginal role when compared to explanations in terms of developmental constraint (Goodwin, 1984; Ho and Saunders, 1984). The process structuralists sought to revive the nineteenth-century project of "rational taxonomy": a classification of biological forms in terms of the generative principles by which form is constructed. The fact that an organism has a particular form is primarily explained by its place in this system. In support of their position, the process structuralists were able to offer striking examples of this kind of explanation. There are only a few patterns of phylotaxis – the successive arrangement of radial parts in a growing plant – and these patterns are typically conserved within lineages of plant species. A general mathematical description of these patterns is available and models of growth that obey this mathematical description are biologically plausible (Mitchison, 1977). If correct, this is an impressively general explanation of many biological traits in many species. The process structuralists also presented methodological arguments. Scientific explanations should appeal to laws of nature, not historical accidents. Explanations of form in terms of the mechanisms of growth are simply better explanations than those that rely on natural selection (Goodwin and Webster, 1996). Most developmentalists have been less extreme than the process structuralists. They do not deny the importance of natural selection, but insist that the course of evolution cannot be understood in terms of selection alone, only in terms of the interaction of selection with the constraints imposed on phenotypic change by development. At the other extreme, some biologists have argued that constraints can only ever be temporary, since evolution can reconstruct the developmental system of the organism so as to achieve whatever outcome is selectively optimal. Darwin himself expressed something like this sentiment when he remarked that his theory embraced both traditional forms of biological explanation, the "unity of type" and the "conditions of existence," but that the conditions of existence was the "higher law" because it explains the origin of the types (Darwin, 1964, p. 206). But there are many highly conserved features of biological lineages that are not plausibly explained by stabilizing selection, such as the fused head-segments of crustaceans mentioned above or the relative position of bones in the tetrapod limb. Something must explain the fact that these features have not been affected by random genetic drift and developmental constraint is an obvious candidate. William Wimsatt (1986, 1999) has offered a highly general argument for the view that developmental constraints will be harder for selection to remove than to construct. It is widely accepted that the ability of natural selection to create complex adaptation depends on the ability to create those adaptations cumulatively, adding features one at a time. Wimsatt argues that new adaptations will be constructed by utilizing existing developmental structures in the organism, so that the ability to develop the new feature is left dependent on the continued existence of the older features. Wimsatt calls this process "developmental entrenchment" and argues that it will lead to features of the organism becoming progressively less open to selective modification in their own right as additional features are built "on top."

Another argument for the adaptationist perspective concedes the role of development as a *cause* of form, but questions its value as an *explanation* of form. One of the primary aims of biology is to explain the fact that organisms are well adapted for their conditions of life (Dawkins, 1986; Dennett, 1995). Naturally, there is a developmental explanation of how each organism is constructed, but this cannot explain the fact that organisms are well adapted. How could the developmental structure of organisms ensure in and of itself that organisms are well suited to the demands of their environment?

Of course, large quantities of evolutionary change may be non-adaptive, in which case these alternative theories may well be important in parts of evolution, but only in the boring parts of evolution, not the parts concerned with what is special about life as opposed to non-life (Dawkins, 1986, p. 303 – *Some process structuralist targets of this remark are identified by name on p. 307.*)

Peter Godfrey-Smith (1999) has christened this view "explanatory adaptationism" to distinguish it from the "empirical adaptationist" view that almost every feature of organisms has an adaptive explanation. For the explanatory adaptationist, the problem with developmental explanations is not that they are false, but that they explain the wrong thing.

Ronald Amundson (1994) has argued that adaptationists and developmentalists are to a significant extent talking past one another because they have very different concepts of developmental constraint. In developmental biology, a developmental constraint explains why certain phenotypes do not occur, either generally or in some particular group of organisms. The fact that a feature conforms to a developmental constraint in this sense is consistent with it being perfectly adapted to its environment. In the study of adaptation, however, developmental constraints are postulated to explain why organisms are unable to construct the optimally adaptive phenotype. This, second understanding of constraint is manifested in another of Godfrey-Smith's categories: "methodological adaptationism." This is view that the best way to reveal developmental constraints is to build optimality models and look at how nature deviates from what is optimally adaptive. In this sense, constraint and adaptation are opposed to one another by definition. Like Godfrey-Smith's distinction between empirical and explanatory adaptationism, Amundson's distinction between constraints on form and constraints on adaptation goes some way to explain why the debate between adaptationists and developmentalists has produced more heat than light. But even after these conceptual clarifications, there remain genuine empirical differences between the two views, as Amundson himself makes clear. The underlying empirical issue is how much of the space of possible biological forms ("morphospace") is ruled out by the fact that organisms built using the fundamental techniques shared by the earth's biota cannot develop in that way. One way to represent this disagreement is by different predictions about what would happen to a population of organisms in the absence of selection. The adaptationist "null hypothesis" is that random variation would spread the population evenly through an increasingly large region of morphospace. The developmentalist "null hypothesis" is that even without selection organisms would be found clustered in some regions of morphospace and excluded from others because of developmental constraints on the production of variants (Alberch, 1982). Developing this theme, Paul Griffiths (1996) has argued that what appear to be conceptual or methodological differences between process structuralists and extreme adaptationists may in reality be manifestations of this empirical disagreement. The empirical disagreement produces conflicting intuitions about whether development or natural selection is more explanatory because a request for explanation presumes a contrast between the state of affairs to be explained and other possible states of affairs (Van Fraassen, 1991). The question "why is this organism *here* in morphospace?" implies the contrast "as opposed to some other region of morphospace." Because process structuralists think most regions of morphospace are developmentally impossible they will see an explanation of how the organism develops its actual form as highly explanatory. By explaining how this form is possible, it contrasts it with the forms that are impossible. The adaptationist assumes that almost all forms are developmentally possible, so learning that the actual form is possible does not explain the contrast between this form and the adjacent forms.

Biocomplexity and self-organisation

Support for the idea that selection is not the only factor determining biological form was provided in the 1990s by the new sciences of complexity (Burian and Richardson, 1990; Bechtel and Richardson, 1993). Some complex systems possess an intrinsic tendency to occupy highly ordered states, so selection is not the only possible source of order in living systems; see also Riedl (1977). Stuart Kauffman's (1993) simulations of networks of "genetic" elements suggested that basic biological phenomena such as autocatalytic cycles required for the origin of life or the array of cell-types required for the emergence of multi-cellular life are highly probable outcomes of random variation in complex chemical or, later, genetic networks. This is in striking contrast to the traditional view that such complex outcomes are highly improbable and must be explained by cumulative selection of many, much smaller increases in order. Kauffman's simulations also suggested that selection is relatively ineffective when the "genetic" elements are strongly interconnected so that the activity of one depends on that of many others, something that is probably true of actual genes. Because Kauffman's work suggests that order may be generated without selection, and that selection may not be able to overcome the intrinsic tendencies of systems, he has sometimes been seen as providing support for the process structuralist position (Goodwin et al., 1993). But other elements of Kauffman's work do not lend themselves to this interpretation. Self-organization and selection can reinforce one another: self-organisation can enrich the input to selection and selection can "tune" developmental parameters to encourage the production of complex variants (Depew and Weber, 1995). In recent years, even highly adaptationist authors such as Daniel Dennett (1995) have made use of Kauffman's work.

Current Status of Problems

Genetic information

There is an "interactionist consensus" in the life sciences that all traits are dependent on both genetic and environmental factors in development (Sterelny and Griffiths, 1999, pp. 13–17). The consensus emerged from early twentieth-century critiques of the concept of instinct and from parallel critiques of the concept of innateness in early ethology. But this is consistent with the view that genes cause development in a radically different way from other, "environmental" factors. Genes are widely believed to contain a program that guides development and to contain information about the evolved traits of the organism. Despite the ubiquity of talk of genetic information in molecular and developmental biology, the predominant view in recent philosophical work on this topic has been that "genetic information" and "genetic program" have a precise meaning only in the context of the relationship between DNA sequence, RNA sequence, and protein structure (Sarkar, 1996; Griffiths and Knight, 1998; Godfrey-Smith, 1999; Kitcher, 2001). In their broader applications, these ideas are merely picturesque ways to talk about correlation and causation.

The obvious way to explicate information talk in biology is via information theory. Information in this sense is the systematic dependence of a signal on a source, a dependence that is created by a set of channel conditions. In the case of development, the genes are normally taken to be the source, the life-cycle of the organism is the signal and the channel conditions are all the other resources needed for the life-cycle to unfold. But it is a fundamental feature of information theory that the role of source and channel condition can be reversed. A source/channel distinction is imposed on a causal system by an observer. The source is one channel condition whose current state the signal is being used to investigate. If all other resources are held constant, a life-cycle can give us information about the genes, but if the genes are held constant, a life-cycle can give us information about whichever other resource we decided to let vary. So far as causal information goes, every resource whose state affects development is a source of developmental information (Johnston, 1987; Gray, 1992; Griffiths and Gray, 1994; Oyama, 2000a).

The fact that causal information conforms to this "parity thesis" is now quite widely recognized (Godfrey-Smith, 1999; Sterelny and Griffiths, 1999; Maynard Smith, 2000; Kitcher, 2001). A common response has been to analyze genetic information using teleosemantics, the philosophical program of reducing meaning to biological function (teleology) and then reducing biological function to natural selection (Millikan, 1984; Papineau, 1987). In his version of the teleosemantic approach, John Maynard Smith compares natural selection to computer programming using the "genetic algorithm" technique. The genetic algorithm programmer randomly varies the code of a computer program and selects variants for their performance. In the same way, natural selection randomly varies the genes of organisms and selects those organisms for their fitness. Just as the function of the selected computer program is to perform the task for which it was selected, the biological function of successful genes is to produce the developmental outcomes in virtue of which they were selected. Such genes are intentionally directed onto, or about, those effects. The defective hemoglobin gene in some human populations, which has been selected because it sometimes confers resistance to malaria, carries teleosemantic information about malaria resistance. However, teleosemantic information is fundamentally unsuited to the aim of avoiding parity. The most fully developed teleosemantic account of developmental information is the "extended replicator theory" (Sterelny et al., 1996; Sterelny, 2000a), which recognizes from the outset that teleosemantic information exists in both genetic and in some non-genetic developmental causes. Griffiths and Gray (1997) argue that teleosemantic information exists in a much wider range of developmental causes. Teleosemantic information exists in any inheritance system that is a product of evolution, including epigenetic inheritance systems. The term "epigenetic inheritance system" is used to denote biological mechanisms which produces resemblances between parents and offspring and which works in parallel with the inheritance of nuclear and mitochondrial DNA (Jablonka and Szathmáry, 1995). Every organism inherits a great deal besides its DNA. To develop normally, the egg cell must contain features such as: basal bodies and microtubule organising centres, correct cytoplasmic chemical gradients, DNA methylation patterns, membranes and organelles, as well as DNA. Changes in these other resources can cause heritable variation that appears in all the cells descended from that egg cell. Differences in methylation, for example, are important in tissue differentiation during the lifetime of a single organism, but they can also pass between the generations. Methylation patterns are often applied to the DNA in a sperm or egg by the parent organism. DNA methylation inheritance has excited a great deal of interest because of it is easy to see how it could play a role in conventional, micro-evolutionary change. Wider forms of epigenetic inheritance include the inheritance of symbiotic microorganisms, habitat and host imprinting, and the care of offspring. All these mechanisms are candidates for evolutionary explanation – they did not come about by accident. This means that the physical traces by which these inheritance mechanisms influence the next generation have biological functions and thus, on the teleosemantic approach, that these traces contain information. The widest form of epigenetic inheritance is "niche construction." Many features of an organism's niche exist only because of the effects of previous generations of that species on the local environment (Laland et al., 2001). However, despite the evolutionary importance of niche construction, the *collectively* constructed features of a species' niche are not adaptations of the *individual* organism, and hence probably cannot be assigned teleosemantic information content.

Genetic program

The concept of the genetic program has proved as controversial as that of genetic information (Keller, 1995). Its critics have questioned whether development is more program-like than any other law governed physical process. There is a sense in which the planets compute their courses around the sun, integrating the forces that act on them to determine the trajectory they will follow. If the idea of a genetic program comes to no more than this, then it is of little scientific value. Some historians of molecular biology have argued that the history of the genetic program concept in molecular biology is one of retreat from literal hypothesis to guiding metaphor to mere tool for popularization (de Chadarevian, 1998); see also Sarkar (1996). In contrast, Alexander Rosenberg (1997) has defended the view that the study of development is the study of how the embryo is "computed" from the genes and proteins contained in the egg cell. Rosenberg's argument is that striking recent successes in developmental molecular biology have concerned genes which switch other genes on or off in hierarchical cascades of gene activation. What, he asks, could be a more powerful vindication of the idea that the genes contain a self-executing program for development? Keller (1999) has rejected this interpretation of the science, arguing that gene activation in the developing embryo is precisely *not* like the unfolding of a stored program, but instead like distributed computing, in which processes are reliably executed by local interactions in networks of simple elements. The mathematician Henri Atlan adds another perspective to this debate, arguing that if there is a program for development in any sense analogous to programs in computer science, then the program is not in the genome. Atlan argues that a rigorous deployment of the analogy identifies DNA sequences with the data accessed at various times whilst a program is running. The program itself is running on the cellular mechanisms that transcribe and process DNA (Atlan and Koppel, 1990).

Developmental systems theory (DST)

DST is an alternative account of the relationship between genes and other factors in development. It has its roots in a longstanding tradition of dissatisfaction with the concepts of instinct, innateness, genetic information and genetic program amongst workers in comparative psychology and developmental psychobiology (Gottlieb, 2001). When used with care, ideas of instinct, innateness, genetic program and genetic information constitute a kind of "methodological preformationism" in which biological form is treated *as if* it was transmitted intact to the next generation so as to avoid the need to deal with the complexities of development. Very often, however, these concepts are treated as if they were substantial explanatory constructs, leading to the illusion that no developmental explanation is needed for traits that are "innate," "hardwired" or "in the genes"! In place of these ideas, DST argues for a thorough-going epigenetic account of development. Biological form is not transmitted intact, or as an intact representation of that form, but must be reconstructed in each generation by interaction between physical causes. Moreover, there is no one element that controls development or prefigures its outcomes. The term "developmental system" refers to the system of physical resources that interact to produce the life-cycle of a particular evolving lineage. A lineage is redefined as a causally connected sequence of similar individual life-cycles and inheritance is redefined as the reliable reproduction of developmental resources down lineages. This definition includes all the mechanisms of epigenetic inheritance, as well as niche construction and the mere reliable persistence of features upon which the developmental system can draw. Natural selection becomes the differential reproduction of heritable variants of developmental systems due to relative improvements in their functioning, a process which leads to change over time in the composition of populations of developmental systems (Griffiths and Gray, 2001).

The book that drew the developmental systems tradition together and gave it a definitive name was Susan Oyama's (2000a) *The Ontogeny of Information: Developmental Systems and Evolution*, first published in 1985 at around the same time as several of Oyama's important papers (Oyama, 2000b). Philosophers of biology began to discuss these new ideas in the 1990s, some aiming to develop and extend Oyama's approach (Moss, 1992; Griffiths and Gray, 1994), Griffiths and Gray (1997) and others to critically evaluate it. Cor van der Weele (1999) has argued that the criticisms of contemporary neo-Darwinism offered by DST are almost completely orthogonal to those of the process structuralists. DST could potentially treat developmental resources atomistically and rely on selection as the primary explanation of biological form. In reality, however, most DST authors have been sympathetic to the idea that developmental constraints and emergent developmental organization are real and play a role in evolution. Kim Sterelny and others (Sterelny et al., 1996) have accepted some of the critical points made by DST, but argued that these do not justify abandoning the replicator in favor of the developmental system as the unit of evolution. Epigenetic inheritance can be accommodated by enlarging the cast of replicators to include some inherited non-genes. The fact that replicators require a specific context to exert the causal influence can be handled in a manner similar to earlier critiques of the dependence of single genes on their genetic contexts (Sterelny and Kitcher, 1988). Schaffner has argued that most work in molecular developmental biology conforms to DST strictures about the distributed control of development and the context-sensitivity of genetic and other causes. He also argues that a certain instrumental privileging of genetic causes is a justifiable part of research practice (Schaffner, 1998).

The most thorough presentation of DST and its application to date is by Oyama et al. (2001), a volume that also contains critical contributions by some of the authors mentioned here.

Analyses of gene concept

Controversies about the role of genes in development in evolution have generated controversies about the definition of the gene. These have not been sterile debates over the "right" definition. The debates have concerned how genes are actually defined by various kinds of biologists, what this indicates about their thinking and whether genes so defined can bear the theoretical weight placed upon them. An excellent introduction to recent debates over the concept of the gene is given by Beurton et al. (2000). There has been a great deal of criticism of the evolutionary gene concept of George C. Williams according to which a gene is *any* sequence of DNA "which segregates and recombines with appreciable frequency" (Williams 1966, p. 24). Many authors in the philosophy of evolutionary biology have discussed whether change over time in populations of evolutionary genes can explain change at the phenotypic level (Sterelny and Griffiths, 1999, pp. 77–93). In the philosophy of developmental and molecular biology, however, the central issue has been the relationship between genes and phenotypes. The classical molecular gene concept, which emerged in the 1960s and is still orthodox in textbook presentations of genetics defines a gene as a stretch of DNA that expresses a particular polypeptide via transcription and translation. This identifies an individual gene by a particular, minimal "phenotype" to which it gives rise. As mentioned above, Waters still defends something close to this concept of the gene as both central to and adequate for the practice of molecular biology and his account has been criticized by Eva Neumann-Held (Waters, 1994a,b; Neumann-Held, 1998). Griffiths and Neumann-Held (1999) have argued that the development of gene concepts from the turn of the century to the present day has been driven by the twin desires to find a structural unit in the DNA itself and to have that unit make some constant contribution to development. They argue that current knowledge about the multiple functions of many genes makes this difficult if not impossible and suggest (but do not endorse) identifying a sequence of DNA with a unique norm of reaction of gene products across cellular contexts. Their own proposal is to identify a specific gene with a DNA sequence plus the context needed to pin down a single gene product in the manner of the classical molecular concept.

Other authors have argued that two distinct notions of gene play a role in molecular biology: "structural" genes that code for polypeptides used to make structural proteins and "regulatory" or "developmental" genes involved in developmental signaling (Morange, 2000). The most famous examples of developmental genes are the homeobox genes – highly conserved sequences that are involved in segmentation in arthropods and in forming the axes of the vertebrate embryo. Developmental genes have become the favored example of both the friends and enemies of the genetic program concept (Gilbert, 2000). Those critical of the concept take the facts about developmental genes to show that the same sequence can have a radically different effect in a different context. Advocates of the program concept are impressed by how much of the developmental process can be "controlled" by a few genes.

Lenny Moss (2002) has criticized both Waters' analysis and the analysis of Neumann-Held and Griffiths, and argued that the very same genes are both multipotential in the manner of the "developmental gene" and, in another context, defined by a determinate phenotypic effect. Moss proposes that the whole range of uses of the gene concept in contemporary biology can be reduced to two competing conceptualizations of the gene that, he argues, were implicit from the earliest days of genetics. The first way of conceiving of a gene, which Moss calls "Gene-P," is a manifestation of the instrumental preformationist research strategy discussed above. In research contexts in which scientists are interested in establishing or exploiting gene-phene correlations, it makes sense to treat genes as if *as if* they were defined by their association with a certain phenotypic outcome. Blue eyes occur if a gene involved in the synthesis of the brown protein is damaged in some way. What makes a DNA sequence a gene for blue eyes is not any particular sequence nor any knowledge of the developmental pathway that leads to blue eyes but only the fact that the presence of this gene can be used to predict blue eyes. That example comes from classical Mendelian genetics, but contemporary molecular genetics also makes use of the Gene-P concept. BRCA1, the gene for breast cancer, is treated as a Gene-P. Moss's other gene concept (Gene-D) is defined by its molecular sequence. Gene-D is a developmental resource that can make any of a multitude of different contributions to development in different contexts. Moss uses the example of the N-CAM gene, the gene that produces the so-called "neural cell adhesion molecule." The N-CAM gene is a specific nucleic acid sequences from which any of 100 different isoforms of the N-CAM protein may potentially be derived. This protein is expressed in different tissues at different developmental stages in many different forms.

> So where a Gene-P is defined strictly on the basis of its instrumental utility in predicting a phenotypic outcome and is most often based upon the absence of some normal sequence, a Gene-D is a specific developmental resource, defined by its specific molecular sequence and thereby functional template capacity and yet it is indeterminate with respect to ultimate phenotypic outcomes (Moss, 2001, p. 88).

Moss argues that many uses of molecular findings that have been criticized by, for example, the developmental systems theory, arise from taking findings that make sense using the Gene-D concept and interpreting them as if they involved the Gene-P concept. For example, Moss would see it as inappropriate to describe one of the classical developmental genes – sequences used in the control of gene expression in many parts of many distantly related species – as a "gene for" the large section of the phenotype of one of those species in whose development it is implicated.

Evolutionary developmental biology

One of the most exciting trends in recent biology has been the emergence of "evolutionary developmental biology" – the integrated study of evolution and development (Raff and Raff, 1987; Hall, 1992; Raff, 1996). Evolutionary developmental biology simultaneously explores the impact of development on the evolutionary process and the evolution of development. A common philosophical interpretation of this trend in biology is that the "molecular revolution" has "opened the black box" created as part of the modern synthesis. What were previously two kinds of empirical work that led to very different and conflicting pictures of life – evolutionary genetics and developmental biology – can now be empirically integrated so as to yield a single picture (Burian, 1997). Waddington's notion of developmental canalization, for example, has been interpreted as the result of the ubiquity of negative and positive feedback loops in the regulation of gene expression (Freeman, 2000). The developmental concept of a "morpho-genetic field" has been reinterpreted as an emergent phenomena resulting from gene regulation (Gilbert et al., 1996).

A central issue in the older debate between developmentalists and adaptationists was the extent to which phenotypes are holistic entities in which change in one part affects every other. Part of Gould and Lewontin's critique of adaptationism was that it assumes an implausibly atomistic phenotype. Many traits of organisms, they argued, cannot be optimized by selection because they are developmentally linked to other traits. In reply, adaptationists accused their critics of having an implausibly holistic conception of the phenotype. After all, the documented examples of natural selection, to say nothing of artificial selection, demonstrate that many traits can be altered without causing any dramatic reorganization of the phenotype. The argument, mentioned above, that developmental constraints are created by evolution and can therefore be dissolved by evolution was also used to support the adaptationist position. Work in developmental evolutionary biology has helped to make this debate more tractable and progressive. A key concept in evolutionary developmental biology is "developmental modularity." A developmental module is a set of developmental processes that strongly interact with one another and interact only weakly with processes outside the module (Müller and Wagner, 1991). Modules can be the result of the same pattern of connectivity holding within the genome, so that the developmental module corresponds to a "genetic module." Alternatively, developmental modularity can be an emergent phenomena resulting, for example, from the emergence of physical boundaries in the embryo. Existing knowledge in developmental molecular biology strongly suggests that development is modular and models of the evolution of development suggest that selection will favor the emergence of modularity (Wagner et al., 1997).

The concept of developmental modularity can be used to reexamine some of the older issues concerning developmental constraints. Developmental modules represent a natural partition of the phenotype in units whose evolution can proceed relatively independently. An accurate model of evolutionary dynamics must incorporate the fact that the evolving phenotype is neither atomistic nor holistic, but modular. It is far from obvious that this fact should be interpreted as showing the importance of what Amundson (1994) has termed "constraints on adaptation." If developmental modules are the real biological characters of which organisms are composed then saying that selection is constrained by having to act on modules is nearly as odd as saying that it is constrained by having to act on features of the phenotype. Philosophers of biology are starting to rethink issues in evolutionary theory in terms of the modularity concept and the results promise to be of the highest interest (Brandon, 1999; Sterelny, 2000a,b).

Future Work

The debate over the role of information concepts in biology is in full swing at present and likely to continue. The renewed contact between the philosophies of evolutionary and developmental biology is also likely to occupy many writers for some time to come. One developmental concept that seems likely to be revisited after some years of neglect is that of innateness. In developmental biology "innateness" seems as charmingly old-fashioned a theoretical construct as "instinct" and equally peripheral to any actual account of gene regulation or morphogenesis. In behavioral ecology, some authors regard the innateness concept as irretrievably confused and a term that all serious scientific workers should eschew (Bateson, 1991) while others claim that the popular demand to know if something is "in our genes" is best construed as a question about whether a trait is an adaptation (Symons, 1992, p. 141). In cognitive psychology, however, whether a trait is innate in its traditional sense – coming in some sense from "inside" rather than the "outside" – is still a key question, and the subject of heated debate (Cowie, 1999). Some philosophers of biology have tried to bring work in developmental biology to bear on the psychological debate (Ariew, 1999) and, judging by recent conference presentations, more work of this kind can be expected.

References

Alberch, P. (1982): "Developmental Constraints in Evolutionary Processes," in J. T. Bonner (ed.), *Evolution and Development*, New York: Springer Verlag, 313–32.

Amundson, R. (1994): "Two Concepts of Constraint: Adaptationism and the Challenge from Developmental Biology," *Philosophy of Science*, 61(4), 556–78.

Ariew, A. (1999): "Innateness is Canalization: In Defense of a Developmental Account of Innateness," in V. G. Hardcastle (ed.), *Where Biology Meets Psychology: Philosophical Essays*, Cambridge, Mass.: MIT Press, 117–38.

Atlan, H. and Koppel, M. (1990): "The Cellular Computer DNA: Program or Data?" *Bulletin of Mathematical Biology*, 52(3), 335–48.

Bateson, P. (1991): "Are There Principles of Behavioural Development?"in P. Bateson (ed.), *The Development and Integration of Behaviour*, Cambridge: Cambridge University Press, 19–39.

Bechtel, W. and Richardson, R. (1993): *Discovering Complexity*. Princeton: Princeton University Press.

Beurton, P., Falk, R. and Rhineberger, H.-J. (2000): *The Concept of the Gene in Development and Evolution*. Cambridge, Cambridge University Press.

Brandon, R. N. (1999): "The Units of Selection Revisited: The Modules of Selection," *Biology and Philosophy*, 14(2), 167–80.

Burian, R. M. (1997): "On Conflicts between Genetic and Developmental Viewpoints – And Their Attempted Resolution in Molecular Biology," in M. L. Dalla Chiara, K. Doets,

D. Mundici and J. van Bentham (eds.), *Structures and Norms in Science*, Dordrecht: Kluwer, 243–64.

Burian, R. and Richardson, R. (1990): *Form and Order in Evolutionary Biology: Stuart Kauffman's Transformation of Theoretical Biology*. Proceedings of the Philosophy of Science Association. East Lansing, Michigan: Philosophy of Science Association.

Cowie, F. (1999): *What's Within? Nativism Reconsidered*. Oxford: Oxford University Press.

Darwin, C. (1964): *On The Origin of Species: A Facsimile of the First Edition*. Cambridge, MA: Harvard University Press.

Dawkins, R. (1986): *The Blind Watchmaker*. London: Longman.

de Chadarevian, S. (1998): "Of Worms and Programs: *Caenorhabitis Elegans* and the Study of Development," *Studies in History and Philosophy of the Biological and Biomedical Sciences*, 29(1), 81–105.

Dennett, D. C. (1995): *Darwin's Dangerous Idea*. New York: Simon and Schuster.

Depew, D. J. and Weber, B. H. (1995): *Darwinism Evolving: Systems Dynamics and the Genealogy of Natural Selection*. Cambridge, Mass.: Bradford Books/MIT Press.

Eldredge, N. and Gould, S. J. (1972): Punctuated Equilibria: An Alternative to Phyletic Gradualism," in T. J. M. Schopf (ed.), *Models in Paleobiology*, San Francisco: Freeman, Cooper & Co., 82–115.

Freeman, M. (2000): "Feedback Control of Intercellular Signaling in Development," *Nature*, 408, 313–19.

Gilbert, S. C. (2000): "Genes Classical and Genes Developmental: The Different uses of Genes in Evolutionary Syntheses," in P. Beurton, R. Falk and H.-J. Rheinberger (eds.), *The Concept of the Gene in Development and Evolution*, Cambridge: Cambridge University Press, 178–92.

Gilbert, S. F., Opitz, J. M. and Raff, R. A. (1996): "Resynthesising Evolutionary and Developmental Biology," *Developmental Biology*, 173, 357–72.

Godfrey-Smith, P. (1999): "Adaptationism and the Power of Selection," *Biology and Philosophy*, 14(2), 181–94.

Goodwin, B. C. (1984): "Changing from an Evolutionary to a Generative Paradigm in Evolutionary Biology," in J. W. Pollard (ed.), *Evolutionary Theory*, New York: Wiley and Sons, 99–120.

Goodwin, B. C. and Webster, G. (1996): *Form and Transformation: Generative and Relational Principles in Biology*. Cambridge: Cambridge University Press.

Goodwin, B. C., Kauffman, S. A. and Murray, J. D. (1993): "Is Morphogenesis an Intrinsically Robust Process?" *Journal of Theoretical Biology*, 163, 135–44.

Gottlieb, G. (2001): "A Developmental Psychobiological Systems View: Early Formulation and Current Status," in S. Oyama, P. E. Griffiths and R. D. Gray (eds.), *Cycles of Contingency: Developmental Systems and Evolution*, Cambridge, Mass.: MIT Press, 41–54.

Gould, S. J. and Eldredge, N. (1977): "Punctuated Equilibria: The Tempo and Mode of Evolution Reconsidered," *Paleobiology*, 3, 115–51.

Gould, S. J. and Lewontin, R. (1979): "The Spandrels of San Marco and the Panglossian Paradigm," *Proceedings of the Royal Society of London*, B 205, 581–98.

Gray, R. D. (1992): "Death of the Gene: Developmental Systems Strike Back," in P. Griffiths (ed.), *Trees of Life*, Kluwer: Dordrecht, 165–210.

Griffiths, P. E. (1996): "Darwinism, Process Structuralism and Natural Kinds," *Philosophy of Science*, 63(3), Supplement, S1–S9.

Griffiths, P. E. and Gray, R. D. (1994): "Developmental Systems and Evolutionary Explanation," in D. L. Hull and M. Ruse (eds.), *The Philosophy of Biology*, Oxford: Oxford University Press, 117–45.

Griffiths, P. E. and Gray, R. D. (1997): "Replicator II: Judgment Day," *Biology and Philosophy*, 12(4), 471–92.

Griffiths, P. E. and Gray, R. D. (2001): "Darwinism and Developmental Systems," in S. Oyama, P. E. Griffiths and R. D. Gray (eds.), *Cycles of Contingency: Developmental Systems and Evolution*, Cambridge, Mass.: MIT Press, 195–218.

Griffiths, P. E. and Knight, R. D. (1998): "What is the Developmentalist Challenge?" *Philosophy of Science*, 65(2), 253–58.

Griffiths, P. E. and Neumann-Held, E. (1999): "The Many Faces of the Gene," *BioScience*, 49(8), 656–62.

Hall, B. K. (1992): *Evolutionary Developmental Biology*. New York: Chapman and Hall.

Hamburger, V. (1980): "Embryology and the Modern Synthesis in Evolutionary Theory," in E. Mayr and W. B. Provine (eds.), *The Evolutionary Synthesis: Perspectives on the Unification of Biology; with a New Preface by Ernst Mayr*, Cambridge, Mass.: Harvard University Press, 97–112.

Ho, M.-W. and Saunders, P. (eds.) (1984): *Beyond Neo-Darwinism: An Introduction to the New Evolutionary Paradigm*. Orlando, FL: Academic Press.

Hull, D. (1974): *Philosophy of Biological Science*. New Jersey: Prentice-Hall Inc.

Jablonka, E. and Szathmáry, E. (1995): "The Evolution of Information Storage and Heredity," *Trends in Ecology and Evolution*, 10(5), 206–11.

Johnston, T. D. (1987): "The Persistence of Dichotomies in the Study of Behavioural Development," *Developmental Review*, 7, 149–82.

Kauffman, S. A. (1993): *The Origins of Order: Self-Organisation and Selection in Evolution*. New York: Oxford University Press.

Keller, E. F. (1995): *Refiguring Life: Metaphors of Twentieth Century Biology*. New York: Columbia University Press.

Keller, E. F. (1999): "Understanding Development," *Biology and Philosophy*, 14(3), 321–30.

Kitcher, P. (1984): "1953 and All That: A Tale of Two Sciences," *Philosophical Review*, 93, 335–73.

Kitcher, P. (2001): "Battling the Undead: How (and how not) to Resist Genetic Determinism," in R. Singh, K. Krimbas, D. Paul and J. Beatty (eds.), *Thinking about Evolution: Historical, Philosophical and Political Perspectives (Festschrift for Richard Lewontin)*, Cambridge: Cambridge University Press, 396–414.

Laland, K. N., Odling-Smee, F. J. and Feldman, M. W. (2001): "Niche Construction, Ecological Inheritance, and Cycles of Contingency in Evolution," in S. Oyama, P. E. Griffiths and R. D. Gray (eds.), *Cycles of Contingency: Developmental Systems and Evolution*, Cambridge, Mass.: MIT Press, 117–26.

Maynard Smith, J. (2000): "The Concept of Information in Biology," *Philosophy of Science*, 67(2), 177–94.

Maynard Smith, J., Burian, R., Kauffmann, S., Alberch, P., Campbell, J., Goodwin, B., Lande, R., Raup, D. and Wolpert, L. (1985): "Developmental Constraints and Evolution," *Quarterly Review of Biology*, 60(3), 265–87.

Millikan, R. G. (1984): *Language, Thought and Other Biological Categories*. Cambridge, Mass.: M.I.T. Press.

Mitchison, J. G. (1977): "Phylotaxis and the Fibonnaci Series," *Science*, 196, 270–75.

Morange, M. (2000): "The Developmental Gene Concept: History and Limits," in P. Beurton, R. Falk and H.-J. Rheinberger (eds.), *The Concept of the Gene in Development and Evolution*, Cambridge: Cambridge University Press, 193–215.

Moss, L. (1992): "A Kernel of Truth? On the Reality of the Genetic Program," *Philosophy of Science Association Proceedings*, 1, 335–48.

Moss, L. (2001): "Deconstructing the Gene and Reconstructing Molecular Develomental Systems," in S. Oyama, P. E. Griffiths, R. D. Gray (eds.), *Cycles of Contingency: Developmental Systems and Evolution*, Cambridge, Mass.: MIT Press, 85–97.

Moss, L. (2002): *What Genes Can't Do*. Cambridge, Mass.: MIT Press, forthcoming.

Müller, G. B. and Wagner, G. P. (1991): "Novelty in Evolution: Restructuring the Concept," *Annual Review of Ecology and Systematics*, 22, 229–56.

Nagel, E. (1961): *The Structure of Science: Problems in the Logic of Scientific Explanation*. London: Routledge & Kegan Paul.

Neumann-Held, E. M. (1998): "The Gene is Dead – Long Live the Gene: Conceptualising the Gene the Constructionist Way," in P. Koslowski (eds.), *Sociobiology and Bioeconomics. The Theory of Evolution in Biological and Economic Theory*, Berlin: Springer-Verlag, 105–37.

Oyama, S. (2000a): *The Ontogeny of Information: Developmental Systems and Evolution*. Durham, North Carolina: Duke University Press.

Oyama, S. (2000b): *Evolution's Eye: A Systems View of the Biology–Culture Divide*. Durham, North Carolina: Duke University Press.

Oyama, S., Griffiths, P. E. and Gray, R. D. (eds.) (2001): *Cycles of Contingency: Developmental Systems and Evolution*. Cambridge, Mass.: MIT Press.

Papineau, D. (1987): *Reality and Representation*. New York: Blackwells.

Raff, R. (1996): *The Shape of Life: Genes, Development and the Evolution of Animal Form*. Chicago: University of Chicago Press.

Raff, R. A. and Raff, E. C. (eds.) (1987): *Development as an Evolutionary Process*. New York: Alan R. Liss. Inc.

Riedl, R. (1977): *Order in Living Systems: A Systems Analysis of Evolution*. New York and London: Wiley.

Rosenberg, A. (1997): "Reductionism Redux: Computing the Embryo," *Biology and Philosophy*, 12, 445–70.

Sarkar, S. (1996): "Biological Information: A Sceptical Look at Some Central Dogmas of Molecular Biology," in S. Sarkar (ed.), *The Philosophy and History of Molecular Biology: New Perspectives*, Dordrecht: Kluwer Academic Publishers, 183, 187–232.

Sarkar, S. (1998): *Genetics and Reductionism*. Cambridge: Cambridge University Press.

Schaffner, K. (1967): "Approaches to Reduction," *Philosophy of Science*, 34, 137–47.

Schaffner, K. (1969): "The Watson–Crick Model and Reductionism," *British Journal for the Philosophy of Science*, 20, 325–48.

Schaffner, K. (1993): *Discovery and Explanation in Biology and Medicine*. Chicago and London: University of Chicago Press.

Schaffner, K. (1998): "Genes, Behavior and Developmental Emergentism: One Process, Indivisible?" *Philosophy of Science*, 65(2), 209–52.

Sterelny, K. (2000a): "The 'Genetic Program' Program: A Commentary on Maynard Smith on Information in Biology," *Philosophy of Science*, 67(2), 195–201.

Sterelny, K. (2000b): "Development, Evolution and Adaptation," *Philosophy of Science*, 67(3), Supplement, S369–S387.

Sterelny, K. and Griffiths, P. E. (1999): *Sex and Death: An Introduction to the Philosophy of Biology*. Chicago: University of Chicago Press.

Sterelny, K. and Kitcher, P. (1988): "The Return of the Gene," *Journal of Philosophy*, 85(7), 339–61.

Sterelny, K., Dickison, M. and Smith, K. C. (1996): "The Extended Replicator," *Biology and Philosophy*, 11(3), 377–403.

Symons, D. (1992): "On the Use and Misuse of Darwinism in the Study of Human Behavior," in J. H. Barkow, L. Cosmides and J. Tooby (eds.), *The Adapted Mind: Evolutionary Psychology and the Generation of Culture*, Oxford: Oxford University Press, 137–59.

Van der Weele, C. (1999): *Images of Development: Environmental Causes in Ontogeny*. Buffalo, NY: State University of New York Press.

Van Fraassen, B. C. (1991): "The Pragmatics of Explanation," in R. Boyd, P. Gasper and J. D. Trout (eds.), *The Philosophy of Science*, Cambridge, Mass.: MIT Press: 317–27.

Waddington, C. H. (1957): *The Strategy of the Genes: A Discussion of Some Aspects of Theoretical Biology*. London: Ruskin House/George Allen and Unwin Ltd.

Wagner, G. P., Booth, G. and Homayoun, B. C. (1997): "A Population Genetic Theory of Canalization," *Evolution*, 51(2), 329–47.

Waters, C. K. (1994a): "Why the Antireductionist Consensus Won't Survive the Case of Classical Mendelian Genetics," in E. Sober (ed.), *Conceptual Issues in Evolutionary Biology, 2nd edn*, Cambridge, Mass.: MIT Press, 401–17.

Waters, C. K. (1994b): "Genes Made Molecular," *Philosophy of Science*, 61, 163–85.

Williams, G. C. (1966): *Adaptation and Natural Selection*. Princeton: Princeton University Press.

Wimsatt, W. C. (1976): "Reductive Explanation: A Functional Account," in R. S. Cohen (ed.), *Proceedings of the Philosophy of Science Association, 1974, vol. 2*. East Lansing, Michigan: Philosophy of Science Association, 617–710.

Wimsatt, W. C. (1986): "Developmental Constraints, Generative Entrenchment and the Innate-Aquired Distinction," in W. Bechtel (ed.), *Integrating Scientific Disciplines*, Dordrecht: Martinus Nijhoff, 185–208.

Wimsatt, W. C. (1999): "Generativity, Entrenchment, Evolution, and Innateness: Philosophy, Evolutionary Biology, and Conceptual Foundations of Science," in V. G. Hardcastle (ed.), *Where Biology Meets Psychology: Philosophical Essays*, Cambridge, Mass.: MIT Press: 139–79.

Chapter 13

Cognitive Science

Rick Grush

Introduction

Philosophy interfaces with cognitive science in three distinct, but related, areas. First, there is the usual set of issues that fall under the heading of philosophy of science (explanation, reduction, etc.), applied to the special case of cognitive science. Second, there is the endeavor of taking results from cognitive science as bearing upon traditional philosophical questions about the mind, such as the nature of mental representation, consciousness, free will, perception, emotions, memory, etc. Third, there is what might be called *theoretical cognitive science*, which is the attempt to construct the foundational theoretical framework and tools needed to get a science of the physical basis of the mind off the ground – a task which naturally has one foot in cognitive science and the other in philosophy. In this article, I will largely ignore concerns of the first two sorts. As for the first, other entries in this volume cover topics such as explanation and reduction in detail. As for the second, little of interest has emerged from this research agenda, at least so far. I will focus on the third topic, the theoretical foundations of cognitive science. This article will begin with a discussion of behaviorism and the cognitive revolution which overturned it, thus setting the agenda for much of what is now philosophy of cognitive science. My discussion of this will focus on three topics: cognitive architecture, content assignation, and the "counter-revolution" of embodied/embedded cognition and dynamical systems theoretic approaches to cognitive science. I will close with some more broadly cast speculations about future directions in the field.

Historical Background: Behaviorism and the Cognitive Revolution

It will be useful to start with the behaviorism of the early part of the twentieth century. As part of a broad intellectual movement aimed at making inquiry into the nature of the world systematic and reliable, it was believed that a science should admit only publicly observable entities, states, operations, and theoretical entities that could be readily reduced to, or cashed out in terms of, publicly observable entities, states, or operations. Various names were associated with this movement – empiricism, operationalism, verificationism, logical positivism. This movement clashed with much of what was traditionally believed about the mind, which was long thought to be the repository of states and operations which were, almost by definition, private and hence not publicly observable. Since thoughts, feelings, dreams, and the like were not part of what could be observed by the public, so much the worse for them as legitimate objects of scientific inquiry. Rather, behaviorism, as the then kosher psychological theory, officially recognized only stimuli, responses, and entities which could be readily reduced to them, such as strengths of connections between a stimulus and a response. (For *loci classici* of behaviorism, see Watson (1913, 1930) and Skinner (1938, 1957).)

On the heels of these scientific biases (what has been called *psychological behaviorism*) came philosophical biases (sometimes called *philosophical* or *analytical behaviorism*) to the effect that mental entities were either fictional, or that, contrary to what might be thought, that their status as private was fictional – putative mental states, such as "expecting rain" being really no more than complex patterns of overt behavior lacking a private mental cause (Ryle, 1949). This ontological puritanism covered all mental entities and states, including

(i)   qualitative states, such as impressions of pain or red
(ii) contentful states such as thoughts and desires, and
(iii) mental operations such as reasoning and planning.

As such, behaviorism offered answers to the central questions of philosophy and mind, questions concerning the status of the mental and its relation to the physical. The answers were that there was nothing distinctively mental – such terms and expressions as are to be found in everyday discourse about the mind and mental states either fail to refer, or refer to complex sets or patterns of overt physical states and processes. For example, Quine (1960) argues that the "meanings" of linguistic expressions, rather than being mental or non-physical entities, are (to a good first approximation) simply sets of stimulus conditions.

Behaviorism itself is perhaps most interesting as an object lesson in just how implausible a view can be adopted by scientists and philosophers and receive the status of orthodoxy. It is of historical interest in that it was the context for the cognitive revolution which overturned it and which provides the current theoretical backdrop of most cognitive science, psychology and philosophy of psychology and cognitive science. The cognitive revolution in essence was the realization that any adequate theory of human and animal mentality would need to posit representational states between sensory stimulus and behavioral response – at least for a great many domains of behavior. These states would be theoretical, and not simply reducible to constructs of observables.

The cognitive revolution brought about a renewed legitimacy of talk and theorizing about some types of mental or cognitive states, specifically, content-bearing states such as beliefs, desires, or more generally states which were about things, or carried information about things, and over which operations (such as inference) could be performed so as to solve problems and plan. The other class of mental states rejected by behaviorism – qualitative sense impressions – did not get resurrected by the cognitive revolution. The revolution was brought about primarily by three influences, the first psychological, the second linguistic, and the third mathematical.

First psychology. In the middle third of the twentieth century, Edward Tolman and a great many collaborators and followers demonstrated complex maze navigation behavior in rats that resisted explanation in terms of stimulus-response mechanisms, but seemed, rather, to suggest that the rats built up complex representational states, or *cognitive maps*, while exploring, and then used these representational structures to solve novel navigation problems in novel ways; see, for example, Tolman (1948). In the early 1970s, O'Keefe and Dostrovsky (1971) found so-called *place cells* in the rat hippocampus – cells which fired when and only when the rat was in a given location. O'Keefe and Dostrovky appeared to find Tolman's maps in the brain. (A great irony lies in Tolman's work. Early in his career Tolman took his rat maze navigation investigations to be squarely within the behaviorist tradition. He once, in an attempt to sum up his faith in the behaviorist methodology, wrote "everything important in psychology . . . can be investigated in essence through the continued experimental and theoretical analysis of the determiners of rat behavior at a choice point in a maze" (1938, p. 34). Given the results that his work led to, one can read an unintended element of prophetic truth in these words.)

In linguistics, Noam Chomsky provided powerful arguments to the effect that no purely stimulus-driven mechanisms could possibly learn the structure of natural language, and that rather, language learning seemed to require at least some innate cognitive representational structures which circumscribed possible grammars that were then selected from by exposure to linguistic data (Chomsky, 1957, 1959). In fact, Chomsky (1959) was explicitly directed against Skinner's behaviorist theory of language.

In mathematics, the theory of computation developed by Turing (1936) and others provided a theoretical framework for describing how states and processes interposed between input and output might be organized so as to execute a wide range of tasks and solve a wide range of problems. There were no fairly direct neural correlates found of the entities posited by Chomsky's linguistic theories or the burgeoning computational theory of cognition (as there was in the case of the cognitive maps). The framework of McCulloch and Pitts (1943) attempted to show how neuron-like units acting as AND and OR gates, etc., could be arranged so as to carry out complex computations. And while evidence that real neurons behave in this way was not forthcoming, it at least provided some hope for physiological vindication of such theories.

The cognitive revolution made common currency of the view that complex behavior is, in large part at least, controlled by inner representational states. These representations carry contents – they are about things – and they are operated on by processes in such a way that the system can solve problems or make plans. These two topics, the nature of the processes or "cognitive architecture," and the contents carried by representational states, have attracted the bulk of the interest in the philosophy of cognitive science.

Current Topics

Cognitive architecture

Perhaps the first philosophical issue broached by the cognitive revolution was the issue of architecture. Now that inner representational states were given a new lease, the questions of what processes operated on them and what they were came to the fore. The following two subsections briefly discuss the two primary trends on this topic in the last three or so decades.

"Classical" cognitive science and artificial intelligence The rise of computer science made available a way of thinking about the mind which has had great influence. The idea was that the mind is like a program, and the brain is the hardware (or "wetware") on which this program runs; see, for example, Turing (1950) and Newell and Simon (1976). The computer model provides an architecture according to which the states of the cognitive system are, in the first instance, representational states with conceptual contents corresponding to entities like names, predicates, quantifiers, etc., in natural language (Fodor, 1975). Combinations of these yield representations with propositional contents. And the processes which operate over these representations are primarily inferential, and learning conceived of as a matter of hypothesis formulation and testing.

This idea has a number of *prima facie* advantages. First, it renders the mind ontologically unmysterious, for the mind is merely a certain *functional organization* of matter. Second, it seems to secure a lasting role for psychology in the face of threats to the effect that neuroscience will tell us all we need to know about the mind, and at the same time to tell us what the correct tools are for theorizing about the mind. This is because if the mind is like a computer program, then the brain is more or less irrelevant to understanding it. In the same way that one can learn all about, e.g., a certain word processing program regardless of what kind of computer it happens to be running on (amount of memory, type of processor, operating system, etc.), so too, the details of the mind are independent of implementation. Studying a computer will not tell you anything about the programs one might run on it. Rather, we study the input–output operation of the mind, how it behaves when it breaks down in various ways, and on the basis of this we learn about the program that the brain is running.

This trend in cognitive psychology and computer science ushered in a trend in philosophy of mind: *functionalism*. According to functionalism, the mind was not some mysterious entity, but was merely a functional organization of matter; see, for example, various of the essays in Putnam (1975). Not only did functionalism supply the above-mentioned "software" theory of mind's contentful states such as beliefs, but it also provided tools to give an account of qualitative states – something that the development of the computer model of the mind, which was functionalism's inspiration, was quite unconcerned with. The idea was that a qualitative state, like the state of being in pain, was merely a functionally specified state of the mind, a state with the right sorts of connections to input, output, and other interposed states. And not only qualitative states, but content-bearing states could, it was thought, also be defined in functional terms. This approach to assigning content to cognitive states will be discussed in a later section (page 278).

So the hope was that an account, inspired by computer science, of the mind and its states – both qualitative and content-bearing – would finally solve the perplexing problems of mind that had baffled philosophers for so long.

One of the problems that immediately beset functionalist accounts of mentality was the observation that if true, then anything which had the correct functional organization would be a mind and be in qualitative and contentful states – anything, including big arrangements of bottles connected by string, tinkertoy constructions, or large water-pipe and valve systems. The proposal that if one arranges a big collection of cans on strings in the right way that the whole mess would feel pain seemed to many philosophers to constitute a *reductio* of the position; for this and other criticisms, see Block (1978). Another well-known objection comes from John Searle (1980) who argues that a computer program, or an appropriately programmed computer, designed to process natural language – that is, something with the right functional organization – is not sufficient for really understanding language, since someone could manually run through the program and successfully process the linguistic input while having no understanding at all of the language in question. The conclusion reached is that genuine human understanding is not, in fact, just a matter of our mental implementation of the right program; the mind is not just a functional organization of matter.

Connectionism　Connectionism as a method of solving problems and as a theoretical stance in cognitive science has been around in one form or another at least since the middle of the twentieth century. But it was clearly the publication of McClelland and Rumelhart's *Parallel Distributed Processing* volumes – (McClelland and Rumelhart, 1986); for the philosophical *locus classicus*, see Churchland (1989) – that thrust the framework into the spotlight in philosophy, cognitive science and neuroscience. The basic idea of connectionism (I here give a brief description of only one, but perhaps the best known, connectionist scheme) is to process information by representing it numerically (as a set of numerical values, aka a *vector*) and passing it through sets of interconnected units in parallel – in particular through the web of connections between these units. In effect, it takes a vector as input, pushes it through a matrix that represents the weights of the connections, and provides as output another vector. By changing the efficacy of the connections that make up the web, the system can be configured so as to implement a broad range of functions, manifested here as vector–vector mappings. Furthermore, simple schemes exist for getting such systems to learn how to solve problems by trial and error, as opposed to the need for explicit hand-programming present in traditional artificial intelligence.

In addition to the learning aspects of connectionist nets, one of the advantages often claimed by proponents of connectionism is *biological plausibility*. The claim is that the connectionist units function roughly like neurons, the connections between them are analogous to axons and dendrites, and the connection weights are analogous to the efficacy one neuron has in making another fire. These broad analogies aside, assessing the accuracy of the claim to biological plausibility is not so straightforward. Many connectionist networks are simply not candidates for biological plausibility at all for any number of reasons. For instance, some assign entire sentential contents to single units, and it is implausible that a single neuron has a propositional content associated with its activity. To take one more example, the most powerful learning algorithms, including back-propagation, by which these nets learn require that signals travel two ways along the same connection – and this seems not to be something that can happen in biological neurons.

On the other hand, other connectionist models have a high degree of biological plausibility, either because they employ learning algorithms that need only mechanisms which real neurons are known to exhibit, or because they are specifically designed to mimic some known neural system, or (often) both.

The modeling successes of connectionism have been impressive, but not complete. Assessing overall merits is difficult because of the range of models and applications in both connectionist and traditional AI modeling, but to a first approximation traditional models do much better at so-called high-level processes, such as planning, reasoning, and language processing, while connectionist models do much better at so-called low-level processes, such as perception and motor control.

Philosophical issues in cognitive architecture   The two hottest philosophical topics in the 1990s concerning cognitive architecture centered on language and putatively language-like cognitive states. The first was a revisiting of an issue that was first broached in the debate with the behaviorists – the ability to learn to process certain kinds of linguistic structures on the basis of exposure to linguistic data. It was claimed that since connectionist schemes learned via exposure to data, that they would be subject to the same sorts of limitations that killed off the behaviorists – namely an inability to account for the linguistic competence we in fact have. One prominent example is the ability to process dependency relations that span a dependent clause, such as "The boy who likes the girls runs away" in which the singular 'runs' goes with the singular 'boy', even though it is next to the plural 'girls'. In one of the most cited papers in psycholinguistics, Elman (1991) managed to train a connectionist network to successfully process such embedded clauses, casting into doubt the arguments to the effect that mere exposure to data would be insufficient. The status of this debate is difficult to assess, though, as Elman's model still managed only a rather modest task, and it is not at all clear that similar connectionist models would be able to account for more complex patterns. The jury is out.

The second issue, first voiced by Fodor and Pylyshyn (1988), was the systematicity of cognition. It is an empirical fact, they claimed, that any creature able to entertain the thought *Rab* will *ipso facto* be able to entertain the thought *Rba* – for instance, being able to think "John loves Mary" implies the ability to think "Mary loves John." You simply do not, the argument claims, have any cognitive systems that could have thoughts of the first sort without the ability to have thoughts of the second. Given this, they argue, the cognitive architecture must be comprised of symbols that can be recombined in ways analogous to the names and predicates of first-order predicate logic. A number of points might be questioned, such as the initial assumption to the effect that this systematicity is actually an empirical fact (but this seems not too implausible), and the inference from recombinability to an architecture defined over syntactic tokens analogous to first-order predicate logic predicates and names. This inference is surely shaky, as there are kinds of structure other than logical structure. (See Smolensky (1988) for a defense of connectionism.)

Content assignation

The second of the major topics in the philosophy of cognitive science is content assignation. This question is much more a philosophical enterprise than questions of cognitive architecture, at least judging by the people who publish in the area. The problem is this: We know that there are representational states (this is the core of the cognitive revolution), and that the vehicles of these representational states are presumably neural states. But what is it in virtue of which these physical states carry the content they do?

Informational, causal, covariational accounts   The first answer to this question we will consider is the one often implicitly assumed by most people working in cognitive science and neuroscience. On this view, a physical state P means or represents some content C (*u* is *F*, say) iff P co-varies with C (the *F*-ness of *a*), and hence carries information about C. For instance, neuroscientists believe themselves to be finding cells which represent faces or specific shapes when they record the activity of such cells and find some which fire strongly when, and only when, the stimulus (face, shape, color, etc.) is present in the visual field. Since a typical situation in which two things covary is when one of them is the exclusive cause of the other, the three descriptions *informational, covariational, causal*, are closely related, though they are not synonymous (Dretske, 1981, 1988; Fodor, 1987, 1990).

So, the basic idea is that neural or cognitive states represent those things that cause them. This seems innocent enough, but problems arise almost immediately. One is the problem of distility – it is true that a certain shaped stimulus in my visual field will cause a certain set of neurons to fire in my visual cortex. But it is also true that, in this case, it is a pattern of ganglion cell activity in my retina that is causing that pattern of activity in my visual cortex. It is also the case that the experimenter who pushed the button making the shape appear on the screen in front of me caused those cells in my visual cortex to fire. The causal chain back from those cells firing in my brain is continuous and long, and it is not clear how we can single out one element in that chain as being the one that determines content.

Another problem is the disjunction problem (there are actually a few different things that go under the heading of the disjunction problem, but I will discuss only one). Suppose that my brain is such that when a horse is in front of me, a certain cell fires. We might say then, if we can solve the distility problem, that the cell means "horse." But now suppose that, on a dark night or in fog, a cow is in front of me, and this cell fires. We might think that I misidentified the cow as a horse. But on the causal account, since this neural firing can be caused by either horses or cows, it would have a disjunctive meaning: "horse or cow." And hence error is impossible. I will not, on this account, have misidentified the horse as a cow, but correctly identified the cow as a member of the disjunctive type *horse or cow*.

These problems have been addressed, with questionable success, by a variety of means – these include, but are not limited to, appeal to ideal conditions or learning conditions, so as to determine which causes are the ones that really set the content, and which are spurious.

Functional role/conceptual role semantics   The idea here, inspired by the functionalist accounts of mind and mental states, is that a state's content, or meaning, is its *conceptual role* (Field, 1977; Block, 1987). For example, a state, call it #, which is such that, when it interacts with states whose meaning is *5* and *7*, the state meaning *12* is reliably produced, and when interacting with states meaning *2* and *6* a state meaning *8* is reliably produced, would have the functional role of *addition*, and hence would mean *addition*. Thus, the state's meaning as an addition operator is supplied by the function that that state has in the system. Of course, the same is true for all such states, including the ones just now labeled as meaning *5* and *7*, etc. So, in fact, the functional-role meaning of all the states of a system are co-determined in a holistic manner. In its basics, conceptual role accounts are similar to functional role accounts, but place more emphasis on roles in inferences specifically rather than any functional relations.

A major problem for such accounts is that there were purported to be proofs to the effect that for any finite functional system, there would be an infinite number of incompatible yet internally consistent interpretations of its states. (I say purported, because the proofs tell us about very delimited types of system, and it is not clear that all functional role type systems are such that these proofs apply to them.) So, to take the example above to the next level of sophistication, if the states '$ # %' yield '!', while '* # @' yield '&', then perhaps $ means 5, # means *addition*, % means 7, and ! means 12, while * means 2, @ means 6 and & means 8. Alternately, # could mean *multiplication*, while ! means 35 and & means 12. That is, one will always be able to find an infinite number of interpretations for all the states in a system which are internally consistent, but which are inconsistent with each other. Functional role, or so it seemed, did not determine a (single, determinate) meaning for functional states after all.

One attempted rescue maneuver was to combine the functional role and causal accounts (so-called *two-factor theories*), by allowing causal links between some of the system's states and objects or properties in the world to fix the interpretation of these states in such a way as to anchor the interpretation of the other inter-posed states. So the idea is that if the state $ is reliably produced when and only when exactly three objects are in view, then that state will mean 3, and if some other state is reliably produced when and only when a horse is in view, that that state means *horse*. With the meaning of many such states fixed, it will be possible to eliminate many or all of the alternate interpretations and fix just one.

The hope is that by combining functional role semantics and causal accounts, it might be possible to solve the alternate interpretation problem faced by the former, as well as the distility and disjunction problem faced by the latter (the state's functional role will determine a content as being just one of the items along the causal chain, or as being just one of the disjunctively sufficient causes, etc.). The stratagem of going two-factor, or more generally of including as relevant factors things outside the cognitive system proper, is also aimed at solving other problems, such as the fact that in different contexts, states with the same concep-tual role might differ in referent.

Biosemantics A theory of content that is currently very popular (indeed, the spirit of her proposal is now embraced by the original major exponent of the informa-tional theory, Dretske) was first introduced by Ruth Garrett Millikan (1984, 1989). Her biosemantic proposal is that we can fix the content of a state by appeal to the evolutionary history of the mechanisms that support that state. The hope is that this can solve the problems facing the bare causal/informational accounts. The idea is that a neural state means C (even though it might be caused by C or D or E), if the reason for that state's evolutionary selection (or more adequately, for the selection of the mechanisms which support that state and its operation) is that it carried information about C. So, for example, while it might be the case that either flies or random retinal ganglion cell firings can get the neurons in the frog's brain that control the tongue to become active, the explanation for why that mechanism was selected would need to appeal to flies, and not to random retinal ganglion cell firings. We would not have explained why that mechanism was selected for if we mentioned that it became active during random retinal activity, but we would explain why it was selected by appealing to the fact that it carried information about flies.

Despite its current popularity, biosemantics has been the subject of a number of criticisms, including the charge that it depends on questionable evolutionary explanations. Another objection is that biosemantics entails that if there is some natural biological mechanism that can represent C, and we construct an exact physical duplicate, that the duplicate will not be able to represent C. This flies in the face of most cognitive neuroscience, which assumes than an explanation for the representational properties of the brain are a function of its physical constitution. If biosemantics is correct, then only those mechanisms which came about through the right sort of evolutionary processes represent anything. The objection is a serious one and often misunderstood. Consider the following analogy. Structural engineering claims that the weight-supporting properties of a bridge are a function of its physical constitution – the materials involved and their configuration. It doesn't matter if the bridge was built by Smith or Jones: if they have the same physical parts in the same configuration, their weight bearing properties will be the same. It would be an odd sort of claim, one clearly incompatible with what is known in physics, to maintain that Smith and Jones could build physically identical bridges, but that Smith's would carry a large load and Jones' would crumble immediately – as though something magical, beyond the explanatory reach of physics, is transmitted through Smith's fingertips. This is precisely what biosemantics wants us to believe about the representational properties of the brain. Somehow, genuine representational properties are like some mysterious ether, without physical effect (any physical effects could be duplicated without the aid of evolution, after all), that presumably moves with DNA. The objection is not fatal – one could after all just bite the bullet and hold that the physical constitution of the brain does not determine its representational properties –, but the objection shows that there is a serious tension between biosemantics and materialism as normally conceived.

Eliminative materialism (EM)   For some sort of completeness I will now discuss eliminative materialism. This position is often misunderstood to be one which argues against there being any contents at all, claiming that notions of content and representation are merely entities posited by a bad theory of mentality (though perhaps one proponent of EM, Stephen Stich (1983), has defended a position which is fairly close to this). For the most part, those philosophers who identify themselves as eliminative materialists, such as Paul Churchland (1981) and Richard Rorty (1965), have only held that certain kinds of mental contents or mental states are fictional, not that the notion of content or representation or mental state are ill-conceived *tout court*. For instance, Rorty argued against the idea that there were anything like essentially private inner sensations which posed insurmountable obstacles to the explanation of behavior in physical terms. But he never argued against the notion of a mental states *simpliciter*. And Churchland's eliminative materialism is directed not only at such things as private mental qualia, but also against representations with propositional content, such as beliefs. He does not argue against the notion of content *per se*, and, in fact, has provided a number of positive views on what non-propositional content is and how it is carried by physical states.

Counter-revolution

The 1990s witnessed a resurgence of what might be called a counter-revolution to the cognitive revolution. Though there has always been resistance to various of the dogmas of the cognitive revolution, this resistance never became a serious challenge to the orthodoxy. One notable example of this movement from the 1960s is the work of J. J. Gibson (1966), whose theory of ecological perception attempted to show how much of what might have been thought to require sophisticated information processing and memory in the cognitive system could in fact be carried out by simpler mechanisms which

(i) exploited information made available in the environment by various invariances and
(ii) were tuned to organismically relevant affordances.

Though Gibson's work was widely read, it did not have the systematic effect on cognitive science that it might have. It did, however, remain salient enough to exert a heavy influence on the current counter-revolution, and make Gibson one of its heroes. This current trend is perhaps most centrally expressed in the two related movements of embedded/embodied cognition (E/E), and dynamical systems theoretic approach to cognition (DST).

Embedded/embodied (E/E) cognition This movement starts by providing a caricature of the traditional view of cognition. According to this caricature, organisms have sense organs that act as transducers, turning peripheral sensory information into symbols which are passed to a central processor. This central processor then manipulates these symbols together with symbols from stored data structures, and forms a plan or settles on some solution to a problem. At this point the central processor sends a bolus of symbols to output transducers, which control effectors so as to produce some sort of movement or other effect on the body or environment.

Proponents of the E/E movement then argue for, and provide examples to support, the idea that many problems can be solved by simple non-representational mechanisms operating in embodied interaction in a structured environment in which the organism is embedded. For example, rather than maintain sophisticated cognitive maps of its environment for use in navigation, a bee might simply have mechanisms which guide it in certain directions with respect to the clearly visible sun. In conjunction with a simple internal clock, such a humble mechanism can be very powerful, and solve many navigation and homing problems that might have otherwise been thought to require sophisticated internal representational structures.

One of the *loci classici* of this movement is Brooks (1991; see also Beer, 1995; Clark, 1997), in which he describes two robots, Alan and Herbert, which have the task of tooling around the hallways of the lab looking for empty soda cans, and when finding one, clearing it away. They have, however, no powerful central processor which takes in symbolic representations of sensor data and then plans routes or executes can collecting maneuvers. Rather, the robots have a "subsumption architecture" in which the bulk of the work is done by a number of independent systems with close links to their own sensors and with little or most often no manipulation of representations. The simple independent systems often interact closely with the world itself rather than with representations of it – prompting the slogan that 'the world is its own best representation'. These robots are claimed to execute their task in a manner much more robust than that of other robots using more traditional methods.

Dynamical systems theory (DST) At about the same time the digital-computer-inspired cognitive revolution got going, one of the contenders in the game of cognitive architecture was the cybernetics camp (Ashby, 1952; Weiner, 1948); for the contemporary revival, see Port and van Gelder (1995). These researchers were very much inspired by mathematical and technological advances in control theory and dynamical systems theory, one of whose main applications was the autonomous control of vehicles and guided weapons systems (the term "cybernetics" derives from the Greek term for the pilot of a ship). Simple feedback control systems were the prime theoretical tool. To see the appeal, note that a thermostat (a feedback controller) does a very good job of regulating the temperature in a room without any sophisticated inner representations about the thermal properties of the room or the power of the heating and cooling systems. Rather, it simply subtracts a current measure of the temperature from a goal temperature, and does one of three actions depending on whether the result is positive, negative, or zero. Similarly, an autopilot can keep a plane flying straight by making simple comparisons between a few numerically specified goal values, and a few instrument readings, and executing one of a small number of actions based on the mismatch, if any.

The cyberneticists' tools of choice for describing these feedback control mechanisms was the growing mathematical apparatus of dynamical systems theory. Dynamical systems theory is a mathematical apparatus for representing systems and their evolution over time. The systems is represented by a set of state variables. The set of state variables establishes a state space: an abstract space in which each point represents one possible state of the system, and the set of all points represents all possible states of the system. The rules of evolution for the system (the *dynamic*) specify how the system will evolve in time – that is, which point it will move to as a function of which point it is at now. The dynamic thus establishes a set of paths through state space (trajectories) that the system will traverse. Dynamical systems theory supplies tools for discussing such systems and their behavior over time.

Note that feedback control mechanisms work because they are in continuous interaction with (i.e. they receive continuous feedback from) the environment in which they are embedded. This continual feedback can, in many situations, make complicated internal mechanisms unnecessary. This is the conceptual link between feedback control and the E/E movement. The connection to the DST movement is simply the fact that the tools of dynamical systems theory are well suited to describing feedback control systems. In fact, almost all of the examples used by proponents of DST are feedback control systems – a small subclass of the possible dynamical systems.

In any case, to the extent that feedback control mechanisms can solve complex problems, two things seem to follow. First, symbolic systems of the sort posited by Newell and Simon-inspired artificial intelligence are in fact not necessary for solving such problems. Second, closely coupled interaction between the agent's body and the environment may, contrary to the encapsulated central processor view of classical AI, be needed to solve many problems.

The most salient feature of this debate is the extent to which the two sides talk past each other, because each of the two sides adopts a different paradigm for what counts as "cognitive." Classicists take reasoning, playing chess, and processing language as paradigm cases, and the E/E and DST camps take sensorimotor tasks as central. Each research program, to all outward appearances, seems to do the better job of accounting for its preferred "cognitive" tasks.

Objections to the counter-revolution   Assessing the merit of the counter-revolution is not straightforward, but a first gloss on what is right and wrong about it is this: the counter-revolution is right that representations understood as symbols structured along something like first-order predicate logic and manipulated via something like inference rules probably have very limited application in understanding the various aspects of cognition; but the counter-revolution is quite wrong to try to exorcise the notion of representation altogether. Representation is here to stay. How to correctly understand its various manifestations is what is up for grabs. I will say more about this in the final section, but for now, some objections to the counter-revolution.

The most serious objection is that the bulk of the abilities studied by cognitive science are abilities executed, or executable, without any dynamic, embedded interactions with the environment. The Watt governor (a favorite example of the DST camp), or Brooks' robots, do nothing if not hooked up in the right way with the right things to interact with. Human cognition, on the other hand, seems to chug along fine in silent contemplation. Chess players try out moves in their heads before trying them on the board, people plan routes to drive home before getting in their car, and people dream of France while silent and motionless in their beds at night. All of these things, and many more, require a representational story for their explanation. How these representations are best understood is, of course, another story. (For an account that combines the core insights of the DST/EE camp while providing for genuine representations, see Grush (1995, 1997).)

Future Directions

As this section requires guesses as to what the future holds, it will of course reflect my biases to a degree even greater than the previous sections. Reader beware!

First, some bare predictions as to how current issues are likely to resolve. For starters, the cognitive revolution is here to stay because it is, in its essentials, right. Insofar as it pushes anti-representationalism, the counter-revolution is misguided and will be washed out in time. On the other hand, the counter-revolution is right to stress the role of the body, the environment, and real-time activity in cognition and problem solving. The solution will involve rethinking the nature of cognition and representation in such a way as to move away from the idea of the disembodied central processor and toward the idea of representations and processes that are more closely tied to agent-environment interactions, but without denouncing representations. The tools of dynamical systems theory are unlikely to have much lasting impact on our understanding of central features of cognition such as language, thought and reasoning.

As far as topics in cognitive architecture go, it is likely that different tasks such as memory, perception, reasoning, will turn out to involve different sorts of processing at an architectural level. But for the more central cognitive systems, they most certainly involve structured representations which can recombine in ways at least analogous to the behavior of the symbols posited by classical computational cognitive science. However, these representations are most likely operated on by processes which are not at all well-described by the formalism of first-order predicate logic or its extensions, exactly because cognitive representations will be found to have their structural features because of their semantic features – syntax being merely a shadow cast by semantics.

Along these lines, there will be growing appreciation for the theories of cognitive linguistics, and especially Ronald Langacker's Cognitive Grammar (Langacker, 1987, 1991, 2000); see also Talmy (2000) and Fauconnier et al. (1996) not only for their provision of the correct tools for understanding human linguistic competence, but also because of the light they shed on cognitive representation and processing in general. Cognitive Grammar takes the view that linguistic expressions are meaningful in virtue of their parasitism on the meaningfulness of representational structures whose home is in perception and action. I will say a bit more about this below.

As for content assignation, two points can be made with some confidence. The first is that for the purposes of cognitive science and neuroscience, at least for the foreseeable future, the off-the-shelf causal/informational theories will be fine. The second is that a philosophically adequate account of the content-bearing properties of physical/neural states is a long way off, and will almost certainly have no resemblance to any causal/informational or biosemantic account. As to what form the eventual correct account will take, only a few general features can be discerned. It must be an account that explains how a given arrangement of physical/neural entities can create its own representational endowment or potential, without recourse to external objects or states of affairs (such as evolutionary history or causal antecedents), or outside interpreters. And relatedly, it must be an account which explains the objectivity of contents – that is, which does not simply take it as unproblematic that the content carried by a cognitive system is of objects and states of affairs which are understood to be independent of their being represented (this will be discussed more below).

Now to some predictions of a positive nature. A growing trend among the newer generation of those who identify themselves as philosophers of cognitive science is a growing appreciation for traditional topics in philosophy, especially topics whose genesis was in Kant, but which have more current expression in the work of philosophers such as Peter Strawson and Gareth Evans. A major goal of this trend is to understand representational structures – such as the representation of space (both egocentric and allocentric), of oneself as an agent in space, and of objects as permanent denizens of the world which are represented as being independent of being represented; that is, as objective. Such representational schemes take as basic the cognitive system's representation of itself as an embodied agent actively engaged in an environment populated with temporally extended objective processes. For example, Bermudez (1998) provides an account of self-consciousness (understood as self-representation) that relies on non-conceptual representational machinery whose home is in perception and action; Metzinger (1993) provides an account of consciousness and qualia that relies on the brain's own self-representation; Grush (2000, 2002) provides and account of the neural mechanisms of spatial representation, self-representation and objectivity.

Such accounts, if they succeed, could represent a radical rethinking of the nature of representation. The new notion will maintain that truth-conditions are crucial for representational content, but that rather than taking truth-conditions to be the satisfaction of an n-place predicate by n objects, truth-conditions will be reconceived as located, structured, objective processes. *Located* in the sense that they are conceived as being spatially and temporally related to the conceiver (unlocated processes being a degenerate case); structured in the sense that the processes are conceived as involving the interaction of multiple entities (objects and properties being a limiting case); *objective* in the sense that these processes are conceived as being independent entities in a world that is independent of the representer (Strawson, 1959, ch. 2); and *processes* being temporally extended (temporally punctate states of affairs, such as the possession of a property by an object, being again a degenerate case). Such representational structures will be largely learned from actual embodied interaction in an environment, and will reflect this perspective in their content – that is, it will be the environment-as-interacted-with that is represented in the first instance, not the environment as it is in itself, whatever that might mean.

In addition to this trend, a number of others can be discerned. I will mention only two. First, the study of emotions, their nature, their connection to reasoning, and their neural substrate, is a growing area of research (Damasio, 1994; Griffiths, 1997). Second, the nature of the "theory of mind" issue, while it has been the focus of attention for some small groups, appears to be taking on more currency, in philosophy, psychology and neuroscience generally; see, for example, Carruthers and Smith (1996). The "theory of mind" phenomenon can be illustrated as follows. Imagine a child watching a person A hide something S in the kitchen and then leave. B then comes in, and moves S from the kitchen to the living room. A returns, and the child is asked where A will look for S. Before a certain age, children answer that A will look in the living room. This is, after all, where S is. After a certain age, children realize that A's actions are mediated by A's representation of the environment and not the environment itself – they realize that other people have representational minds. This issue is of import for a number of philosophical questions (including the philosophy of mind and action), psychology, and for understanding a number of phenomena, such as autism and various psychopathologies.

References

Ashby, W. R. (1952): *Design for a Brain*. New York: Wiley.

Beer, R. (1995): "A Dynamical Systems Perspective on Agent–Environment Interaction," *Artificial Intelligence*, 72, 173–215.

Bermudez, J. L. (1998): *The Paradox of Self-Consciousness*. Cambridge, MA: MIT Press.

Block, N. (1978): "Troubles with Functionalism," in W. Savage (ed.), *Minnesota Studies in the Philosophy of Science: vol. IX*. Minneapolis: University of Minnesota Press, 261–325.

Block, N. (1987): "Functional Role and Truth Conditions," *Proceedings of the Aristotelian Society*, LXI, 157–81.

Brooks, R. (1991): "Intelligence without Representation," *Artificial Intelligence*, 47, 139–59.

Carruthers, P. and Smith, P. (eds.) (1996): *Theories of Theories of Mind*. Cambridge: Cambridge University Press.

Chomsky, N. (1957): *Syntactic Structures*. Paris: Mouton.

Chomsky, N. (1959): "Review of Skinner's *Verbal Behavior*," *Language*, 35, 26–58.

Churchland, P. (1981): "Eliminative Materialism and the Propositional Attitudes," *Journal of Philosophy*, 78(2), 67–90.

Churchland, P. (1989): *A Neurocomputational Perspective: The Nature of Mind and the Structure of Science*. Cambridge, MA: MIT Press.

Clark, A. (1997): *Being There: Putting Brain, Body and World Back Together Again*. Cambridge MA: MIT Press.

Damasio, A. (1994): *Descartes' Error*. New York: Avon Books.

Dretske, F. (1981): *Knowledge and the Flow of Information*. Cambridge, MA: MIT Press.

Dretske, F. (1988): *Explaining Behavior*. Cambridge, MA: MIT Press.

Elman, J. (1991): "Distributed Representations, Simple Recurrent Networks, and Grammatical Structure," *Machine Learning*, 7, 195–225.

Fauconnier, G., Sweetser, E. and Smith Sweet, E. (eds.) (1996): *Spaces, Worlds and Grammar*. Chicago: University of Chicago Press.

Field, H. (1977): "Logic, Meaning and Conceptual Role," *Journal of Philosophy*, 69, 379–408.

Fodor, J. (1975): *The Language of Thought*. Cambridge, MA: Harvard University Press.

Fodor, J. (1987): *Psychosemantics*. Cambridge, MA: MIT Press.

Fodor, J. (1990): *A Theory of Content and Other Essays*. Cambridge MA: MIT Press.

Fodor, J. and Pylyshyn, Z. (1988): "Connectionism and Cognitive Architecture: A Critical Analysis," *Cognition*, 28, 3–71.

Gibson, J. J. (1966): *The Senses Considered as Perceptual Systems*. Boston: Houghton Mifflin.

Griffiths, P. (1997): *What The Emotions Really Are*. Chicago: University of Chicago Press.

Grush, R. (1995): "Emulation and Cognition," PhD thesis, UC San Diego, UMI.

Grush, R. (1997): "The Architecture of Representation," *Philosophical Psychology*, 10(1), 5–25. Reprinted in W. Bechtel, P. Mandik, J. Mundale and R. Stufflebeam (eds.), *Philosophy and the Neurosciences: A Reader*, Oxford: Basil Blackwell, forthcoming.

Grush, R. (2000): "Self, World and Space: The Meaning and Mechanisms of Ego- and Allocentric Spatial Representation," *Brain and Mind*, 1(1), 59–92.

Grush, R. (2002): *The Machinery of Mindedness*, in preparation.

Langacker, R. (1987, 1991): *Foundations of Cognitive Grammar* (2 vols). Stanford: Stanford University Press.

Langacker, R. (2000): *Grammar and Conceptualization*. The Hague: Walter de Gruyter.

McClelland, J. and Rumelhart, D. (eds.) (1986): *Parallel Distributed Processing: Explorations in the Microstructure of Cognition* (2 vols). Cambridge, MA: MIT Press.

McCulloch, W. S. and Pitts, W. H. (1943): "A Logical Calculus of the Ideas Immanent in Nervous Activity," *Bulletin of Mathematical Biophysics*, 5, 115–33.

Metzinger, T. (1993): *Subjekt und Selbstmodel*. Paderborn: Mentis.

Millikan, R. G. (1984): *Language, Thought, and Other Biological Categories*. Cambridge, MA: MIT Press.

Millikan, R. G. (1989): "Biosemantics," *Journal of Philosophy*, 86(6), 281–97.

Newell, A. and Simon, H. (1976): "Computer Science as Empirical Inquiry: Symbols and Search," *Communications of the Association for Computing Machinery*, 19, 113–26.

O'Keefe, J. and Dostrovsky, J. (1971): "The Hippocampus as a Spatial Map: Preliminary Evidence from Unit Activity in the Freely Moving Rat," *Brain Research*, 34, 171–5.

Port, R. and van Gelder, T. (1995): *Mind as Motion: Explorations in the Dynamics of Cognition*. Cambridge, MA: MIT Press.

Putnam, H. (1975): *Mind, Language and Reality: Philosophical Papers, vol. 2*. Cambridge: Cambridge University Press.

Quine, W. V. O. (1960): *Word and Object*. Cambridge, MA: MIT Press.

Rorty, R. (1965): "Mind–Body Identity, Privacy, and Categories," *Review of Metaphysics*, 19(1), 24–54.

Ryle, G. (1949): *The Concept of Mind*. London: Hutchinson and Company, Ltd.

Searle, J. (1980): "Minds, Brains and Programs," *Behavioral and Brain Sciences*, 3, 417–24.

Skinner, B. F. (1938): *The Behavior of Organisms: An Experimental Analysis*. New York: Appleton-Century.

Skinner, B. F. (1957): *Verbal Behavior*. Englewood Cliffs, NJ: Prentice-Hall.

Smolensky, P. (1988): "On the Proper Treatment of Connectionism," *Behavioral and Brain Sciences*, 11, 1–74.

Stich, S. (1983): *From Folk Psychology to Cognitive Science: The Case Against Belief*. Cambridge MA: MIT Press.

Strawson, P. F. (1959): *Individuals*. New York: Routledge.

Talmy, L. (2000): *Toward a Cognitive Semantics* (2 vols). Cambridge, MA: MIT Press.

Tolman, E. C. (1938): "The Determiners of Behavior at a Choice Point," *Psychological Review*, 45, 1–41.

Tolman, E. C. (1948): "Cognitive Maps in Rats and Men," *Psychological Review*, 55, 189–208.

Turing, A. M. (1936): "On Computable Numbers, with an Application to the Entscheidungs-Problem," *Proceedings of the London Mathematical Society 2nd Series*, 42, 230–65.

Turing, A. M. (1950): "Computing Machinery and Intelligence," *Mind*, 59, 433–460.

Watson, J. (1913): "Psychology as a Behaviorist Views It," *Psychological Review*, 20, 158–77.

Watson, J. (1930): *Behaviorism, revised edn*. New York: Norton.

Weiner, N. (1948): *Cybernetics: Or, Control and Communication in the Animal Machine*. New York: Wiley.

Chapter 14

Social Sciences

Harold Kincaid

Introduction

Philosophy of the social sciences is a live area of research, in large part because foundational philosophical issues are still up for grabs in the social sciences themselves.[1] This survey argues that philosophy of social science is undergoing a transition, one reflecting larger trends in philosophy and philosophy of science. Those trends are away from *a priori* conceptual analysis and towards some sort of naturalized epistemology that makes the actual practice of scientific research central.

The past literature in the field has concerned largely what the social sciences can or cannot do, what they must be like or never could be. Those views were typically defended on broad conceptual grounds. Interesting positions resulted and issues were clarified. However, these arguments claimed more for purely philosophical considerations than they can deliver. Hence more recent work approaches the issues in much closer contact with the actual practice of social research. Conceptual analysis still has a place, namely, in clarifying debates in and about the social sciences. Yet there are good reasons to think that the key issues are intimately connected with empirical issues in the social sciences themselves.

Survey of the Past Literature

The issues that have dominated the philosophy of the social sciences during the last century have their origins in the seminal work of Weber and Durkheim. Durkheim (1965) defended sociology as an autonomous discipline independent from psychology. Sociology was autonomous because there is a realm of "social facts" that exists, in some sense, independently of the individual. It is those facts that sociology investigates. Moreover, those social facts can influence individuals in ways they do not realize. Hence investigating social facts cannot be done entirely in terms of individuals' professed self conceptions; social science has to go beyond the individual subject's own categories to propose theoretical concepts that explain social processes.

Weber (1968) took a very different position. To him, talk of social facts over and above individuals bordered on a simple-minded reification. Social explanation is above all the explanation of the behavior of individuals. Explaining the behavior of individuals requires a special method of sympathetic understanding, commonly labeled with the German term *Verstehen*. That is because, unlike the natural sciences, the social sciences study beings who attribute meaning to the world. Weber nonetheless thought an objective social science possible.

Weber also had various ideas about how the social sciences explain. Weber thought social explanation must deal with what he called "ideal types." He put this notion to various uses,[2] but its central function was to handle the complexity of the social world. Real social processes involve very many interacting causal factors; real social processes show significant diversity across individuals, times, and places. Ideal types attempt to abstract from that complexity. The laws of supply and demand were one such abstraction or idealization on Weber's view. He believed that they worked in the same way and with the same legitimacy on his view as idealizations in the natural science.

Durkheim and Weber raise here, in incipient form, most of the issues that have comprised the philosophy of social science:

- *Naturalism vs. antinaturalism*   Do the social sciences proceed according to the same methods and criteria of adequacy as the natural sciences?
- *Interpretivism vs. causalism*   Does explaining social phenomena require reference to meanings in a way that makes the social sciences fundamentally different from the natural sciences or are social phenomena amenable to ordinary causal explanation?
- *Instrumentalism vs. realism*   Can, should, and/or must the models of social scientists be taken as literal descriptions of social reality rather than as illuminating stories with some other less direct connection to reality?
- *Holism vs. individualism*   Can and/or must the social sciences explain in terms of irreducible social entities or must all adequate explanations be in terms of individuals?

Of course, how these questions interrelate is not clear. Nor is their meaning entirely obvious in these brief formulations. I thus want to spell out further the issues involved by looking at the relevant literature from the relatively recent past.

The most common arguments against naturalism fall into one of two categories: those that turn on the meaningful nature of human behavior, and those that do not. Let's focus first on arguments that do not depend on meaning. Some of the most influential turned on metaphysical facts about social phenomena that allegedly preclude scientific study. For example, two claims have been made:

1   Social systems are not closed systems – they are inherently subject to influences from biological and physical forces. This means social explanations are radically incomplete in a way that explanations in the natural sciences are not (Davidson, 1994, Taylor, 1971).

2   Social phenomena are only loosely connected to physical phenomena, for the same social process can be realized in indefinitely many different physical instantiations. That means social phenomena are too open ended to behave in the lawlike ways required for scientific study (Searle, 1984).

Another influential antinaturalist argument is that the social sciences are inherently nonexperimental, thus precluding rigorous scientific study. This argument concerns the nature of social scientific inquiry, not the nature of society directly as in the previous arguments. Yet metaphysical claims about society are indirectly relevant here as well. If we want to argue that experimentation in the social sciences is not just difficult but inevitably doomed, then it is probably something about the phenomena – the open nature of social systems, the unique nature of human agents, etc. – that grounds that necessity.

We can turn next to arguments against naturalism from the meaningful nature of human behavior. This brings us to the debate over interpretivism. At issue, recall, is roughly whether human behavior is amenable to causal explanation. The alleged obstacle is the meaningful nature of human behavior – the fact that humans interpret the acts of others and of themselves, that human utterances have semantic content, and so on.

Various routes have been proposed for reaching the antinaturalist conclusion from the premise that human behavior is meaningful:

1   One argument is that the social sciences must explain behavior in terms of reasons. But reasons and their associated behavior cannot stand in a causal relation. That I expressed rage because I was angry is true by definition. That A causes B however can never be true by definition. So reasons cannot be causes (Melden, 1961).

2   A related argument offered by many is that the social sciences do not explain behavior but action. Actions are not just bodily movements but behavior with a meaning. So behavior always has to be interpreted; as a result, the social sciences cannot have "brute data" like the natural sciences. Social science is a hermeneutic enterprise like literary interpretation (Taylor, 1971).

3   Others have pointed to the alleged fact that human action is constituted by rules or norms (Winch, 1958). Drawing inspiration from Wittgenstein, they argue that understanding social behavior is like understanding the moves in a game – to know what was done and why is to see the point of the act, the norms covering it, and so on. This means we must understand the act from the subject's viewpoint and that the point of social science is understanding, not causal explanation as in the natural sciences.

4   Another persuasive argument points to the place of assumptions about rationality in interpretation. To interpret another is to make sense of them. To make sense of them I use the principle of charity – I assume they do what is rational by my lights. But this means there are *a priori* constraints of quasi-evaluative nature that guide interpretation, unlike explanation in the natural sciences (Davidson, 1984, Hollis, 1996).

The next set of issues on our list concerns realism and instrumentalism. The realism-antirealism debate in the social sciences has taken its main impetus from questions about the role of idealizations and simplifications in social explanation. Given its widespread use of such devices, economics has not surprisingly been the chief focus of this debate. The problem is this: economic theories make enormous simplifying assumptions – for example, that agents have all relevant information, that there are markets for every good at every time, and the like. Yet these assumptions are literally false. So how can such theories accurately describe how the economic realm works?

The most persuasive answer among economists was given by the Nobel laureate Friedman (1953). He claimed that the realism of assumptions does not matter. What does matter is predictive success. In short, the point of theories is to predict. This is a classical instrumentalist view of science.

Similar antirealist sympathies lie behind the doctrine known as operationalism as it is manifest in the social sciences. In the social sciences, operationalism asserts that theoretical terms have meaning only through the criteria used to measure them. From this doctrine, it is not a long step to the conclusion that theories are really a fifth wheel – their purpose is to help predict and collate, not to describe an independent reality. Behaviorists in psychology espoused such a view when they claimed that mental states were unneeded; economists advocate a similiar position in the theory of revealed preference that replaces utility maximizing with consistently ordered observable choices.

A final set of issues involves questions about explanation in the social sciences. Do explanations in the social sciences proceed in the same way as those in the natural sciences? If not, naturalism seems at risk. The past literature has focused on this topic in at least two ways: in debates over explanations invoking functions, and in the holism–individualism debate.

The debate over functions arises because the social sciences employ some odd looking explanations. They explain various social practices as existing because they promote benefical goals or purposes. Yet they do not claim that these practices were designed or intended by anybody.

Such functional explanations historically have been rife in the social sciences. Marxists (Miliband, 1969) claimed that the state exists in order to promote the interests of the ruling class; Durkheim (1933) claimed that the division of labor exists in order to promote social solidarity. Yet social scientists proceeded without any account of how these explanations worked or why they were legitimate.

In the 1950s, philosophers began to raise questions about functional explanation and attempted to subsume them under the nomonological-deductive model of explanation. Hempel (1965) analyzed functional explanations in terms of what was necessary for successful functioning, asking if laws could be identified tying specific traits to success. He argued that no such laws existed because of "functional equivalents" – other social practices that could serve the same function – and thus that functional explanations were largely inadequate.

We saw earlier that the individualism–holism debate goes back to the origins of the social sciences. Beginning in the 1950s, it became a lively issue for philosophers in addition to social scientists. Individualists claimed the scientific mantle of materialism and reductionism for their side. Perhaps the best known advocate along these lines was Watkins. He appealed to metaphysical facts: "the ultimate constituents of the social world are individuals" and "social events are brought about by people." These "metaphysical commonplaces" as he called them have the "methodological implication that large-scale phenomena should be explained in terms of the situations, dispositions, and beliefs of individuals" (Watkins, 1973, p. 179).

On the holist side, other metaphysical facts were invoked. Following Durkheim, it was argued that social facts have an external existence to individuals (Gellner, 1968) or that causal influences from social facts have ineliminable place (Ruben, 1985). Other holist arguments invoked facts about language: to attribute check writing to an individual presupposes a large set of institutional facts about banks and so on (Mandelbaum, 1973).

Current State of the Debates

Current debates in the philosophy of social science are a mix of issue clarification, conceptual argumentation, and empirically oriented discussions raised by the practice of social research. I begin with current work on naturalism and related issues.

Before looking at specific arguments for and against naturalism, it will be helpful to identify more clearly what is at issue. The question is usually put in terms of methodology: can the social sciences use the methods of the natural sciences? However, this formulation immediately confronts several difficulties. "The methods of the natural sciences" is unclear. At the level of particulars, the different natural sciences do not share the same methods. Biologists rely heavily on gel electrophoresis and astronomers on recordings of light beyond the visible spectrum – at this level, their methods are very different. Moreover, asking about "methods" does not reveal the root issue. We are interested, roughly speaking, in scientific respectability: can the social sciences produce scientific knowledge? Unless methods are foolproof (they are not), determining what methods the social sciences use still does not tell us about their scientific standing.

So the naturalism debate is perhaps better seen as focusing on these two questions:

- Can the social sciences produce good science by the broad standards of the natural sciences?
- Can the social science produce good science *only* by meeting the broad standards of the natural sciences?

Focusing on the *broad* standards of the natural sciences – on general criteria for confirmation and explanation – looks for what is common to the diverse procedures of the natural sciences and also avoids the claim that there is a single criterion either at a time or over time for scientific goodness. The differences between the two questions distinguishes two versions of naturalism, one that says the social sciences can meet natural standards and the other that says they must.

We identified arguments against naturalism from the past literature that turned on the nonexperimental nature of social science, the open nature of social systems, and the loose connection between the social and the physical. (Questions about meaning will be taken up separately later.) It is doubtful that these alleged obstacles rule out good social science. Perhaps these traits differentiate the social sciences from, say, particle physics. Yet they hardly distinguish the social sciences from evolutionary biology or ecology, for example. Darwin's evidence for evolution by natural selection was nonexperimental and current evidence largely remains in that category. Biological systems are open. Biological categories, likewise, have a loose tie to the physical. For example, there are millions of physical structures that realize the notion of an "antibody"; fitness, the fundamental notion of evolutionary biology, likewise has a similar open-ended connection to the physical. Assuming that these areas of biology produce good science, the above traits cannot be principled obstacles to successful social science.

However, these problems can be real in practice. One way such problems turn up, in practice, is that most claims in the social sciences are qualified *ceteris paribus* – or, other things being equal or assuming nothing else interferes. *Ceteris paribus* clauses are a problem for naturalism because they raise doubts about whether social science claims can be confirmed and can explain. Frequently, what else must be equal cannot be fully stated. That means when data contradict the hypothesis at issue, we do not know where to place blame – on the hypothesis or on the unknown confounding factors. Giving credit is equally problematic, for positive results might be the consequence of other things not being equal. It looks like *ceteris paribus* claims are untestable and vacuous. Doubts also arise about social *explanations* invoking *ceteris paribus* clauses. An assertion qualified *ceteris paribus* describes how things would be if all else was equal. But how do claims about how things would be explain how things in fact are?

Various solutions to these problems have been proposed. One solution we saw above came from Friedman. His response to unrealistic assumptions was to deny that the point of theories was to do any more than predict. So, it did not matter if social science claims were qualified *ceteris paribus* so long as they predicted the data.

This instrumentalism might be justified on other grounds, but unless it is, it is no solution to the *ceteris paribus* problem. Social scientists claim to explain, to cite causes, and to make policy recommendations. Economists, for example, assert that various macroeconomic variables – the rate of interest, the money supply, etc – influence employment and growth. If the theories economists use to explain these phenomena make false assumptions, then we have reason to doubt those explanations. Successful predictions may be lucky accidents that will not hold up in new circumstances; they may result from spurious correlations. The only way to rule out this situation is to provide evidence that the postulated mechanisms are the true ones – and that means finding evidence somehow that other things are equal.

Another response to the *ceteris paribus* problem has been to spell out truth conditions – to determine what states of affairs would make a claims qualified *ceteris paribus* true. Numerous proposal have been offered, with Hausman (1981, 1992) being the most direct application to the issue in the social sciences. According to Hausman, the phrase "*ceteris paribus*" has a fixed meaning, but what proposition it picks out depends on the context. The claim "All *A*s are *B*s, *ceteris paribus*" is true in context *C* just in case *C* picks out a property *P* such that "Everything that is both *A* and *P* is *B*" is true.

Giving truth conditions may help remove doubts that claims qualified *ceteris paribus* are meaningful. Yet stating truth conditions and having good evidence that those conditions obtain are entirely different things, and it is the latter problem that remains for the social sciences. Using Hausman's analysis, how are we to identify property *P*? If the *ceteris paribus* clause is unspecified, then how are we to know if *P* is instantiated? Perhaps *P* is only infrequently seen in the contexts we observe. The basic problem remains.

A second approach comes from Cartwright (1989). She provides truth conditions as well, with *ceteris paribus* laws referring to capacities. Her truth conditions face the same problem mentioned above, namely, stating truth conditions still leaves us with the problem whether we every know they are satisfied. However, Cartwright's account has interesting implications for the natural sciences. In her view, neither the social sciences nor the physical sciences produce strict laws about observables. So the fact that social science claims are qualified *ceteris paribus* cannot show a fundamental divide.

There are compelling objections to Cartwright's arguments against strict laws (Earman and Roberts, 1999). Nonetheless, it surely is true that some seemingly quite respectable natural science faces the *ceteris paribus* issue. So, arguably, the key question concerns what strategies there are for dealing with the complexity that leads to *ceteris paribus* claims and whether those strategies are open to the social sciences.

Two questions are at issue: is the claim qualified *ceteris paribus* plausible or true, and does it explain in the circumstances in question? Is it true that, other things being equal, supply and demand determine price? And, is it true in this circumstance where other things are not equal that supply and demand are causally influencing price? Questions of the first sort can be approached by

(a) finding circumstances where other things are equal and confirming the claim directly
(b) showing that as circumstances approach those described by the *ceteris paribus* clauses, the claim becomes more accurate, and
(c) identifying a mechanism that underlies the alleged claim.

Furthermore, questions about both confirming and explaining can be tackled, for example, by

(d) showing that the claim holds even though *ceteris paribus* condition is not or is not known to be met and
(e) showing – either empirically or in models – that the claim holds over diverse contexts, particularly contexts that vary the suspected complicating variables (Hausman, 1981; Kincaid, 1996).

Thus the *ceteris paribus* problem is arguably no inherent obstacle to good social science. It is a problem that can be handled. When and where the social sciences actually do so requires detailed engagement with actual social research. Once again, philosophy of social science issues are continuous with empirical ones.

One specific area where the *ceteris paribus* problem has been seriously studied is economics. Hausman (1992) has defended unrealistic assumptions in economics on the following grounds: Economic data are often of mixed quality and the assumptions behind econometric tests often not met. Moreover, the basic laws of mainstream neoclassical economics – such as the claim that consumers will always prefer more to less for the same price – are *a priori* plausible. So continuing to assert such claims in the face of sometimes conflicting data is reasonable. Economists do become dogmatic, however, when they refuse to consider the prospect that economic causes are not dominant over noneconomic ones. In Hausman's terminology, an inexact science of economics may be defensible but not a separate one.

Hausman's arguments here have met substantial criticisms despite the numerous virtues of the book in which those arguments appear (Maki, 1996, Kincaid, 1996). One compelling worry is that Hausman's defense is still at too high a level of abstraction. It can, indeed, sometime, be reasonable to hold on to initially plausible generalizations in the face of conflicting but questionable evidence. But it is not clear that the contrary evidence in economics is always so questionable or that economic postulates so *a priori* plausible.

Returning now to the debate over naturalism, we have next to consider interpretivism. The idea that human behavior is meaningful conceals many different ideas, and it is helpful to sort them out. At least the following kinds of "meaning" might reasonably be at issue:

Doxastic meaning: what a subject believes
Intentional meaning: what a subject intends, desires, etc.
Linguistic meaning: how the verbal behavior of the subject is to be translated
Symbolic meaning: what the behavior of the subject symbolizes
Normative meaning: what norms the behavior of the subjects reflects or
    embodies

While no doubt there are interconnection between these various senses of "meaning," they are importantly different.

Perhaps the most influential argument from the past based on meaning was the type advanced by Taylor (1971). He argued that there are no brute or bare data in the human sciences and that every categorization of human behavior rests on prior understandings. This argument is regularly repeated by those in the social sciences who want to draw a sharp distinction between quantitative and qualitative methods. Yet this argument ignores the main developments in philosophy of science since the positivists. Given the holism of testing, the experimenter's regress, the place of skill in testing, and other such phenomena widely identified in the natural sciences, "brute data" without "prior understandings" are not to be found there either.

The holism of testing is also relevant to two other traditional interpretivist type arguments, namely, those concerning rationality and the subject's categories. Rationality is thought to make the social sciences different because the interpreter must make normative assumptions about what is rational in order to attribute beliefs to subjects and must assume that the subject is on the whole rational. This principle of charity was treated as an *a priori* methodological rule. The holism of testing would suggest that methodological rules are ultimately justified by their fit with experience and can likewise be rejected when they do not. From this perspective, the principle of charity is a contingent, defeasible rule. We should assume that individuals are rational only to the extent that we have evidence that they are; we should assume our standards of rationality are universal only to the extent that the empirical evidence warrants that claim. Of course, we must make assumptions about rationality to get interpretation off the ground. But the holism of testing tells us that no science gathers data without prior theory. Thus assumptions about rationality are no different than other theoretical assumptions. This general response to the rationality debate has been ably defended by Henderson (1991, 1993). There are nonetheless still dissenters (Hollis, 1996).

The holism of testing also bears upon arguments over the subject's self-conceptions. How the subject views the world is one piece of evidence about his or her behavior, but a piece that must be evaluated on the basis of everything else that we know. Subjects can be entirely mistaken about the causes of their behavior as a large body of work shows (Liska, 1975). To ignore the subject's understanding of the situation is no doubt to court trouble. Yet only experience can tell us when and where the subject's self-understanding provides the best explanation of social phenomena.

We saw, in the last section, that earlier writers also argued that social science explanation essentially involved appeal to reasons, that reasons cannot be causes, and therefore that the social sciences cannot provide causal explanation. That argument has been decisively criticized by Davidson (1980) and others. The problem with the argument is that it rests crucially on the assumption that causal claims cannot be expressed by tautologies or analytic truths since causal relations are contingent. But this confuses the necessity of statements with necessity in the world. "John's angry behavior was caused by his anger" is presumably a tautology but the causal relation presumably occurs between a mental state and John's behavior; the claim that "such and such neural events caused John's behavior" is not a tautology. In fact, every causal claim can be turned into a tautology by the simple device of changing "*A* caused *B*" into "the cause of *B* caused *B*".

Though the reasons can not be causes argument fails, distant relatives of that argument are still taken seriously. Rosenberg has argued in numerous places that the social sciences have not progressed because they rest on belief-desire psychology (1980, 1988). Belief-desire explanations constitute a small closed circle not amenable to independent testing and refinement. Attributing beliefs requires determining what people desire and how strongly, yet to do this we must already know what they believe.

One plausible response to that argument (Henderson, 1991, 1993) is that Rosenberg's problem is real but not inevitable. Actual psychological explanations involve many more different states than belief and desire, which are generic terms. Short- and long-term memory, reasoning heuristics, classification prototypes, and motivation states are examples of the more specific items that feature in psychological explanations. These diverse states can ground separate and independent claims, breaking the simple belief-desire circle and thus allowing for real testing. Once again, it is an empirical issue when and where this problem is real and when and where it can be successfully overcome.

There is another response to Rosenberg's skeptical arguments that actually also defuses most other antinaturalist arguments based on meaning. Even if belief-desire psychology was inherently flawed, that is an obstacle to good science in the social sciences only if all social science must invoke belief-desire psychology. But to assume that is to take a position on another fundamental issue, namely, on the individualism–holism debate. Much social science seems to operate without any particular assumptions about the explanations of individual behavior at all. Sociologists Hannan and Freeman (1989), for example, develop an account of the distribution of kinds of organizations based on competitive environments. Their account is entirely at the level of organizations and other social entities. So, they are not committed, in any obvious way, to a belief-desire account of individual behavior or any other account for that matter.

There are further ongoing controversies associated with the interpretivist tradition that are of great relevance to the practice of some social science. The last two types of meaning listed above – symbolic and normative – are widely invoked, particularly in cultural anthropology. Symbolic meanings are attributed to various practices and objects by social scientists, yet those meanings may not be recognized by the subjects themselves. Meanings that are meanings to no one has seemed problematic (Skorupski, 1976), yet "representations" without any conscious cognizer seem ubiquitous in cognitive science. There remains much work to do in clarifying such claims and in determining if and how they might be warranted.

Norms are likewise widely invoked yet controversial. Economists find them suspicious unless tied to underlying mechanisms of self interest and much work has gone into investigating whether they can be accounted for in that way (Coleman, 1990). Norms are also suspect as independent explainers to both social scientists and philosophers. There is extensive debate among philosophers about the place of rules or norms, where the main worry is commitment to an unending regress (rules needed to apply rules, etc.) (Kripke, 1982). Sociologists raise related doubts which they put in terms of the background institutions needed for norms to function (Hilbert, 1992).

Let's turn now from naturalism issues to questions concerning scientific realism. There has been an extensive debate over scientific realism in the social sciences, though as with debates over realism in general, there are often quite diverse theses at stake. There are two different strands of argument calling themselves "realists" in the recent philosophy of social science literature. One strand appeals to the views and arguments of scientific realists such as Boyd (1983), applying them to the social sciences (Trout, 1998). Here the main claim is that current mature social science produces theories that are at least approximately true – their generalizations pick out real causal processes. The other strand (Bhaskar, 1978, 1979; Pawson, 1989; Cartwright, 1989) makes similar claims but adds on a restriction concerning what it is to pick out "real causal processes." The restraint is that theories refer to natures, capacities, or generative causal mechanisms. (I shall call this variant "essentialist realism.")

On the antirealist side, social constructionist accounts are offered. Perhaps receiving the greatest attention is the work of McCloskey (1985), an economist. He argues that a careful look at actual social science research will show that it is not guided by standard accounts of scientific method but by rhetoric – by what is socially persuasive. This view embodies the sort of social constructionism defended, for example, by Rorty (1987) who holds that the only constraints on science are "conversational." This approach has led to some interesting case studies of actual social science research.

Obviously, more issues and arguments are at stake in these debates than can be assessed here. Two points are worth raising, however. As noted above, there are some serious criticisms of the essentialist realist claim that explaining social research (or natural science research) requires a commitment to capacities or natures

(Papineau, 1991; Earman and Roberts, 1999). Yet the doctrine is broad enough that some important points remain even if one rejects the commitment to capacities or natures. Social scientists have picked up on the idea that they are committed to "generative mechanisms." Here they are rejecting instrumentalist ideas which have shaped social scientific research. Much social research proceeds by running regressions or other statistical tests on a set of variables without any theoretical rationale or without any theoretical account of their interrelation. The variables and their corresponding data are often drawn from ordinary language notions and lack any clear theory of what is being measured and how. Asking for "generative mechanisms" is thus rejecting social science research that claims to stick merely to the observable and to eschew theoretical commitment. In short, this version of realism rejects traditional instrumentalism, a doctrine which underwent withering criticism in philosophy of science some time ago. Thus the social science realists are in this way on the mark.

Another point about the realism debates worth noting concerns the global nature of the arguments on both sides (Kincaid, 2000). The standard version of realism argues that the best explanation for the methodological successes of the social sciences is that the relevant claims are at least approximately true. This argument takes some general characteristic of social scientific practice and infers from that trait that the science in question must be right. In this sense, it is a global argument. Similarly, antirealist arguments go from the claim that all social science involves persuasion, for example, to the claim that it is a social construct. Entire domains of investigation are evaluated in one fell swoop.

This global approach to realism issues is suspect. If the *best* explanation of the methodological practices of the social sciences is their approximate truth, then we must rule out other possible explanations. But there is good empirical evidence that other explanations cannot always be ruled out. For example, significance testing is widely used in the social sciences to determine what is believable and what is not. Yet there are good reasons to think that practice is quite misguided (Cohen, 1994). Apparently, the social processes of the disciplines nonetheless ensure that the practice continues. Similarly, careful study of research on various "psychological disorders" shows the ways standard methods can be used to construct – literally – syndromes (Hacking, 1995). And there are deep foundational issues concerning the requirements for measurement in the social sciences (e.g. additivity) that have arguably been largely ignored rather than solved (Michell, 1999).

Antirealist arguments about the social sciences are likewise implausibly global. As I just suggested, it is plausible there that some social research superficially employs standard scientific procedures while outcomes are really determined by the social interests of investigators. However, showing that this can happen is far from showing that it always does or must.

I turn next to debates over the nature of explanation in the social sciences, beginning with the individualism–holism controversy. The individualism–holism issue has been plagued by debates over unclear theses, and some progress can be made simply by sorting out the relevant issues, particular with the help of debates over reductionism in philosophy of psychology and biology. It is useful to distinguish between claims about ontology, theory reduction, explanation, and confirmation. Within these categories, there are diverse claims as well. Some of the more central individualist claims by category are:

- *Ontological*
  Societies are composed of individuals.
  Societies do not act independently of individuals.
  Social entities do not exist.
- *Theory reduction*
  Any social theory is, in principle, reducible to a theory referring entirely to individuals.
- *Explanation*
  Theories referring only to individuals can fully explain all social phenomena.
  Individualist mechanisms are a necessary condition for social explanation.
- *Confirmation*
  No social theory without individualist mechanisms can be well confirmed.
  Searching for individualist theories is the best route to successful social science.

All these theses have their past and current defenders. The reductionist version of individualism was asserted by Watkins and is currently advocated at times by Elster (1985, p. 5). The first two ontological claims are frequently cited in support. Claims about mechanisms have been urged by Elster (1989), Little (1989) and numerous others. Contemporary holists similarly deny some or all of these claims. Ruben (1985) and Gilbert (1992), for example, deny the ontological claims that social entities do not exist and ground their arguments at least in part in denying that individual theories can explain.

Let's begin with the reductionist version of individualism which, as we will see below, plays an important role in evaluating other individualist theses. Reasonable requirements for reduction are a set of bridge laws providing a lawlike connection between terms of the theory to be reduced and the reducing theory such that the explanations of the reducing or more fundamental theory can explain everything that the reduced theory can. Connections between terms need not be semantic or definitional truths. Connections between terms that presuppose the theory to be reduced (as would be the case if we equated "anger" with "anger expressing behavior" in reduction of the mental to the behavioral) are not sufficient.

The most common argument given in support of the reductionist version of individualism appeals to the metaphysical facts Watkins' cites. Societies are made up of individuals and do not act independently of them. Thus, surely, any theory of social phenomena is reducible to some theory of individual behavior (in principle – actually having the theory on the books may be too much to ask).

This argument fails. One main reason has been discussed not only in the philosophy of social science literature but in the philosophy of psychology and biology literature as well (Fodor, 1974; Hellman and Thompson, 1975; Kincaid, 1990). There it has been argued that mental states or biological states are composed of and do not act independently of (they supervene on) facts about physical states. Yet reducing biology or psychology to physics or chemistry may be impossible. For example, biological terms like "antibody" have millions of different physical realizations, and there may be no way to identify those realizations in purely chemical terms.

So, too, with social explanations. They may employ social categories – for example, "corporation" or "institution" – that can be brought about by indefinitely many collections of individual behavior. Social terms, like mental and biological ones, might be multiply realizable. The needed bridge laws might not exist. This shows that Watkins' two metaphysical truths do not *ensure* that reduction is possible.

Those truths do not entail that we can reduce social theories to individualist ones for another reason: even if we could find lawlike connections between social terms and individual descriptions, those connections might not suffice to show that individualist theories fully explain. For example, if our accounts of individual behavior invoked social processes themselves, reduction would fail.

The upshot here is that the individualism-holism debate in its reductionist guise is really an empirical issue (Kincaid, 1997). To decide if reduction is possible, then we must ask

(a)   if terms referring to social entities are multiply realizable, and
(b)   if accounts of individual behavior make no essential reference to irreducible social terms.

Answering the first question might lead us to ask the empirical question whether there are social processes that are "blind" to individual level detail? For example, if corporations compete and it is corporate profitability that matters, then we might think multiple realizations likely – there may be many different ways to organize individuals into a profitable corporation. Answering the second question would lead us to ask whether accounts of individual behavior can proceed entirely without invoking social structure. Microeconomic explanations often take individual preferences, the distribution of wealth and income, and the existence of property rights as given. Yet those things may be precisely the kind of factors we would want to explain by pre-existing institutional structures.

Put as an empirical issue, the individualism–holism debate is no longer an all-or-nothing issue. If individualism asserts that all social explanation is reducible, then it only takes one strong case of multiple realizations to refute it. Yet the individualist position might (or might not) be plausible for many other domains of social inquiry.

Let's turn briefly to other theses in the individualism–holism debate. Arguably, the ontological claim that social entities are real entities turns on the outcome of the reductionist version of the debate, for one standard criterion for ontological commitment is whether a theory makes essential reference to an entity. The claim that individualist theories can fully explain likewise stands or falls with the reductionist theses. If individualist theories completely explain, that means they have a way to capture social explanations – assuming there are such – in their own terms. And that puts us back to the requirements for reduction.

A further version of individualism requires individualist mechanisms (Elster, 1989; Little, 1989). This thesis is of dubious merit as a general proposal. Mechanisms can be specified at many levels of detail, so the demand is ambiguous from the start. Moreover, we confirm many ordinary explanatory causal claims without providing mechanisms: I can be quite confident that the baseball broke the window without providing the molecular details involved. Whether mechanisms are needed seems to depend on

1   how well confirmed our hypothesis is at the macrolevel
2   how extensive our knowledge is at some microlevel, and
3   the extent to which our macrolevel account presupposes details about mechanisms at that level.

Sometimes mechanisms are important, sometimes they are not.

Another major topic focusing on explanation concerns the status of functional explanations. A first question concerns what we want in an account of functional explanations. Much of the literature on functional explanation seeks a conceptual analysis – a set of necessary and sufficient conditions capturing our ordinary usage of "functions" or "exists in order to" (Wright, 1976). That project is a holdover from past philosophical traditions and one probably not worth pursuing. There is no reason to think there is a single ordinary concept to capture in the first place, little reason to think the concept is captured in necessary and sufficient conditions rather than, say, a prototype, and little reason to want such an analysis – what would being able to successfully predict philosopher's intuitions tell us about the science at issue?

A second, more empirical project would be to clarify functional explanations in practice – to show how they work, how they relate to standard forms of explanation, and what it takes to confirm them. An important first step in this direction came from Cohen (1978). According to Cohen, functional explanations are a subtype of what he calls "consequence explanations." Consequence explanations account for causes by appeals to their affects. A consequence explanation is successful when we can show that A's disposition to have a certain effect is connected in a lawful way with A's existing. In this way Cohen tried to show how functional explanations were an instance of nomological deductive explanations. "A exists in order to do B" explains by showing how to that A exists from a relevant law.

Nomological-deductive accounts are, however, defective in well-known ways. Lawful connections may be due to spurious correlations from common causes.

This suggests making the connection explicitly causal. With this emendation, some functional explanations of the form "*A* exists in order to *B*" make the following claims:

1 that *A* causes *B*, and
2 that *A* persists because it causes *B*.

The first claim is an ordinary causal assertion. The second claim asserts that *A*'s causing *B* results in *A*'s continued existence.

Are explanations of this sort suspect? Current worries are of at least three kinds: that "consequence-etiological" accounts like that given above face various counterexamples (Borse, 1984; Bigelow and Pargetter, 1987), that such explanations cannot be confirmed because they do not cite mechanisms (Elster, 1983; Little, 1989), and that such accounts are unfalsifiable because identifying benefits of a given practice is too easy (Hallpike, 1986). Worries of the first sort assume that the goal is to provide a conceptual analysis. I argued above that the important task is not to find necessary and sufficient conditions that cover all ordinary language usages, so those worries can be put aside – though I think most can be answered nonetheless (Kincaid, 1996). The other two cannot be so easily dismissed.

There are two ways to argue that mechanisms are necessary in functional explanation: as a claim about explanation, in general, or as a claim about only functional explanations. We have already seen that a general demand for mechanisms is dubious. Is there something unique to functional explanations that makes mechanisms necessary nonetheless? The worry roughly is that there is no connection between positive effects and persistence of a social practice. Yet the two causal claims sketched above for functional explanations are, in principle, no different than causal claims put forth by evolutionary biology. Finch beak size and shape apparently exists to promote fitness: the beak traits contribute to more offspring, and having more offspring causes the persistence of the trait. The underlying genetic mechanism may not be known, but that is no more problematic than not knowing the quantum mechanical details underlying the ball smashing the window. Something similar seems true of the demand for social mechanisms. So long as we can show that practice *A* causes *B* and persists because it does so, we have shown enough.

There are, however, many reasons to be suspicious of functional explanations in the actual practice of the social sciences. The evidence given that the two conditions actually hold is often thin. Frequently optimality arguments are given – arguments roughly of the form "trait *x* would be optimal, thus *x* persists because of its effects" – and they can easily be Panglossian exercises (Kincaid, 1995). But such problems seem to be problems of practice rather than principle; they are, after all, problems biologists must confront as well social scientists (Dupré, 1987).

Problems to Come

Important future issues in the philosophy of the social sciences turn on developing the naturalized empirical approach focusing on the actual practice of social research. The previous section has already pointed out a number of unresolved questions that will require a careful look at what social research actually shows. In this section, I briefly survey other issues that arise from investigating and evaluating the actual practice of social research.

Much interesting work in the philosophy of science has resulted from engagement with social constructionist accounts of various episodes in the natural sciences. Philosophy of social science could likewise profit from such an engagement, though interestingly the sociologists of knowledge have not as often turned their gaze on themselves. Here, the task is to identify the collective practices of social research and to use that information in explaining and evaluating the social sciences. Some interesting work in this direction has been done for economics (McCloskey, 1985; Sent, 1997), on the behaviorial sciences (Longino, 1990; Hacking, 1995), on the category of race (Root, 2000), on the status of social science concepts in general (Dupré, 1993), and in feminist critiques of specific social sciences (Wylie, 1996, 1997).

Pursuing such questions also naturally leads to investigating inference in the social sciences – describing how and with what success social scientists move from data to hypothesis. There are numerous issues of this sort where important work is and/or needs to be done. Here are some examples:

- *The extent to which various philosophical models of science fit the social sciences*
  Work has been done assessing how well bootstrapping models apply to archeology (Wylie, 1986) and Lakatosian models to economics (Hands, 1993). A variety of Bayesian analyses of the natural sciences have been produced, but few for the social sciences. Given the central place of frequentist statistical inference in the social sciences, they are a seemingly ripe area for such investigation.
- *Issues in causal modeling*
  There is an important literature – see Pearl (2000) for summary – investigating the prospects for distinguishing causal models given correlational data and other constraints inspired by work in artificial intelligence. While the issues here are not specific to the social sciences, this work has important potential implications for how causal models in the social sciences are evaluated.
- *The force of simulation*
  Simulation is a growing practice in social research. Yet it presents numerous puzzles concerning its point and force. Are simulations supposed to tell us how the world works? If so, they raise the *ceteris paribus* problem in a new way. When are simulations compelling evidence despite their unrealistic assumptions? What other purposes than confirmation can they serve and how?

- *The legitimacy of data mining*
  Data mining is the repeated use of statistical techniques in search of results that support a given model. The practice is widespread in the social sciences, particularly economics. Yet there is strong disagreement over its legitimacy (Backhouse and Morgan, 2000).

- *The nature of measurement*
  The social sciences provide apparently quantitative data in a number of areas. Yet quantitative measurement seems to rest on a number of assumptions about the thing being measured – that it is amenable to an additive scale for example. It is not clear such requirements always hold for social phenomena, though social scientists proceed nonetheless (Michell, 1999). Is there any justification for doing so? Can data derived from measures that do not meet strict requirements of quantitative measurement nonetheless be probative? How and where?

- *Nonstatistical inference practices*
  Among social scientists there is a growing literature on the use of cases and comparative analysis to draw conclusions from observational data (Ragin, 1987; Ragin and Becker, 1992). The strengths and weaknesses of these approaches, their relation to standard models of confirmation, and other such questions are open areas of inquiry.

- *Evaluating counterfactuals*
  Counterfactual claims are widespread in the social sciences (Hawthorn, 1991), though there is very little attention to what their assertion requires. Some discussion occurs in the literature on causal modeling (Pearl, 2000), but those are special circumstances with explicitly defined functional relations. What can be said about areas where such constraints are not available?

- *The place of game theory*
  Game-theoretical models dominate parts of economics and are increasingly important in political science and elsewhere (Bicchieri, 1993). These applications raise in a concrete way the *ceteris paribus* problem, for the models in question invoke strong idealizations and abstractions (Kincaid, 2001). Are those models well confirmed and explanatory nonetheless? When and where? Particularly of relevance to philosophers of science are the assumptions such models make about rationality, both rational action and rational inference by agents.

- *The place of experimental methods*
  Experimental methods are increasingly important in the social sciences, especially economics (Kagel and Roth, 1995). Exactly how such methods work and with what force in the social sciences remains to be determined. Old debates about the ecological relevance of psychological experiments resurface again, this time about whether artificial experimental results tell us anything about real markets, for example.

The social sciences will no doubt continue to raise challenging issues in the philosophy of science.

Notes

1 Philosophy of the social sciences is a broad area and it is not possible to do justice to the entire field anymore than a single article could do so for the philosophy of the natural sciences. This survey is necessarily selective, based on what topics seem to have received the most attention and no doubt on the interests and knowledge of the author. Not discussed in any detail are the following topics: issues in the philosophy of history concerning laws, objectivity, narrative, and the like (Martin, 1989; Roth, 1994); the fact/value distinction in the social sciences (Root, 1993); issues concerning the indeterminacy of translation as it relates to debates in the social sciences (Roth, 1987; Bohman, 1991; Kincaid, 1996); ontology and metaphysics of groups (Ruben, 1985; Gilbert, 1992); game theory and decision theory (Bicchieri, 1993), and no doubt other areas as well.
2 See Gordon (1991) for a discussion.

References

Backhouse, R. and Morgan, M. S. (2000): "Introduction: Is Data Mining a Methodological Problem?" *Journal of Economic Methodology*, 7(2), July, 178–81.

Bhaskar, R. (1978): *A Realist Theory of Science*. Brighton: Harvester.

Bhaskar, R. (1979): *The Possibility of Naturalism*. Brighton: Harvester.

Bicchieri, C. (1993): *Rationality and Coordination*. Cambridge: Cambridge University Press.

Bigelow, J. and Pargetter, R. (1987): "Functions," *Journal of Philosophy*, 84, 181–96.

Bohman, J. (1991): *New Philosophy of Social Science*. Cambridge: MIT Press.

Borse, C. (1984): "Wright on Functions," in E. Sober (ed.), *Conceptual Issues in Evolutionary Biology*, Cambridge: MIT Press, 369–86.

Boyd, R. (1983): "On the Current Status of Scientific Realism," *Erkenntnis*, 19, 45–90.

Cartwright, N. (1989): *Nature's Capacities and Their Measurement*. Oxford: Clarendon Press.

Cohen, G. A. (1978): *Karl Marx's Theory of History: A Defence*. Princeton: Princeton University Press.

Cohen, J. (1994): "The Earth Is Round (p < .05)," *American Psychologist*, 49, 997–1003.

Coleman, J. (1990): *Foundations of Social Theory*. Cambridge: Harvard University Press.

Davidson, D. (1980): "Actions, Reasons, and Causes," in D. Davidson, *Essays on Action and Events*, Oxford: Clarendon Press, 1–19.

Davidson, D. (1984): "Thought and Talk," in D. Davidson, *Inquiries into Truth and Interpretation*, Oxford: Clarendon Press, 155–70.

Davidson, D. (1994): "Psychology as Philosophy," in Martin and McIntyre (eds.), 79–81.

Dupré, J. (1987): *The Latest on the Best: Essays on Evolution and Optimality*. Cambridge: MIT Press.

Dupré, J. (1993): *The Disorder of Things*. Cambridge: Harvard University Press.

Durkheim, E. (1933): *The Division of Labor in Society*. New York: Macmillan.

Durkheim, E. (1965): *The Rules of the Sociological Method*. New York: Free Press.

Earman, J. and Roberts, J. (1999): "'Ceteris Paribus,' There is No Problem of Provisos," *Synthese*, 118(3), 439–78.

Elster, J. (1983): *Explaining Technical Change*. Cambridge: Cambridge University Press.

Elster, J. (1985): *Masking Sense of Marx*. Cambridge: Cambridge University Press.

Elster, J. (1989): *Nuts and Bolts for the Social Sciences*. Cambridge: Cambridge University Press.

Fodor, J. A. (1974): "Special Sciences (Or: Disunity of Science as a Working Hypothesis)," *Synthese*, 28, 97–115.

Friedman, M. (1953): "The Methodology of Positive Economics," in M. Friedman, *Essays in Positive Economics*, Chicago: University of Chicago Press, 3–43.

Gellner, E. (1968): "Holism vs. Individualism," in M. Brodbeck (ed.), *Readings in the Philosophy of the Social Sciences*, New York: Macmillan, 254–69.

Gilbert, M. (1992): *On Social Facts*. Princeton: Princeton University Press.

Gordon, S. (1991): *The History and Philosophy of the Social Sciences*. London: Routledge.

Hacking, I. (1995): *Rewriting the Soul: Multiple Personalities and the Sciences of Memory*. Princeton: Princeton University Press.

Hallpike, C. (1986): *The Principles of Social Evolution*. Oxford: Oxford University Press.

Hands, W. (1993): *Testing, Rationality, and Progress*. Lanham: Rowman and Littlefield.

Hannan, M. and Freeman, J. (1989): *Organizational Ecology*. Cambridge: Harvard University Press.

Hausman, D. (1981): *Capital, Profits, and Prices*. New York: Columbia University Press.

Hausman, D. (1992): *The Inexact and Separate Science of Economics*. Cambridge: Cambridge University Press.

Hawthorn, G. (1991): *Plausible Worlds*. Cambridge: Cambridge University Press.

Hellman, G. and Thompson, F. W. (1975): "Physicalism: Ontology, Determination, and Reduction," *Journal of Philosophy*, 72, 551–64.

Hempel, C. (1965): *Aspects of Scientific Explanation and Other Essays in the Philosophy of Science*. New York: Free Press.

Henderson, D. (1991): "On the Testability of Psychological Generalizations," *Philosophy of Science*, 58, 596–607.

Henderson, D. (1993): *Interpretation and Explanation in the Human Sciences*. Albany: SUNY Press.

Hilbert, R. A. (1992): *The Classical Roots of Ethnomethodology: Durkheim, Weber, and Garfinkel*. Chapel Hill: University of North Carolina Press.

Hollis, M. (1996): *Reason in Action*. Cambridge: Cambridge University Press.

Kagel, J. and Roth, A. (1995): *Handbook of Experimental Economics*. Cambridge: Cambridge University Press.

Kincaid, H. (1990): "Molecular Biology and the Unity of Science," *Philosophy of Science*, 57, 575–93.

Kincaid, H. (1995): "Optimality Arguments and the Theory of the Firm," in Little (1995, pp. 211–36).

Kincaid, H. (1996): *Philosophical Foundations of the Social Sciences: Analyzing Controversies in Social Research*. Cambridge: Cambridge University Press.

Kincaid, H. (1997): *Individualism and the Unity of Science: Essays on Reduction, Explanation, and the Special Sciences*. Lanham, MD: Rowman and Littlefield.

Kincaid, H. (2000): "Global Arguments and Local Realism about the Social Sciencs," *Philosophy of Science*, 67, S667–9.

Kincaid, H. (2001): "Assessing Game-Theoretic Accounts in the Social Sciences," in P. Gardenfors, K. Kijania-Placek and J. Wolenski (eds.), *Proceedings of the Congress on Logic, Methodology and the Philosophy of Science*, Dordrecht: Kluwer, forthcoming.

Kripke, S. A. (1982): *Wittgenstein on Rules and Private Language: An Elementary Exposition*. Cambridge, MA: Harvard University Press.

Liska, A. (1975): *The Consistency Controversy: Readings on the Impact of Attitude on Behavior*. New York: Wiley and Sons.

Little, D. (1989): *Understanding Peasant China*. New Haven: Yale University Press.

Little, D. (ed.) (1995): *On the Reliability of Economic Models*. Dordrecht: Kluwer.

Longino, H. E. (1990): *Science as Social Knowledge*. Princeton: Princeton University Press.

Maki, U. (1996): "Two Portraits of Economics," *Journal of Economic Methodology*, 3, 1–39.

Mandelbaum, M. (1973): "Societal Facts," in J. O'Neill (ed.), *Modes of Individualism and Collectivism*, London: Heineman, 221–34.

Martin, M. and McIntyre, L. (eds.) (1994): *Readings in the Philosophy of the Social Sciences*. Cambridge: MIT Press.

Martin, R. (1989): *The Past Within US: An Empirical Approach to the Philosophy of History*. Princeton: Princeton University Press.

McCloskey, D. (1985): *The Rhetoric of Economics*. Madison, Wisconsin: The University of Wisconsin Press.

Melden, A. (1961): *Free Action*. London: Routledge.

Michell, J. (1999): *Measurement in Psychology: A Critical History of a Methodological Concept*. Cambridge: Cambridge University Press.

Miliband, R. (1969): *The State in Capitalist Society*. New York: Basic Books.

Papineau, D. (1991): "Correlations and Causes," *The British Journal for the Philosophy of Science*, 42, 397–413.

Pawson, R. (1989): *A Measure for Measures: A Manifesto for Empirical Sociology*. London: Routledge.

Pearl, J. (2000): *Causality: Models, Reasoning, and Evidence*. Cambridge: Cambridge University Press.

Ragin, C. (1987): *The Comparative Method: Moving Beyond Qualitative and Quantitative Methods?* Berkeley: University of California Press.

Ragin, C. and Becker, H. (1992): *What Is A Case?* Cambridge: Cambridge University Press.

Root, M. (1993): *Philosophy of Social Science*. Cambridge: Blackwell.

Root, M. (2000): "How We Divide the World," *Philosophy of Science*, 67, S628–40.

Rorty, R. (1987): "Science as Solidarity," in J. Nelson (ed.), *The Rhetoric of the Human Sciences*, Madison: University of Wisconsin Press, 78–92.

Rosenberg, A. (1980): *Sociobiology and the Preemption of Social Science*. Baltimore: Johns Hopkins University Press.

Rosenberg, A. (1988): *Philosophy of Social Science*. Boulder: Westview Press.

Roth, P. (1987): *Meaning and Method in the Social Sciences*. Ithaca: Cornell University Press.

Roth, P. (1994): "Narrative Explanations: The Case of History," in Martin and McIntyre (1994, pp. 701–13).

Ruben, D. (1985): *The Metaphysics of the Social World*. London: Routledge and Kegan Paul.

Searle, J. (1984): *Minds, Brains, and Behavior*. Cambridge: Harvard University Press.

Sent, E.-M. (1997): *The Evolving Rationality of Rational Expectations: An Assessment of Thomas Sargent's Achievements*. Cambridge: Cambridge University Press.

Skorupski, J. (1976): *Symbol and Theory.* Cambridge: Cambridge University Press.

Taylor, C. (1971): "Interpretation and the Sciences of Man," *Review of Metaphysics*, 25, 3–51.

Trout, J. D. (1998): *Measuring the Intentional World: Realism, Naturalism, and Quantitative Methods in the Behavioral Sciences.* Oxford: Oxford University Press.

Watkins, J. (1973): "Methodological Individualism: A Reply," in J. O'Neill (ed.), *Modes of Individualism and Collectivism*, London: Heineman, 179–85.

Weber, M. (1968): *The Methodology of the Social Sciences*, E. Shils and H. Finch (eds.), New York: Free Press.

Winch, P. (1958): *The Idea of a Social Science and Its Relation to Philosophy.* London: Routledge and Kegan Paul.

Wright, L. (1976): *Teleological Explanations.* Berkeley: University of California Press.

Wylie, A. (1986): "Bootstrapping in Un-Natural Sciences: An Archaeological Case," in A. Fine and P. Machamer (eds.), *PSA 1986, vol. I*, East Lansing Michigan: Philosophy of Science Association, 314–22.

Wylie, A. (1996): "The Constitution of Archaeological Evidence: Gender Politics and Science," in P. Galison and D. J. Stump (eds.), *The Disunity of Science: Boundaries, Contexts, and Power*, Stanford: Stanford University Press, 311–43.

Wylie, A. (1997): "The Engendering of Archaeology: Refiguring Feminist Science Studies," *Osiris*, 12, 80–99.

Chapter 15

Feminist Philosophy of Science[1]

Lynn Hankinson Nelson

Highlights of Past Literature

Feminist philosophy of science is located at the intersections of the philosophy of science and feminist science scholarship, and like these, constitutes a diverse tradition of inquiry marked by significant development in the last two decades.[2] The problems and approaches that characterize this tradition, as in other traditions in the philosophy of science, reflect developments in the sciences and science scholarship and are best understood in their light.

It is common to describe the trajectory of feminist interest in, and engagement with, the sciences since the 1970s in terms of an evolution: from an initial emphasis on the sociology of science, with particular attention to factors contributing to women's relative under-representation and less powerful positions in it; to critiques offered by feminist scientists of androcentric research problems, methods, and theories in their disciplines, and their development of constructive alternatives to them; to more broadly focused investigations of relationships between social processes internal and external to science, and its directions and content; to, eventually, critiques and analyses of theories about science, including those developed in the philosophy of science, and the development of approaches informed by feminist science scholarship.

Such descriptions serve as a useful introduction to some of the developments that led to the emergence of feminist philosophy of science, and to some of the empirical research that helped to shape its core questions and problems. But as feminists point out, they are also misleading.[3] For one thing, they suggest discrete "stages" of feminist inquiry, characterized by distinguishable emphases and problems. But feminist scientists and science scholars are among those who have found it difficult to maintain the conventional boundary between the epistemology of science and a number of social processes that characterize or impact on science. For example, the presence and role of androcentric assumptions in shaping research questions, methodological assumptions, and hypotheses in a number of research programs, have been taken by feminists as evidence of relationships between the scope of evidence available to scientists and social processes and non-epistemic values that characterize the broader communities in which science is undertaken; see for example Bleier (1984), Harding (1986), Longino (1990), Nelson (1996) and Wylie (1997a).[4]

The precise nature of the relationships in question, their implications for conventional views about science, and the normative policies to be recommended in their light, remain matters of ongoing investigation and analysis in feminist philosophy of science, as they do in other traditions. But, together with developments in the philosophy of science and science studies disciplines, such relationships explain why the "levels" of analysis earlier outlined have evolved apace in feminist science scholarship and informed each other.

A second problem with descriptions such as that earlier outlined is that they fail to acknowledge the significant impact that developments in the philosophy of science had on the core questions and approaches of feminist philosophy of science. As I later explore, and recent reviews by Sandra Harding and Alison Wylie detail, feminist philosophers have used and expanded on the challenges to logical positivism and empiricism and to key features in the work of Carnap, Hempel, and Nagel that emerged in the discipline beginning in the late 1950s – including evidence that observation is theory-laden and theories underdetermined by available evidence, and arguments for various forms of holism (Harding, 2000; Wylie, 2000). Like their colleagues, feminist philosophers of science have sought to understand the implications of these developments for questions of traditional importance to the philosophy of science, including those concerning the nature of objectivity and of evidential relations, and the role of epistemic and other values in scientific practice, as well as "What constitutes 'good' science and 'good' philosophy of science?" Further, and explored in some detail below, feminists have made substantive use of constructive approaches that emerged in the broader discipline in response to these challenges.

A review of significant past contributions to feminist philosophy of science, however necessarily selective, must begin with feminist science critiques in the 1970s and 1980s. Not only did these critiques identify and raise issues that would help shape core questions in the research tradition, but the first analyses aptly described as engaging the philosophy of science from feminist perspectives were offered by practicing scientists.

Among the earliest feminist science critiques were those focused on the sociology of science. Although the formal barriers to women's participation in the sciences had been removed, feminists identified a host of informal barriers that continued to contribute to women's relative under-representation and less prestigious positions in the sciences overall, and to their particularly small representation in the physical sciences. These barriers (many of which were also recognized by professional science associations) included discriminatory admissions practices in undergraduate and graduate programs; less access to technological aids, financial aid, and grants, for female graduate students; a lower "reinvestment potential" of women's credentials relative to similar credentials among men; and discrimination in hiring and placement.[5] In this and later decades, feminists' analyses of social processes characterizing and/or impacting on the sciences would also explore barriers based on race, ethnicity, and culture (Collins, 1991; Harding, 1986, 1991), and social arrangements "internal" to science – including divisions in cognitive authority within research programs, and prestige hierarchies within and among the sciences (Addelson, 1983).

Conventionally, the first two sets of issues are viewed as unrelated to the epistemology of science, and thus of little or no interest to the philosophy of science. But feminists are among those who argue that one cannot simply assume that factors conventionally regarded as "external" to science, including the social identities and contextual values of scientists, have no impact on the directions or content of scientific research. As I earlier noted, feminist critiques of the "content" of various sciences arguably constitute evidence of just such relationships. Nor, feminists have argued, can one simply assume that social processes internal to science – for example, peer review and funding mechanisms – ensure that the most promising hypotheses and research programs are eventually funded and pursued. Research undertaken in various science studies disciplines indicates that prestige hierarchies and conservatism also have a role in determining such outcomes (Addelson, 1983; Harding, 1986; Longino, 1990; Nelson, 1990).

In the early 1970s, feminist scientists and historians of science turned to the content of science. Feminists in the social sciences and psychology identified androcentrism in the goals, research questions, methods, organizing principles, and theories in their disciplines. In anthropology, sociology, history, economics, and political science, feminists criticized methodological approaches to and accounts of social life that emphasized men's behavior and activities as defining of the so-called public sphere and "culture" and that associated women (explicitly or implicitly) with the so-called private sphere and reproductive activities, in turn treated as "natural" and invariant. Feminist social scientists also criticized the central questions of mainstream research in their disciplines as largely ignoring issues of concern to women, including gender discrimination in the workplace and violence against women; see, for example Wylie (1996a) for an overview of these critiques.

The problem, feminist social scientists argued, was not simply that accounts of social life so characterized are empirically inadequate, ignoring or distorting as they do the productive and diverse nature of women's activities in specific historical and cultural contexts, their variability along the axes of race and class, and phenomena such as domestic violence. As problematic were the associations of men with culture and production and of women with nature and reproduction. By theoretically dichotomizing these connections, they obscure the actual and significant inter-relationships between the domains or spheres. For example, mainstream accounts of the economic structure of twentieth-century capitalism did not address how women's unpaid labor in the so-called private sphere sustained key features of that structure (Hartmann, 1981). Similarly, feminist anthropologists pointed out that many ethnographic studies of "hunter-gatherer" societies only focused on the hunting activities of men, e.g., contributions to Rosaldo and Lamphere (1974). Groundbreaking analyses of androcentric research questions, methods, and hypotheses were offered by feminist sociologists (Smith, 1977), historians (Kelly-Gadol, 1976), and anthropologists (Rosaldo, 1980), including androcentric accounts of human evolution (Slocum, 1975; Tanner and Zihlman, 1976).[6]

In psychology, feminists identified three general problems: influential models of psychological development and psychological maturity were based on empirical research limited to boys and men; there were hypotheses that the trajectory of women's psychological development was "truncated" or "deviant" because it did not fit such models – see, for example Gilligan (1982) for discussion of both problems; and research devoted to finding physical explanations (e.g., sex differences in hemispheric lateralization) for alleged sex differences in mathematical and spatial abilities, was frequently characterized by androcentric assumptions and circular reasoning (Bleier, 1984; Star, 1979).

It would be hard to over-estimate the significance or impact of these critiques for the emergence of feminist philosophy of science. They identified the role of assumptions shaped by current social and political context (e.g., androcentric assumptions) in research that was mainstream and credible, their important consequences for widely-accepted hypotheses and theories, and the role of scientific research in reinforcing social and political practices. Moreover, they often included or led to the development of constructive alternatives in terms of research focuses and hypotheses and research methodologies; see, for example Stanley and Wise (1983), Fonow and Cook (1991), Reinharz (1992) and Wylie (1996b). Finally, they would serve as important resources for feminist investigations of androcentrism in the life sciences and bio-behavioral sciences, to which I now turn.[7]

It is not surprising that early feminist critiques of research in the biological and bio-behavioral sciences often focused on research that sought to establish and/or to explain perceived differences between women and mean on the basis of biology; that imposed gender stereotypes (e.g., male dominance and aggression, and female passivity) on the behavior and social organization of non-human species; and that used the alleged universality of sex differences in behaviors, across cultures and across species, as *prima facie* evidence of their biological origins. Feminist scientists offered detailed analyses of the ways in which androcentrism and/or sexism shaped research questions and problems, hypotheses, and the interpretation of research results in endocrinology, empirical psychology, Sociobiology, evolutionary biology, primatology, and animal sociology. Representative collections of such critiques include the four-volume series titled *Genes and Gender*, the first volume of which was published in 1978 (Tobach and Rosoff, 1978), and the anthologies *Women Look at Biology Looking at Women* (Hubbard et al., 1979), *Biological Woman – The Convenient Myth* (Hubbard et al., 1982), and *Feminist Approaches to Science* (Bleier, 1988).

By the mid 1980s, the critiques offered by feminist scientists had expanded to include the role of sexual dimorphism and androcentrism in biological research not directly related to explaining sex differences (Hrdy, 1981; Longino and Doell, 1983); the emphasis in the bio-behavioral sciences on dominance relationships; and models of biological processes assuming linear and hierarchical causal relationships (Hubbard, 1982; Keller, 1983, 1985). Again, constructive alternatives were proposed to the assumptions and models so criticized: more complex models of social interactions and organization among other species (Haraway, 1986), more complex models of specific biological processes (Bleier, 1984; Fausto-Sterling, 1985; Longino and Doell, 1983), and models of natural relationships emphasizing "order" rather than law-like relationships (Keller, 1985).[8]

The critiques I have summarized raised several important questions. Were the cases involving androcentrism aptly characterized as "bad science"? Was there evidence, for example, that researchers were consciously manipulating data, consciously proposing hypotheses to support traditional assumptions and practices, and/or just less bright than their feminist colleagues? Were such cases idiosyncratic and, thus, without implications for long-standing views about scientific objectivity, method, and so forth? If, as many feminist scientists came to believe, the answer to each question is in most cases "no," then feminist critiques suggested that "science as usual" is a more human, and culturally bound, activity than previously assumed. So, too, the ability of feminist scientists to identify androcentrism in their disciplines and the role of other emphases with obvious relationships to cultural context (e.g., the emphasis on hierarchical relationships) and to propose at least viable (if not more viable) alternatives, suggested relationships between the social/cultural identities of scientists, including the contextual values they bring to their research, and the directions and content of their research. Many feminists take similar conclusions to be appropriate in light of research in various science studies disciplines that suggests relationships between race, ethnicity, and culture, and the directions and content of scientific research; see Harding (1991) for an overview.

Analyses exploring the implications of these issues for the epistemology of science were offered in, among other places, the works by Ruth Bleier, Anne Fausto-Sterling, Ruth Hubbard, and Evelyn Fox Keller cited above. These scientists considered the relationship between theories and observation; the nature of "scientific facts"; continuities between so-called common-sense assumptions and contextual values on the one hand, and the directions and content of scientific research on the other; and the implications of feminist science critique for objectivity – both as an attribute of individual knowers and of knowledge claims. Although all emphasized that the research feminist criticized, and feminists' interests in proposing alternatives in keeping with feminist political goals, constituted evidence that science and scientists were deeply influenced by their social and political contexts, none advocated relativism. All made appeal to conventional epistemic values such as empirical adequacy in their criticisms of specific research problems and hypotheses, and in their proposals for alternative approaches.

At the same time, many feminist scientists, including Evelyn Fox Keller in the following passage, were acutely aware of some tensions and potential dangers. Concluding on the basis of the various lines of critique mentioned above that science is just "politics by other means," would undermine the empirical and normative force of those critiques. And, in more general terms, most feminist scientists recognized that determining "how things are" is crucial in choosing effective courses of action for improving women's lives.

> Joining feminist thought to other social studies of science brings the promise of radically new insights, but it also adds to the existing intellectual danger a political threat. The intellectual danger resides in viewing science as pure social product; science then dissolves into ideology and objectivity loses all intrinsic meaning. In the resulting cultural relativism, an emancipatory function of modern science is negated and the arbitration of truth recedes into the political domain (Keller, 1982, p. 117).

To be sure, the appeals feminist scientists made to research results and other data, in the context of challenging androcentric hypotheses and theories, attested to an assumption that there is a world that constrains theorizing. But it was also clear that the findings and emergence of feminist science critiques would require the development of more complicated and nuanced understandings of such constraints and of science's complex relationships to its social contexts.

As Alison Wylie notes, the work of feminist philosophers of science is continuous with that of feminist scientists in several respects. Some have contributed to the analysis of the role of androcentrism in specific research programs and sciences, and most seek to address the empirical and normative questions initially posed by feminist scientists, albeit in ways frequently informed by developments in the philosophy of science (Wylie, 2000). This work is extensive and, again, my discussion of it is selective, focusing on work representative of major themes and developments.

The first anthology in philosophy to include articles devoted to explorations of the implications of feminist science critique for the philosophy of science was Sandra Harding's and Merrill B. Hintikka's collection, *Discovering Reality: Feminist Perspectives on Epistemology, Metaphysics, Methodology, and Philosophy of Science* (1983). Several contributors explored the questions of why feminist scientists and science scholars were able to recognize problems in mainstream research not previously recognized, and what kind of epistemology might best provide a framework for "justifying" the critiques and alternatives proposed by feminists in the sciences and other disciplines. Harding and Harstock each offered what would become influential and controversial versions of feminist standpoint theory, a framework which traces its roots in Marxism; and other contributors offered analyses of science informed by psychoanalytic theory and postmodernism (Keller and Grontkowski, 1983; Flax, 1983).

Arguably, the single work of the 1980s that would have the most long-term impact in feminist philosophy of science (both in the sense that its central arguments, accepted by some and criticized by others, would provide a general framework for work in the next decade) was Sandra Harding's *The Science Question in Feminism* (1986). Harding engaged in an extensive analysis of the epistemologies feminists were adopting or seeking to develop to explain and/or to justify feminist claims within and about science. She identified three epistemological frameworks available to and drawn upon by feminist scientists and science scholars: feminist empiricism, feminist standpoint theory, and feminist postmodernism. Noting that each framework represents a hybrid – combining (or attempting to combine) feminists' questions, concerns, and findings on the one hand, with those of an older, non-feminist tradition on the other – Harding explored how advocates of each had to wrestle with a "parent" tradition not designed to accommodate or explain the issues with which feminists were concerned.

Harding insisted on the contingency of her analysis, maintaining that each framework was inherently unstable and likely to evolve in response to developments in the others. She also urged ambivalence towards accepting one or the other framework over the others, suggesting that feminists could not afford to close the door on the future science projects that feminist empiricism or standpoint theory might eventually yield. But her analysis also suggested that feminist empiricism was the least promising of the three approaches. While the Marxist and postmodernist traditions traced their roots to "emancipatory" projects, Harding argued, empiricism is inherently conservative, of necessity ruling out the significance of movements of social liberation in explanations of scientific progress. More specifically, Harding attributed several "dogmas" to twentieth-century empiricism – including commitments to individualism and to a distinction between the contexts of discovery and justification – that would make it difficult to construct a viable explanation of the emergence of feminist science scholarship or to justify feminists claims within and about the sciences. Harding also characterized then-current feminist empiricist approaches as themselves inherently conservative, viewing the problems identified by feminist scientists to be "social biases correctable by stricter adherence to the existing methodological norms of scientific inquiry" rather than as reflective of problems with the norms themselves (Harding, 1986, p. 24).

In contrast, Harding argued that feminist standpoint theory insists on the relevance of social location to what it is possible for individuals to know. In brief and oversimplified terms, standpoint theorists argue that knowers are "situated," their vantage points made possible and shaped by concrete material, socio-political, historical, and cultural situations, and that not all epistemic situations are equally advantageous. As Harding has more recently summarized this argument, those in dominant positions within social, political, and/or cultural hierarchies are at an epistemic disadvantage: their activities and material/social situation "both organize and set limits on what [they] can understand about themselves and the world around them" (Harding, 1993, p. 54) such that "the real relations of humans with each other and with the natural world are not visible" (Harstock, 1983; Smith, 1977). The lives and experiences of those not so advantaged, on the other hand, provide the source and grounds for claims more likely to be veridical. Building from these general assumptions, divisions by gender in labor and power can be expected to have epistemological consequences. In Harstock's (1983) formulation of an argument calling for a specifically feminist standpoint epistemology, the traditional division of labor by gender can explain why feminist knowledge claims are superior to those they criticize. Feminist perspectives are "achieved standpoints," made possible by such divisions, and by inherent contradictions between the realities of women's lived lives and the dominant ideology. Finally, Harding used feminist postmodernist approaches, as explored and explicated by Donna Haraway and Jane Flax, among others, to identify challenges to both feminist empiricism and standpoint epistemology, particularly the tenability of "universalizing claims" about the power of reason, science, and the "subject/self" (p. 28).

Although some of the specific arguments just summarized would turn out to be influential, particularly in the social sciences, they would also be challenged by feminist philosophers, including those interested in developing empiricist approaches to science. Indeed, Harding herself, in both this and later work, would identify important problems facing standpoint theory, including what some took to be its tendency toward essentialism and the lack of a convincing argument that feminists and/or women enjoyed an epistemic advantage. At the same time, Harding's analysis, together with debates within feminist theory about essentialism (concerning gender among other categories), the viability of epistemology, and the inextricable relationships between gender and other social relations, would contribute to increased attention to the "situadedness" of knowers and knowledge claims and to the contingent and pragmatic aspects of scientific practice.

Donna Haraway's notion of "partial [particularly and specifically embodied] vision" is representative of feminists' interests in recovering the embodied, socially-enabled, and otherwise concretely-situated aspects of knowing and knowledge-making. Often, efforts to analyze these features of knowing were explicitly linked to concerns with responsibility and with objectivity, and in this regard the following passage by Haraway is also representative.

> not so perversely, objectivity turns out to be about particular and specific embodi-
> ment, and definitely not about the false vision promising transcendence of all limits
> and responsibility. . . . Feminist objectivity is about limited location and situated
> knowledge, not about transcendence and splitting of subject and object. In this way
> we might become answerable for what we learn how to see (Haraway, [1988] 1991,
> p. 190).

Other important contributions to feminist philosophy of science in the 1980s included articles in a two-volume special issue of *Hypatia* devoted to "Feminism and Science" edited by Nancy Tuana (1987, 1988). In addition to critiques of specific research programs and theories, a number of contributors explored the more general implications of feminist science scholarship, including questions concerning the justification of feminist claims (e.g., Linda Alcoff, Helen E. Longino, and Elizabeth Potter). Alcoff's and Potter's articles focused on questions then very much at the heart of feminist theorizing: "On what grounds do we challenge non-feminist and particularly androcentric assumptions and theories?" and: "On what grounds do we argue for the superiority of feminist assumptions and theories?" As Alcoff explored, feminists, even those critical of specific and general aspects of science, were (and continue to be) concerned that feminist critiques and analyses of science do not, in Keller's terms, "[recede] into the political domain," i.e., that they not be or become pure ideology. In advocating that feminists explore the question of what model of theory-choice might be aptly attributed to feminist scientists, Alcoff's discussion is also representative in its use of arguments by mainstream philosophers of science (in this case, Pierre Duhem, Thomas Kuhn, Hilary Putnam, and W. V. Quine) that challenged positivist conceptions of theory choice as "value-neutral" and/or "empirically determined," and that supported one or another version of holism; see also, Longino and Potter, in the same volumes.

I have noted that questions concerning how to justify, as well as how to explain, feminist claims within and about the sciences were at the heart of a good deal of feminist philosophy of science in the 1980s. There were also explorations of the question of whether there could or might eventually be "a feminist science," and what such a science might be like. Despite the influence in the field of appeals to aspects of Marxism, of Continental philosophy, and of postmodernism, explorations of these several questions often presupposed the conventional view that knowledge is propositional and that, if feasible, a feminist science would be at least partly characterized by its "content." The last contribution to the past literature I note here is Helen E. Longino's "Can There Be a Feminist Science?" ([1987] 1989). In contrast to the approaches just mentioned, Longino focused on the question of what it means "to do science as a feminist," rather than that of what the "content" of a "feminist science" might be, and advocated "a process-based approach to characterizing feminist science [rather than a content-based approach]" (p. 45). In this and subsequent work, Longino proposed that an answer to the question, "What does it mean to do science as a feminist?" would involve identifying the "contextual" (i.e., non-epistemic) values as well as epistemic values that feminists bring to their research, and that guide their choices of research problems, theories, and so forth. Her proposals for emphasizing science as practice rather than product, and for exploring the role of contextual values in scientific practice, would provide a general framework for her future work and influence that of others. Perhaps of most significance, Longino would later propose that feminist philosophy of science (she used the phrase 'feminist epistemology' in the work in question) is also best understood and undertaken as a way of doing the philosophy of science (Longino, 1994). I return to these arguments in the next section.

In more general terms, the emphasis on science as a social and value-infused set of practices and on the epistemological significance of the social processes that characterize it, would join explorations of the situatedness and contingency of epistemic practices and scientific knowledge, to become central themes in feminist philosophy of science in the late 1980s and 1990s. In this research, feminists' concerns with understanding the nature and degree of the constraints imposed by the world would continue to hold center stage.

Current Work

As I earlier noted, Harding's 1986 taxonomy of "feminist epistemologies" was influential in more than one sense. Whether one agreed with her characterizations of each of these approaches, her analysis of each in light of the others served to highlight some important strengths and some important weaknesses of feminist approaches informed by the three broader traditions. Standpoint epistemologists have continued to develop and refine the theory in light of their own and others criticisms of its early formulations.[9] And efforts by feminists in the 1990s to develop versions of feminist empiricism would critically engage, sometimes explicitly and sometimes implicitly, Harding's accounts and criticisms of empiricism and of feminist empiricism.

Two early such efforts were Helen E. Longino's *Science as Social Knowledge* (1990) and my *Who Knows: From Quine to Feminist Empiricism* (Nelson, 1990). Although different in the approaches and many of their emphases, they shared a commitment to the empiricist thesis that knowledge is grounded in experience, and theorizing and theories constrained by evidence. In addition, three of their emphases were representative of feminist philosophy of science more broadly: arguments that science communities, rather than scientists *qua* individuals, are the appropriate focus of the philosophy of science; arguments that non-epistemic or contextual values can deeply inform "good" science; and analyses of specific case studies involving gender and science to support these arguments; see, for example Duran (1998) and Wylie (2000).

Longino's argument for recognizing the social nature of scientific practice, and the role within that practice of contextual values, drew on then well-known arguments in the philosophy of science that evidential relationships were far more complex than earlier empiricist accounts of the logic of testing had acknowledged. Briefly put, Longino argued that what determines whether or not someone will take some fact or alleged fact, x, as evidence for some hypothesis h, "is not a natural (for example, causal) relation between the state of affairs x and that described by h but other beliefs that person has concerning the evidential connection between x and h." States of affairs are taken as evidence, Longino continued, "in light of regularities discovered, believed, or assumed to hold" (Longino, 1990, p. 41). Such assumed regularities, Longino argued, are examples of background assumptions that "[always determine] the evidential relation" (p. 60).

Longino cited several other features of current scientific practice as evidence that scientific inquiry is a social rather than an individual process, scientific knowledge a social rather than an individual achievement. These features include the dependence and interdependence of individual scientists on the conditions, conceptual as well as practical, that make their work possible, and the dependence of the sciences, individually and collectively, on the value attributed to their enterprise by the larger social and political community (pp. 67–9). Further, Longino argued that "the clashing and meshing of a variety of points of view" that produces scientific knowledge typically involves "conceptual criticism": i.e., the adjustment of the background assumptions, mentioned above, in light of which the data are interpreted (p. 72). Because such criticism is an inherently social process, "scientific method" includes more than hypothesis testing, and such testing involves subjecting data, hypotheses, and relevant background assumptions to conceptual scrutiny. It is such interactions among scientists that "modify their observations, theories, hypotheses, and patterns of reasoning" and eventually produce scientific knowledge, rather than "individuals applying a method to the material to be known" (Longino, 1993, pp. 110–11).

In arguing that values are inextricable features of science, Longino engaged in analyses of specific research programs, arguing that the background assumptions functioning to determine evidential relations are themselves of different sorts. Some express "constitutive values" (values which determine the rules governing acceptable scientific practice); other background assumptions express contextual values, values which reflect "the social and cultural environment in which science is done" (Longino, 1990, pp. 4–5). One of Longino's central normative claims was that heterogeneous science communities are more likely to produce heterogeneous background assumptions, which in turn will enable the recognition, scrutiny, and modification of such assumptions.

In *Who Knows*, I built on Quine's arguments for holism and naturalism to argue for an empiricist approach to science that did not include commitments to individualism or to a hard and fast boundary between science and non-epistemic values. I argued that Quine's arguments for holism (and related arguments against foundationalism) yielded a view of the evidence supporting specific hypotheses and theories as being of two kinds: experience (conventionally parsed as observation or observation sentences) and inter-theoretic integration with other accepted theories. I also argued that arguments of Quine's demonstrated that both kinds of evidence are social. The bodies of theory in question are, of course, social achievements – the product of collective efforts to explain and predict experience and features of the world. In maintaining that experience is also fundamentally social, I built in part on Quine's (1981) arguments that the sensory experiences (in the sense of phenomenological experiences) recognized as relevant to a particular knowledge claim are themselves shaped and mediated, indeed made possible, by a larger system of historically and culturally specific theory and practice. Finally, I argued that the broad bodies of theory that constitute part of the evidence for specific theories and claims, include claims and assumptions shaped by social and political context, building in part on Quine's arguments for the interdependence of science and so-called "common sense," but extending the latter to include more than physical-object theory, and on cases from feminist science scholarship. Recognizing the role of non-epistemic values in scientific practice does not, I argued, reduce science to ideology. Rather, it suggests the need for more diverse science communities, communities such that the methodological and metaphysical assumptions functioning as evidence for specific research questions and claims would be subjected to a broader range of scrutiny.

In the intervening years, the arguments just summarized have been criticized within feminist theory and by non-feminist critics. Some argue that the emphasis on the role of community standards in the determination of what ultimately counts as authoritative knowledge threatens (or constitutes) relativism. Similarly, there have been criticisms of our individual accounts of the role of non-epistemic values in scientific practice, and charges that these also threaten or constitute relativism. Some feminists remain unsympathetic to empiricism, and particularly naturalism, and many remain unsympathetic to others of Quine's positions. But efforts to develop a version of empiricism, and/or naturalism, commensurate with feminist scholarship and useful to feminist theorists continue (Campbell, 1994, 1998; Duran, 1998; Nelson, 1996; Tuana, 1996).

In more general terms, analyses of evidential relations that approach scientific practice and knowledge as inherently social, and that seek to understand the role of epistemic and non-epistemic values in scientific practice, characterize a good deal of recent work in feminist philosophy of science. And, arguably, many of the most significant recent contributions to feminist philosophy of science neither locate themselves, nor would they be easy to locate, using Harding's (1986) taxonomy. Harding had also characterized epistemology as a justificatory enterprise; but, by the 1990s, arguments against this view, as offered by naturalized philosophers of science and others, were perceived by many philosophers of science as definitive. And although feminists continue to focus on questions concerning the nature and scope of the evidence that supports scientific theories, there is now less explicit concern with the question of what "justifies" feminists' claims within and about science.

With more emphasis on "explanation" than justification, attention to questions about the explanatory principles that should figure in the philosophy of science characterizes a good deal of current work in feminist philosophy of science, as it does work in the broader discipline; see, for example the articles in Alcoff and Potter (1993). Work has been undertaken to further develop models of evidential relations using semantic theory (Giere, 1996; Longino, 1993), Ian Hacking's work on "evidential independence" (Wylie, 1996b), versions of realism (Barad, 1996; Campbell, 1998), and versions of holism (Nelson, 1996; Potter, 1993). Reflecting developments summarized in the previous section, most such efforts incorporate assumptions about the "situadedness" of knowers and knowledge claims. Most also seek to demonstrate that, contrary to traditional assumptions, recognizing the role of social processes, and the partiality and contingency of all knowledge claims, does not entail relativism. Representative of these assumptions and interests, Ronald Giere advocates "perspectival realism" (Giere, 1996), Karen Barad "agential realism" (Barad, 1996), and I advocate "naturalistic realism" (Nelson, 1996).

As the foregoing suggests, feminist philosophers of science have found valuable resources in the broader discipline. They have contributed to and made use of challenges to post-logical positivism in their analyses of science: including evidence suggesting that observation is theory-laden (Harding, 1986); that individual hypotheses do not, in Quine's words, "face the tribunal of experience individually," but do so as part of broader bodies of theory (Longino and Doell, 1983; Longino, 1990; Nelson, 1990); that theories are underdetermined by available evidence (Alcoff, 1989; Longino, [1987] 1989, 1990; Nelson, 1990); that the more specific epistemic virtues traditionally associated with the objectivity of knowledge claims – empirical accuracy, generality of scope, simplicity, and so forth – are often incapable of simultaneous satisfaction, with contingent and pragmatic factors often determining the priority given to one or more over another (Longino, 1990, 1996; Wylie, 1995); and that clusters of disciplinary commitments (metaphysical, methodological, and so forth) are instrumental in the arrival at agreed-upon understandings of given phenomena and research problems.

Moreover, in considering the implications of these challenges for feminist philosophy of science, and vice versa, feminist philosophers have built on, even as they have sought to further develop or refine, constructive approaches in the philosophy of science that emerged in response to them. In addition to approaches already cited, feminists have made appeal to and expanded on Bas van Fraassen's constructive empiricism – e.g., Longino (1990) – Mary Hesse's Network Model – e.g., Alcoff (1989) and Potter (1989); W. V. Quine's naturalism (Antony, 1994; Campbell, 1998; Duran, 1998; Nelson, 1990, 1996); and various aspects of Thomas Kuhn's work, including his arguments concerning the role of epistemic values in scientific practice (Longino, 1995, 1996; Wylie, 1995). Finally, like naturalized philosophers of science, feminists sympathetic to Quinean naturalism, and some critical of it, have explored the implications of developments in the cognitive sciences, and in other relevant disciplines, in their analyses of science; see, for example Nelson and Nelson, 2001).

There is also work to incorporate, in philosophical explanatory principles, the implications of recent investigations into the nature and role of epistemic and non-epistemic values in scientific practices. Arguably, the most influential and provocative analyses of these issues have been offered by Helen Longino. I earlier noted that Longino proposes that one should approach the question of "What does it mean to do science as a feminist?" in terms of the practices of feminist scientists and, in particular, in terms of the values that guide that practice (i.e., guide their assessments of specific hypotheses and theories, and reflect feminist political commitments and goals). The most traditional of the values Longino attributes to feminist scientists is empirical adequacy. But additional values "come into play" in the assessment of theories, hypotheses, and models, Longino (1994, p. 477) argues, because "empirical adequacy is not a sufficient criterion of theory and hypothesis choice". Those she identifies as guiding the practice of feminist scientists are ontological heterogeneity, complexity of relationships, diffusion of power, applicability to current human needs, and novelty (pp. 477–9). What is specifically feminist about these standards, Longino maintains, is that an effect of each is "to prevent gender from being disappeared . . . each makes gender a relevant axis of investigation" (p. 481). Longino describes "the non-disappearance of gender" as "a bottom line requirement of feminist knowers," its intent "to reveal or prevent the disappearing of the experience and activities of women and/or to prevent the disappearing of gender" (p. 481).

In "Cognitive and Non-cognitive Values in Science: Rethinking the Dichotomy", Longino (1996) uses the juxtaposition of some traditional cognitive virtues (simplicity, external consistency, breadth of scope . . .) with those she attributes to the practice of feminist scientists (ontological heterogeneity, novelty, mutuality of interaction . . .) to argue that the former are at least not purely cognitive and, like those that inform the practice of feminist scientists, presently carry political valence. As in her earlier work, Longino assumes a view of evidential relations that emphasizes the role of social processes, and, building in part on Kuhn's arguments, she concludes that the epistemic weight attributed to a particular theoretical virtue in the choice between methods, theories, or research programs, is determined by local, negotiated, and ideally pluralistic, considerations.

Defended and expanded upon, e.g. in Wylie (1995), Longino's arguments have also been criticized. One line of criticism, offered by both feminists and those highly critical of work in feminist philosophy, is that Longino has failed to demonstrate that the role she attributes to non-cognitive values does not yield relativism (Haack, 1996). Another is that, as they operate in the practice of feminist scientists, the significance of some of the values Longino identifies – and, specifically, the non-disappearance of gender, ontological heterogeneity, and complexity of relationships – is that, in a number of domains (e.g., the biological sciences), theories and research that reflect them are likely to be more empirically adequate (Nelson and Nelson, 1994). That is to say, whatever the political salience of these values, critics argue, they are epistemic and it is important that their role as such be recognized.

The last trend in current work in feminist philosophy of science I will note here is work to expand the investigations of the relationships between science and its social context by incorporating insights from non-Western research. Sandra Harding's work in this regard is substantive and influential. In "Multicultural and Global Feminist Philosophies of Science: Resources and Challenges" (1996) and elsewhere, Harding argues that significant philosophical issues concerning science emerge when themes in multicultural and global feminisms, and in postcolonial science studies, are brought to bear on Northern philosophy of science, including Northern feminist philosophy of science. Among the issues Harding identifies and considers are relationships between androcentrism and Eurocentrism in Northern philosophies of science, the expansion of Northern sciences and technologies to developing countries, and gender relations within global political economies. Viewed from the perspective of postcolonial science studies and multicultural and global feminisms, Harding maintains, Northern sciences can be seen to constitute "local," rather than universally applicable, knowledge system. Attention to the distinctive philosophical issues raised by multicultural and global feminisms, she concludes, can expand the concerns of postcolonial and Northern feminist philosophies of science in ways that support the development of epistemologies and ontologies capable of detecting the androcentrism and Eurocentrism of dominant frameworks in the sciences and in the philosophy of science.

Future Work

As the discussion so far indicates, feminists have found valuable resources in the larger discipline of philosophy of science. Indeed, despite severe criticisms of feminist approaches to science offered by those advocating very traditional approaches to the philosophy of science, constructive engagements between feminists and their colleagues are more common today than a decade ago. Contributors to a recent anthology Jack Nelson and I co-edited (Nelson and Nelson, 1996) reflected a wide range of philosophical positions, and most of their explorations of parallels and tensions between feminist and other approaches to science bore little resemblance to the rhetoric characterizing the so-called science wars. Substantive engagement across philosophical positions and traditions has also characterized several recent conferences devoted to values and science, including the conference "Values and Science" at the University of Pittsburgh in Fall 1998, and the conference of the same title at the University of Alabama in February 2000. And several "mainstream" journals have published special issues devoted to feminist philosophy of science (Nelson, 1995).

This is not to say that deep disagreements do not remain between feminists and their colleagues, or among feminist philosophers, concerning the implications of feminist science scholarship for the philosophy of science. It is to say that there are concurrent developments in mainstream and feminist philosophy of science that make positive engagements more likely. These include the abandonment, by many philosophers, of foundationalism and of the emphasis on justifying science; the emergence of approaches (for example, naturalism and semantic theory) that are not easily characterized as "positivist" or "anti-positivist"; and increased attention to the social processes in which scientific knowledge is generated, and to the ways in which values of various kinds serve a positive function in scientific practice.

Given the benefits of pluralism in the broader discipline, there now seems little reason to work to develop or to hope for "a" feminist theory of science. And the work in feminist philosophy I have cited, and much that I have not, also suggests the significant benefits of bringing different philosophical approaches to feminist engagements with science, and the substantial insights into the sciences that work in a variety of disciplines have provided. Indeed, it now seems clear that any systematic investigation of the sciences requires the tools of a wide range of disciplines, and that to understand this complex family of enterprises it is essential that the contributions of practicing scientists, of social scientists, and of science studies scholars from a range of disciplines, should inform one another.

I have also suggested, at least by implication, the need for further exploration of some specific issues. There is a need for further development of the notions of situatedness and contingency, and the related emphasis on reflexivity, in ways that will support, rather than undermine, the empirical content and normative implications of the important claims feminists have made within and about the sciences. I have suggested that there is also a need for further investigations into the nature of the epistemic and non-epistemic values as they figure in the practices of science, particularly of how non-epistemic values (such as those brought by feminists) can have a positive role, and the need for further investigations of the rationale for and consequences of abandoning the traditional distinction between epistemic and non-epistemic values.

Notes

1 Some work mentioned or discussed in this chapter, particularly early work in feminist philosophy of science, was described as "feminist epistemology." Taking disciplines to be "organized and institutionalized bodies of research focused on a core set of questions" (Burian, 1993, pp. 387–8), the philosophy of science and epistemology are distinct disciplines – the first taking the sciences "proper" as its subject of study, the second knowledge more generally. Here I use "feminist philosophy of science" to discuss feminist analyses of science, including those initially categorized as "feminist epistemology" (e.g., work in the collection *Feminist Epistemologies*, edited by Alcoff and Potter).

2 As will become clear, feminist philosophy of science is an inherently interdisciplinary research program. The analyses offered by feminist philosophers build substantively on the analyses and research of feminist scientists, as well as on work in the philosophy of science. In addition, feminist scientists often engage issues of traditional concern to the philosophy of science. Indeed, as I later outline, arguably the first analyses aptly described as feminist philosophy of science were offered by feminist scientists.

3 See, for example, the introductions to Keller and Longino (1996) and Nelson and Nelson (1996) and Harding (2000) and Wylie (2000).

4 I here presuppose the standard distinction between contextual values and epistemic or constitutive values – between, respectively, factors that are conventionally treated as appropriately "external" to science (non-epistemic interests and values) and those conventionally treated as epistemic and appropriately "internal" to science (e.g., theoretical virtues such as empirical adequacy, generality of scope, and simplicity). Although this distinction is increasingly challenged in science scholarship, it is useful precisely because it has been and remains influential. The conventional boundary between the sociology of science and the epistemology of science is also contested, and my reference to issues involving the sociology of science is not a reference to the research program so named. As Joseph Rouse details (1996), there are important differences between research programs that assume a hard and fast boundary between the epistemology and the sociology of science – including work in the sociology of knowledge and traditional approaches in the philosophy of science – and approaches in feminist

philosophy of science and other research traditions that do not assume this conventional boundary (e.g., versions of "social epistemology" and "social empiricism" as advocated, respectively, by Steve Fuller and Miriam Solomon that do not (Rouse, 1996).

5   The literature is extensive. Representative works include Aldrich (1978), Hornig (1979), Rossiter (1982), *Signs* (1978) and National Academy of Sciences (1983).

6   See Eichler (1988) and Wylie (1996a, 1996b) for excellent overviews of these developments.

7   As Alison Wylie has shown in a number of publications, the development of interest in gender and archaeology followed a somewhat different trajectory from that outlined here. See, for example, Wylie (1997b).

8   Due to space limitations, and the different focuses of feminist critiques of sciences whose subject matter is not overtly concerned with gender, I do not here discuss feminist analyses focusing on the physical sciences. There is, however, a good deal of work in this area. See Barad (1996), chapters in Keller (1985), Potter (1989), Spanier (1995) and Traweek (1988).

9   Wylie (2000) includes an excellent overview of developments in standpoint theory.

References

Addelson, K. P. (1983): "The Man of Professional Wisdom," in Harding and Hintikka (1983, pp. 165–86).

Alcoff, L. M. (1989): "Justifying Feminist Social Science," in Tuana (1989, pp. 85–103).

Alcoff, L. and Potter, E. (eds.) (1993): *Feminist Epistemologies*. New York and London: Routledge.

Aldrich, M. L. (1978): "Women in Science: Review Essay," *Signs*, 4(1), 126–35.

Antony, L. (1994): "Quine as Feminist: The Radical Import of Naturalized Epistemology," in L. Antony and C. Witt (eds.), *A Mind of One's Own: Feminist Essays on Reason and Objectivity*, Boulder: Westview Press, 185–225.

Barad, K. (1996): "Meeting the Universe Halfway: Realism and Social Constructivism Without Contradiction," in Nelson and Nelson (1996, pp. 161–94).

Bleier, R. (1984): *Science and Gender: A Critique of Biology and Its Theories on Women*. New York: Pergamon Press.

Bleier, R. (ed.) (1988): *Feminist Approaches to Science*. New York: Pergamon Press.

Burian, R. M. (1993): "Technique, Task Definition, and the Transition from Genetics to Molecular Genetics," *Journal of the History of Biology*, 26(3), 387–407.

Campbell, R. (1994): "The Virtues of Feminist Empiricism," *Hypatia*, 9, 90–115.

Campbell, R. (1998): *Illusions of Paradox: A Feminist Epistemology Naturalized*. Ithaca: Cornell University Press.

Collins, P. H. (1991): *Black Feminist Thought: Knowledge, Consciousness and the Politics of Empowerment*. New York: Routledge.

Duran, J. (1998): *Philosophies of Science/Feminist Theories*. Boulder: Westview Press.

Eichler, M. (1988): *Nonsexist Research Methods: A Practical Guide*. Boston: Allen and Unwin.

Fausto-Sterling, A. (1985): *Myths of Gender: Biological Theories of Women and Men*. New York: Basic Books.

Flax, J. (1983): "Political Philosophy and the Patriarchal Unconscious," in Harding and Hintikka (1983, pp. 245–81).

Fonow, M. and Cook, J. A. (eds.) (1991): *Beyond Methodology: Feminist Scholarship as Lived Research*. Bloomington: Indiana University Press.

Giere, R. N. (1996): "The Feminism Question in the Philosophy of Science," in Nelson and Nelson (1996, pp. 3–16).

Gilligan, C. (1982): *In a Different Voice: Psychological Theory and Women's Development*. Cambridge, MA: Harvard University Press.

Haack, S. (1996): "Science as Social? – Yes and No," in Nelson and Nelson (1996, pp. 79–94).

Haraway, D. (1986): "Primatology is Politics by Other Means," in R. Bleier (ed.), *Feminist Approaches to Science*, New York: Pergamon Press, 77–118.

Haraway, D. (1988): "Situated Knowledges: The Science Question in Feminism and the Privilege of Partial Perspective," *Feminist Studies*, 14, 575–99. Reprinted with revisions in Haraway D. (ed.) (1991): *Simians, Cyborgs, and Women*, New York and London: Routledge, 183–201.

Harding, S. (1986): *The Science Question in Feminism*. Ithaca: Cornell University Press.

Harding, S. (1991): *Whose Science? Whose Knowledge?* Ithaca: Cornell University Press.

Harding, S. (ed.) (1993): *The "Racial" Economy of Science: Towards a Democratic Future*. Bloomington: Indiana University Press.

Harding, S. (1996): "Multicultural and Global Feminist Philosophies of Science: Resources and Challenges," in Nelson and Nelson (1996, pp. 263–88).

Harding, S. (2000): "Feminist Philosophies of Science," in Joan Callahan (ed.), *The APA Newsletter on Feminism and Philosophy*, 99(2), 190–2.

Harding, S. and Hintikka, M. B. (eds.) (1983): *Discovering Reality: Feminist Perspectives on Metaphysics, Epistemology, Methodology, and Philosophy of Science*. Dordrecht: D. Reidel.

Harstock, N. (1983): "The Feminist Standpoint: Developing the Ground for a Specifically Feminist Historical Materialism," in Harding and Hintikka (1983, pp. 283–310).

Hartmann, H. (1981): "The Family as the Locus of Gender, Class, and Political Struggle: The Example of Housework," *Signs*, 6(3), 366–94.

Hornig, L. S. (1979): *Climbing the Academic Ladder: Doctoral Women Scientists in Academe*. Washington, DC: National Academy of Sciences.

Hrdy, S. B. (1981): *The Woman That Never Evolved*. Cambridge, MA: Harvard University Press.

Hubbard, R. (1982): "Have Only Men Evolved?" in Hubbard et al. (1982, pp. 17–46).

Hubbard, R. and Lowe, M. (eds.) (1979): *Genes and Gender II: Pitfalls in Research on Sex and Gender*. New York: Gordian Press.

Hubbard, R., Henifin, M. S. and Fried, B. (eds.) (1979): *Women Look at Biology Looking at Women: A Collection of Feminist Critiques*. Boston: G. K. Hall.

Hubbard, R., Henifin, M. S. and Fried, B. (eds.) (1982): *Biological Woman – The Convenient Myth*. Cambridge: Schenkman.

Keller, E. F. (1982): "Feminism and Science," in N. Keohane, M. Rosaldo and B. Gelpi (eds.), *Feminist Theory*, Chicago: University of Chicago Press, 113–26.

Keller, E. F. (1983): *A Feeling for the Organism: The Life and Work of Barbara McClintock*. New York: W. H. Freeman.

Keller, E. F. (1985): *Reflections on Gender and Science*. New Haven: Yale University Press.

Keller, E. F. and Grontkowski, C. (1983): "The Mind's Eye," in Harding and Hintikka (1983, pp. 207–24).

Keller, E. F. and Longino, H. E. (eds.) (1996): *Feminism and Science*. Oxford and New York: Oxford University Press.

Kelly-Gadol, J. (1976): "The Social Relations of the Sexes: Methodological Implications of Women's History," *Signs*, 1, 809–23.

Longino, H. E. (1987): "Can There be a Feminist Science?" *Hypatia*, 2, 51–64. Reprinted in Tuana (1989, pp. 45–57).

Longino, H. E. (1990): *Science as Social Knowledge*. Princeton: Princeton University Press.

Longino, H. E. (1993): Subjects, Power, and Knowledge: Description and Prescription in Feminist Philosophies of Science," in Alcoff and Potter (1993, pp. 101–20).

Longino, H. E. (1994): "In Search of Feminist Epistemology," *The Monist*, 77(4), 472–85.

Longino, H. E. (1995): "Gender, Politics, and the Theoretical Virtues," *Synthese*, 104(3), 383–97.

Longino, H. E. (1996): "Cognitive and Non-Cognitive Values in Science: Rethinking the Dichotomy," in Nelson and Nelson (1996, pp. 39–58).

Longino, H. E. and Doell, R. (1983): "Body, Bias and Behavior: A Comparative Analysis of Reasoning in Two Areas of Biological Science," *Signs*, 9(2), 206–27.

National Academy of Sciences (1983): *Climbing the Ladder: An Update on the Status of Doctoral Women Scientists and Engineers*. Washington, DC: National Academy Press.

Nelson, L. H. (1990): *Who Knows: From Quine to Feminist Empiricism*. Philadelphia: Temple University Press.

Nelson, L. H. (ed.) (1995): *Synthese*, 104(3), Special Issue on Feminism and Science.

Nelson, L. H. (1996): "Empiricism Without Dogmas," in Nelson and Nelson (1996, pp. 95–120).

Nelson, L. H. and Nelson, J. (1994): "Feminist Values and Cognitive Virtues," in D. Hull, M. Forbes and R. M. Burian (eds.), *PSA 1994, vol. 2*, East Lansing, MI: Philosophy of Science Association, 120–9.

Nelson, L. H. and Nelson, J. (eds.) (1996): *Feminism, Science, and the Philosophy of Science*. Dordrecht: Kluwer.

Nelson, L. H. and Nelson, J. (eds.) (2001): *Re-Reading the Canon: Feminist Interpretations of W. V. Quine*. University Park Penn: Penn State Press.

Potter, E. (1989): "Modeling the Gender Politics in Science," in Tuana (1989, pp. 132–46).

Potter, E. (1993): "Gender and Epistemic Negotiation," in Alcoff and Potter (1993, pp. 161–86).

Quine, W. V. O. (1981): "Empirical Content," in W. V. O. Quine, *Theories and Things*, Cambridge: Harvard University Press.

Reinharz, S. (1992): *Feminist Methods in Social Research*. Oxford: Oxford University Press.

Rosaldo, M. Z. (1980): "The Use and Abuse of Anthropology: Reflections on Feminism and Cross-Cultural Understanding," *Signs*, 5, 389–417.

Rosaldo, M. Z. and Lamphere, L. (eds.) (1974): *Woman, Culture, and Society*. Stanford: Stanford University Press.

Rossiter, M. (1982): *Women Scientists in America: Struggles and Strategies to 1940*. Baltimore: Johns Hopkins University Press.

Rouse, J. (1996): "Feminism and the Social Construction of Scientific Knowledge," in Nelson and Nelson (1996, pp. 195–216).

Signs: Journal of Women in Culture and Society (1978): 4(1), Special Issue: Women, Science and Society.

Slocum, S. (1975): "Woman the Gatherer: Male Bias in Anthropology," in R. Reiter (ed.), *Toward an Anthropology of Women*, New York: Monthly Review Press, 9–21.

Smith, D. (1977): "Women's Perspective as a Radical Critique of Sociology," *Sociological Inquiry*, 44, 7–13.

Spanier, B. (1995): *Im/Partial Science: Gender Ideology in Molecular Biology*. Bloomington: Indiana University Press.

Stanley, L. and Wise, S. (1983): *Breaking Out: Feminist Consciousness and Feminist Research*. Boston: Routledge and Kegan Paul.

Star, S. L. (1979): "Sex Differences and the Dichotomization of the Brain: Methods, Limits, and Problems in Research on Consciousness," in Hubbard and Lowe (1979, pp. 113–30).

Tanner, N. and Zihlman, A. (1976): "Women in Evolution," *Signs*, 1, 585–608.

Tobach, E. and Rosoff, B. (eds.) (1978): *Genes and Gender I: On Hereditarianism and Women*. New York: Gordian Press.

Traweek, S. (1988): *Beamtimes and Lifetimes: The World of High Energy Physics*. Cambridge, MA: Harvard University Press.

Tuana, N. (ed.) (1987): *Hypatia*, 2(3), Special Issue on Feminism and Science I.

Tuana, N. (ed.) (1988): *Hypatia*, 3(1), Special Issue on Feminism and Science II.

Tuana, N. (ed.) (1989): *Feminism and Science*. Bloomington: Indiana University Press.

Tuana, N. (1996): "Revaluing Science: Starting from the Practicws of Women," in Nelson and Nelson (1996, pp. 17–38).

Wylie, A. (1995): "Doing Philosophy as a Feminist: Longino on the Search for a Feminist Epistemology," *Philosophical Topics*, 23(2), 345–58.

Wylie, A. (1996a): "Fcminism and Social Science," in E. Craig (ed.), *Encyclopedia of Philosophy*, New York and London: Routledge, 191–4.

Wylie, A. (1996b): "The Constitution of Archaeological Evidence: Gender Politics and Science," in P. Galison and D. J. Stump (eds.), *The Disunity of Science*. Stanford: Stanford University Press, 311–43.

Wylie, A. (1997a): "Good Science, Bad Science, or Science as Usual? Feminist Critiques of Science," in L. D. Hager (ed.), *Women in Human Evolution*. New York: Routledge, 29–55.

Wylie, A. (1997b): "The Engendering of Archaeology: Refiguring Feminist Science Studies," *Osiris*, 12, 80–99.

Wylie, A. (2000): "Feminism in Philosophy of Science: Making Sense of Contingency and Constraint," in J. Hornsby and M. Fricker (eds.), *Companion to Feminism and Philosophy*, Cambridge: Cambridge University Press, 166–82.

Index